Lecture Notes in Computer Science 6716

Commenced Publication in 1973
Founding and Former Series Editors:
Gerhard Goos, Juris Hartmanis, and Jan van Leeuwen

Editorial Board

David Hutchison
Lancaster University, UK

Takeo Kanade
Carnegie Mellon University, Pittsburgh, PA, USA

Josef Kittler
University of Surrey, Guildford, UK

Jon M. Kleinberg
Cornell University, Ithaca, NY, USA

Alfred Kobsa
University of California, Irvine, CA, USA

Friedemann Mattern
ETH Zurich, Switzerland

John C. Mitchell
Stanford University, CA, USA

Moni Naor
Weizmann Institute of Science, Rehovot, Israel

Oscar Nierstrasz
University of Bern, Switzerland

C. Pandu Rangan
Indian Institute of Technology, Madras, India

Bernhard Steffen
TU Dortmund University, Germany

Madhu Sudan
Microsoft Research, Cambridge, MA, USA

Demetri Terzopoulos
University of California, Los Angeles, CA, USA

Doug Tygar
University of California, Berkeley, CA, USA

Gerhard Weikum
Max Planck Institute for Informatics, Saarbruecken, Germany

Rafael Muñoz   Andrés Montoyo
Elisabeth Métais (Eds.)

Natural Language Processing and Information Systems

16th International Conference on Applications
of Natural Language to Information Systems, NLDB 2011
Alicante, Spain, June 28-30, 2011
Proceedings

 Springer

Volume Editors

Rafael Muñoz
Andrés Montoyo
University of Alicante, School of Computing, 03080 Alicante, Spain
E-mail: rafael@dlsi.ua.es; montoyo@dlsi.ua.es

Elisabeth Métais
CNAM- Laboratoire Cédric, 292 Rue St. Martin, 75141 Paris Cedex 03, France
E-mail: elisabeth.metais@cnam.fr

ISSN 0302-9743   e-ISSN 1611-3349
ISBN 978-3-642-22326-6   ISBN 978-3-642-22327-3 (eBook)
DOI 10.1007/978-3-642-22327-3
Springer Heidelberg Dordrecht London New York

Library of Congress Control Number: 2011930837

CR Subject Classification (1998): I.2.7, H.3, H.2, I.5, J.3, H.2.8, I.2.6

LNCS Sublibrary: SL 3 – Information Systems and Application, incl. Internet/Web
and HCI

© Springer-Verlag Berlin Heidelberg 2011
This work is subject to copyright. All rights are reserved, whether the whole or part of the material is
concerned, specifically the rights of translation, reprinting, re-use of illustrations, recitation, broadcasting,
reproduction on microfilms or in any other way, and storage in data banks. Duplication of this publication
or parts thereof is permitted only under the provisions of the German Copyright Law of September 9, 1965,
in its current version, and permission for use must always be obtained from Springer. Violations are liable
to prosecution under the German Copyright Law.
The use of general descriptive names, registered names, trademarks, etc. in this publication does not imply,
even in the absence of a specific statement, that such names are exempt from the relevant protective laws
and regulations and therefore free for general use.

Typesetting: Camera-ready by author, data conversion by Scientific Publishing Services, Chennai, India

Printed on acid-free paper

Springer is part of Springer Science+Business Media (www.springer.com)

Preface

The 16th International Conference on Applications of Natural Language to Information Systems (NLDB 2011) was held during June 28–30, 2011, at the University of Alicante, Spain. Since the first NLDB conference in 1995, the main goal has been to provide a forum for the discussion and dissemination of research on the integration of natural language resources within the field of information system engineering.

The development and convergence of computing, telecommunications and information systems has already led to a revolution in the way that we work, communicate with each other, buy goods and use services, and even in the way we entertain and educate ourselves. As the revolution continues, large volumes of information are increasingly stored in a manner which is more natural for users to exploit than the data presentation formats typical of legacy computer systems. Natural language processing (NLP) is crucial to solving these problems and language technologies are indispensable to the success of information systems. We hope that NLDB 2011 was a modest contribution toward this goal.

NLDB 2011 contributed to the increase in the number of goals and the international standing of NLP conferences, largely due to its Program Committee, whose members are renowned researchers in the field of NLP and information system engineering. These high standards which have been set for NLDB can also be measured by the significant number of papers submitted (74) in this edition. Each paper was reviewed by three members of the Program Committee or by their recommended subreviewers. As a result of the review process, 11 articles were accepted as regular papers and another 11 were accepted as short papers. Additional contributions were selected for the Poster and the Doctoral Symposium sessions held at the conference.

Finally, we would like to thank all the reviewers for their involvement and excellent work. We extend these thanks to our invited speakers, Michael Thelwall and Horacio Saggion, for their valuable contribution, which undoubtedly increased the interest in the conference. We would also like to express our gratitude to the people who helped in the organization of the different parts of the conference program. We would especially like to thank Miguel Angel Varó for setting up and maintaing all Web services for the conference.

June 2011

Rafael Muñoz
Andrés Montoyo
Elisabeth Métais

Organization

Conference Chairs

Rafael Muñoz   University of Alicante, Spain
Elisabeth Métais   CEDRIC/CNAM, France

Program Chair

Andrés Montoyo   University of Alicante, Spain

Doctoral Symposium

Paloma Moreda   University of Alicante, Spain
Elena Lloret   University of Alicante, Spain

Poster Session

Alexandra Balahur   University of Alicante, Spain
Jesús M. Hermida   University of Alicante, Spain

Organizing Committee

Alexandra Balahur   University of Alicante, Spain
Ester Boldrini   University of Alicante, Spain
Antonio Ferrández   University of Alicante, Spain
Jesús M. Hermida   University of Alicante, Spain
Fernando Llopis   University of Alicante, Spain
Elena Lloret   University of Alicante, Spain
Patricio Martínez-Barco   University of Alicante, Spain
Andrés Montoyo   University of Alicante, Spain
Paloma Moreda   University of Alicante, Spain
Rafael Muñoz   University of Alicante, Spain
Manuel Palomar   University of Alicante, Spain
Jesús Peral   University of Alicante, Spain

Program Committee

Akhilesh Bajaj   University of Tulsa, USA
Alexander Hinneburg   University of Halle, Germany
Alfredo Cuzzocrea   University of Calabria, Italy
Andreas Hotho   University of Kassel, Germany

Andrés Montoyo   Universidad de Alicante, Spain
Antje Düsterhöft   Hochschule Wismar, Germany
Bernhard Thalheim   Kiel University, Germany
Cedric du Mouza   CNAM, France
Christian Kop   University of Klagenfurt, Austria
Christian Winkler   University of Klagenfurt, Austria
Christina J. Hopfe   Cardiff University, UK
Deryle Lonsdale   Brigham Young Uinversity, USA
Elisabeth Métais   CNAM, France
Epaminondas Kapetanios   University of Westminster, UK
Fabio Rinaldi   University of Zurich, Switzerland
Farid Meziane   Salford University, UK
Frederic Andres   University of Advanced Studies, Japan
Georgia Koutrika   Stanford University, USA
Grigori Sidorov   National Researcher of Mexico, Mexico
Günter Neumann   DFKI, Germany
Günther Fliedl   University of Klagenfurt, Austria
Hae-Chang Rim   Korea University, Korea
Harmain Harmain   United Arab Emirates University, UAE
Heinrich C. Mayr   University of Klagenfurt, Austria
Helmut Horacek   Saarland University, Germany
Hiram Calvo   National Polytechnic Institute, Mexico
Irena Spasic   Manchester Centre for Integrative Systems
                                      Biology, UK
Isabelle Comyn-Wattiau   CNAM, France
Jacky Akoka   CNAM, France
Jana Lewerenz   Capgemini Düsseldorf, Germany
Jian-Yun Nie   Université de Montréal, Canada
Jon Atle Gulla   Norwegian University of Science and
                                      Technology, Norway
Juan Carlos Trujillo   Universidad de Alicante, Spain
Jürgen Rilling   Concordia University, Canada
Karin Harbusch   Universität Koblenz-Landau, Germany
Krishnaprasad Thirunarayan   Wright State University, USA
Leila Kosseim   Concordia University, Canada
Luis Alfonso Ureña   Universidad de Jaén, Spain
Luisa Mich   University of Trento, Italy
Magdalena Wolska   Saarland University, Germany
Manuel Palomar   Universidad de Alicante, Spain
Max Silberztein   Université de Franche-Comté, France
Nadira Lammari   CNAM, France
Odile Piton   Université Paris I Panthé on-Sorbonne, France
Panos Vassiliadis   University of Ioannina, Greece
Paul Johannesson   Stockholm University, Sweden
Paul McFetridge   Simon Fraser University, Canada

Philipp Cimiano   CITEC, University of Bielefeld, Germany
Pit Pichappan   Annamalai University, India
Rafael Muñoz   Universidad de Alicante, Spain
René Witte   Concordia University, Canada
Roger Chiang   University of Cincinnati, USA
Rossi Setchi   Cardiff University, UK
Samira Si-Said Cherfi   CNAM, France
Stéphane Lopes   Université de Versailles, France
Udo Hahn   Friedrich-Schiller-Universität Jena, Germany
Veda Storey   Georgia State University, USA
Vijay Sugumaran   Oakland University Rochester, USA
Yacine Rezgui   University of Salford, UK
Zornitsa Kozareva   Information Science Institute, University
                            of South California, USA
Zoubida Kedad   Université de Versailles, France

Additional Reviewers

Alexandra Balahur   John McCrae
Arturo Montejo   Maria Bergholtz
Doina Tatar   María Teresa Martín Valdivia
Elena Lloret   Óscar Ferrández
Jesús M. Hermida

Sponsoring Institutions

Consellería d'Educació, Generalitat Valenciana, Spain
University of Alicante, Spain

Table of Contents

Full Papers

Short Papers

Posters

Doctoral Symposium

Invited Talks

1 Easy Access to Textual Information: The Role of Natural Language Processing

Horacio Saggion
Universitat Pompeu Fabra
Grupo TALN - Departament of Information and Communication technologies
Spain

In the internet era we face the recurring problem of information overload thus creating a major need for computer tools able to constantly distill, interpret, and organize textual information available on-line effectively. Over the last decade a number of technological advances in the Natural Language Processing field have made it possible to analyze, discover, and machine-interpret huge volumes of online textual information. One clear example of these advances in the field is the nowadays ubiquitous presence of automatic translation tools on the Web. However other NLP technologies have still a fundamental role to play in allowing access to textual information. In this talk I will present an overview of research in the area of Natural Language Processing aiming at facilitating access to textual information online. I will review the role of text summarization, question answering, and information extraction in making textual information more accessible. I will also discuss the issue of access to textual information for people with learning disabilities and present the ongoing work in this area of the Simplex project which aims at producing a text simplification system to facilitate easy access to textual information.

2 Sentiment Strength Detection in the Social Web: From YouTube Arguments to Twitter Praise

Michael Thelwall
University of Wolverhampton
Statistical Cybermetrics Research Group
UK

Sentiment analysis or opinion mining is mainly concerned with the automatic identification of subjectivity and polarity in text. It has been commercially successful because of the potential market research applications for tools that are able to detect customer reactions to products. Recently, new tools have been developed that are able to detect the strength of sentiment in texts with reasonable accuracy across a wide range of social web contexts. The addition of

R. Muñoz et al. (Eds.): NLDB 2011, LNCS 6716, pp. 1–2, 2011.
© Springer-Verlag Berlin Heidelberg 2011

strength information for sentiment promises to allow more fine-grained analyses of bodies of texts. At the same time, there have been a number of large scale non-commercial analyses of texts in the social web that have shed light on wider patterns of sentiment in society or around specific issues. This talk will describe the results from applying a general purpose unsupervised sentiment strength algorithm to two social web contexts: YouTube comments and Twitter posts. The results demonstrate how large scale social trends can be identified relatively easily and how small scale network-based interactions can also be analysed automatically for sentiment.

COMPENDIUM: A Text Summarization System for Generating Abstracts of Research Papers

Elena Lloret[1,*], María Teresa Romá-Ferri[2], and Manuel Palomar[1]

[1] Dept. Lenguajes y Sistemas Informáticos
Universidad de Alicante
Apdo. de correos, 99, E-03080 Alicante, Spain
{elloret,mpalomar}@dlsi.ua.es
[2] Department of Nursing, University of Alicante, Spain
mtr.ferri@ua.es

Abstract. This paper presents COMPENDIUM, a text summarization system, which has achieved good results in extractive summarization. Therefore, our main goal in this research is to extend it, suggesting a new approach for generating abstractive-oriented summaries of research papers. We conduct a preliminary analysis where we compare the extractive version of COMPENDIUM (COMPENDIUM$_E$) with the new abstractive-oriented approach (COMPENDIUM$_{E-A}$). The final summaries are evaluated according to three criteria (content, topic, and user satisfaction) and, from the results obtained, we can conclude that the use of COMPENDIUM is appropriate for producing summaries of research papers automatically, going beyond the simple selection of sentences.

Keywords: Human Language Technologies, NLP Applications, Text Summarization, Information Systems.

1 Introduction

The vast amount of information currently available has fuelled research into systems and tools capable of managing such information in an effective and efficient manner. That is the case of Text Summarization (TS), whose aim is to produce a condensed new text containing a significant portion of the information in the original text(s) [20]. In particular, TS has been shown to be very useful as a stand-alone application [2], as well as in combination with other systems, such as text classification [19].

---

* This research has been supported by the FPI grant (BES-2007-16268) from the Spanish Ministry of Science and Innovation, under the project TEXT-MESS (TIN2006-15265-C06-01) and project grant no. TIN2009-13391-C04-01, both funded by the Spanish Government. It has been also funded by the Valencian Government (grant no. PROMETEO/2009/119 and ACOMP/2011/001). The authors would like to thank Ester Boldrini, Paloma Moreda, Isabel Moreno, Helena Burruezo, Jesús Hermida, Jorge Cruañes, Fernando Peregrino, Héctor Llorens, Felipe Sellés, and Rubén Izquierdo for their help in the manual evaluation of the summaries.

R. Muñoz et al. (Eds.): NLDB 2011, LNCS 6716, pp. 3–14, 2011.
© Springer-Verlag Berlin Heidelberg 2011

Among the wide range of applications of TS, one especially interesting concerns the generation of abstracts of research papers. In the research context, every article published in a conference, journal, etc. must include an abstract written by the author, which is a summary of the main topics and findings addressed in the research presented. These abstracts are very important to provide an idea of what the article is about, and they can be used not only by humans, but also by automatic systems for indexing, searching and retrieving information without having to process the whole document. TS can be very useful for automatically generating such abstracts. However, to carry out this process is very challenging and difficult. This is shown by the fact that, although there have been some attempts to generate abstracts in the recent years [4], [16], most of the current work on TS still focuses on extractive summarization[1] [10], [12], [21]. The main problem associated to extractive summarization is the lack of coherence resulting summaries exhibit, partly due to non-resolved coreference relationships, and the wrong link between sentences.

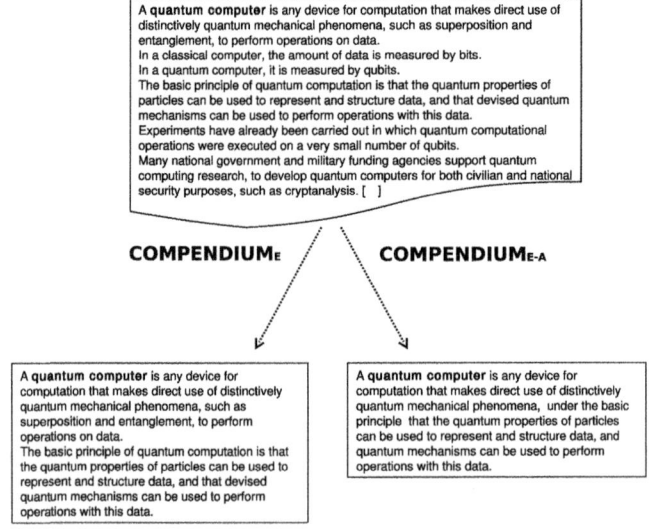

Fig. 1. Intuitive illustration of the COMPENDIUM system

In this paper, our goal is to analyze to what extent TS is useful for a specific application: the automatic generation of research paper abstracts. In particular, we propose two generic TS approaches: one based on a pure extractive summarizer (COMPENDIUM$_E$) and a novel one that combines extractive and abstractive techniques (COMPENDIUM$_{E-A}$). In this preliminary research work, we want to

---

[1] Extractive summarization consists of selecting the most important sentences of a document to form the final summary, whereas abstractive implies also the generation of new information.

study to what extent such approaches are appropriate to serve as a surrogate of the source document or, in contrast, if they can only be used in order to provide users with an idea of what is important in the document. Results show that extractive and abstractive-oriented summaries perform similarly as far as the information they contain, but the latter is more appropriate from a human perspective, when a user satisfaction assessment is carried out. Figure 1 illustrates the differences between an extractive summary (COMPENDIUM$_E$) and an abstractive one (COMPENDIUM$_{E-A}$) in an intuitive way.

The remaining of the paper is structured as follows: Section 2 introduces previous work on TS approaches developed for specific applications. Section 3 describes in detail the data set employed, analysing also the human generated abstracts provided with research papers. Further on, in Section 4 we explain the suggested TS approaches and how we developed them. The experiments carried out, the results obtained together with a discussion for each type of evaluation are provided in Section 5. Finally, the conclusions of the paper together with the future work are outlined in Section 6.

2   Related Work

In this Section, we explain previous work on different approaches that have used TS techniques to address specific tasks (e.g. for generating Wikipedia articles, weather forecast reports, etc.).

Sauper and Barzilay [18] propose an automatic method to generate Wikipedia articles, where specific topic templates, as well as the information to select is learnt using machine learning algorithms. The templates are obtained by means of recurrent patterns for each type of document and domain. For extracting the relevant content, candidate fragments are ranked according to how representative they are with respect to each topic of the template. Other approaches, that also rely on the use of templates to organize and structure the information previously identified, are based on information extraction systems. For instance, in Kumar et al. [9] reports of events are generated from the information of different domains (biomedical, sports, etc.) that is stored in databases. In such research, human-written abstracts are used, on the one hand, to determine the information to include in a summary, and on the other hand, to generate templates. Then, the patterns to fill these templates in are identified in the source texts. Similarly, in Carenini and Cheung [4], patterns in the text are also identified, but since their aim is to generate contrastive summaries, discourse markers indicating contrast such as *"although"*, *"however"*, etc. are also added to make the summary sound more naturally.

Natural Language Generation (NLG) has been also applied for adding new vocabulary and language structures in summaries. In Yu et al. [22] very short summaries are produced from large collections of numerical data. The data is presented in the form of tables, and new text is generated for describing the facts that such data represent. Belz [1] also suggests a TS approach based on NLG, in order to generate weather forecasts reports automatically.

Another specific application of TS concerns the generation of summaries from biographies. The idea behind multi-document biography summarization is to produce a piece of text containing the most relevant aspects of a specific person. Zhou et al. [23] analyzed several machine learning algorithms (*Naïve Bayes*, *Support Vector Machines*, and *Decision Trees*) to classify sentences, distinguishing between those ones containing biographic information (e.g. the date/place of birth) from others that do not.

The generation of technical surveys has also been addressed in the recent years. In Mohammad et al. [14] citations are used to automatically generate technical surveys. They experimented with three types of input (full papers, abstracts and citation texts), and analyzed different already existing summarization systems to create such surveys, for instance LexRank [5]. Among the conclusions drawn from the experiments, it was shown that multi-document technical survey creation benefits considerably from citation texts.

Our research work focus on generating research abstracts, and in particular analyzing the capabilities of different TS techniques for such purposes. This problem has been addressed in previous literature. For instance, Pollock and Zamora [15] used cue words for generating abstracts. This technique consists of determining the relevance of a sentence by means of the phrases or words it contains that could be introduce relevant information, such as *"in conclusion"* or *"the aim of this paper"*. Furthermore, Saggion and Lapalme [17] exploited also this kind of information, by means of a set of patterns that were later combined with the information extracted from the document. In our approach, we do not rely on specific patterns, nor we learn the struture or the content from already existing model abstracts. In contrast, we want to analyze the appropriateness of two generic TS approaches to generate abstracts of research papers in particular for the medical domain, and assess to what extent they would be appropriate for this task.

3   Description of the Data Set: Research Papers

We collected a set of 50 research articles from specialized journals of medicine directly from the Web. Each article contains a human-written abstract which will be used as a model summary. Furthermore, the articles also include an outline, a set of keywords, as well as several figures and tables. The data set was first preprocessed and only the main content of each document was kept for further processing. In other words, the outline, bibliographic entries, keywords, figures and tables were removed. Table 1 shows some properties of the data set after the preprocessing stage. Additionally, it also contains some figures corresponding to the human-written abstracts. As can be seen, the documents are rather long (more than 2,000 words on average), whereas the abstracts are shorter (162 words on average), thus being their compression ratio with respect to the whole article quite high (13%).

3.1   Analysis of the Model Summaries

Before conducting the TS process, an analysis of the human-written abstracts is carried out. The reason why such analysis is done is to quantify and understand

Table 1. Properties of the data set and the human-written abstracts

Avg. number of sentences per document	83.03
Avg. number of words per document	2,060
Avg. number of words per human abstracts	162.7
Compression ratio for abstracts wrt. documents	13%

the nature of human-written abstracts. As we expected, only a small number of abstracts (18%) have an extractive nature, containing at least one identical sentence expressed later in the article. Instead of a pure extractive nature, this can be considered a combination of extractive and abstractive strategies. The percentage of identical sentences ranged between 9% and 60%, depending on the abstract. On the contrary, the 82% of the abstracts have an abstractive nature, but these abstracts have not been created following a pure abstractive strategy either, since important fragments of information are selected identically from the source document, and then they are generalized and connected with others in a coherent way, through discourse markers and linking phrases. As a result of this analysis, one may think that a TS approach that combines extractive with abstractive techniques together can be more appropriate to tackle this task.

The first and the second summaries in Table 2 correspond to a fragment of a medical article and a human abstract, respectively. It is worth noticing that 50% of the sentences in the human abstract are identical (sentences 1 and 2 correspond to sentences 1 and 78 in the original article) and 50% are new sentences. Moreover, as it can be seen the third and the fourth sentence in the human abstract have been generated from relevant pieces of information that appears in the original document (e.g. sentence 3 contains information from sentences 8, 26, 33 and 81).

4   Text Summarization Approach: COMPENDIUM

In this Section, the two suggested approaches for generating summaries are explained. First we described COMPENDIUM$_E$, a pure extractive TS approach (Section 4.1), and then we take this extractive approach as a basis in an attempt to improve the final summaries by integrating abstractive techniques, leading to COMPENDIUM$_{E-A}$ (Section 4.2).

Besides a fragment of a medical article and a human abstract, Table 2 shows two summaries generated with COMPENDIUM$_E$ and COMPENDIUM$_{E-A}$, which are explained in the next subsections. It is worth mentioning that these approaches produce generic summaries, and for generating them neither the keywords of the original article nor the information in the titles or in the abstract have been taken into consideration. As it can be seen, the first two sentences in the summary generated with COMPENDIUM$_E$ are also presented in the human abstract. Moreover, the resulting summary for COMPENDIUM$_{E-A}$ has also some sentences in common with COMPENDIUM$_E$, whereas others have been compressed or merged.

Table 2. Example of different types of summaries and the corresponding source article

Fragment of the original medical article:
1: Histologic examination of lesions plays a key role in the diagnostics of cutaneous lupus erythe-matosus (LE). 2: LE has a broad spectrum of histological signs which are related to the stages of the lesions, but some signs apply to all stages (e.g.mucin deposition). [...] 8: From the histological standpoint, LE can be classified only into early, fully developed, late LE, and special manifesta-tions of LE.[...] 26: If interface dermatitis becomes manifest at it's maximal expression, Rowell's syndrome must be considered. [...] 33: In subtypes of LE with epidermal involvement, damage to keratinocytes is a distinctive sign which is very helpful in establishing the diagnosis. [...] 81: LE may be associated with Kikuchi-Fujimoto disease (histiocytic necrotizing lymphadenitis). [...]
Human Abstract:
1: Histologic examination of lesions plays a key role in the diagnostics of cutaneous lupus ery-thematosus (LE). 2: LE has a broad spectrum of histopathological signs, which are related to the stages of the lesions. 3: In addition to the main subtypes of LE, we report on special manifestations like Rowell's-syndrome and Chilblain LE, and give an account of Kikuchi-Fujimoto disease (histi-ocytic necrotizing lymphadenitis), which may be associated with systemic LE. 4: Furthermore the most considerable histopathologic differential diagnoses are discussed.
Extractive summary generated with COMPENDIUM$_E$:
1: Histologic examination of lesions plays a key role in the diagnostics of cutaneous lupus ery-thematosus LE . 2: LE has a broad spectrum of histological signs which are related to the stages of the lesions, but some signs apply to all stages e.g. mucin deposition . 3: Histologic findings of skin lesions are essentially identical for systemic lupus erythmatosus SLE and cutaneous LE. 4: From the histological standpoint, LE can be classified only into early, fully developed, late LE, and special manifestations of LE. 5: The early histologic findings of LE lesions are sparse superficial perivascular lymphocytic infiltrates, neutrophils and sometimes nuclear dust immediately beneath the dermoepidermal junction. 6: Few individual necrotic keratinocytes and focal vacuolar alter-ation of basal cells may occur. 7: According to Kuhn et al. the presence of even slight epidermal or junctional involvement should exclude LE tumidus.
Abstractive-oriented summary generated with COMPENDIUM$_{E-A}$:
1: LE lesions plays a key role in the diagnostics. 2: LE has a broad spectrum of histological signs which are related to the stages of the lesions, but some signs apply to all stages e.g. mucin deposition . 3: LE lesions are essentially identical for systemic lupus erythmatosus SLE. 4: LE can be classified only into early histologic. 5: LE lesions are sparse superficial perivascular lymphocytic infiltrates neutrophils and sometimes nuclear dust immediately beneath the dermoepidermal junction. 6: Few individual necrotic keratinocytes and focal vacuolar alteration of basal cells may occur. 7: According to Kuhn et al the presence of even slight epidermal or junctional involvement should exclude.

4.1   COMPENDIUM$_E$: Extractive Summarization

This TS approach relies on four main stages for producing extractive summaries:

- **Preprocessing:** first of all, a basic linguistic analysis is carried out in order to prepare the text for further processing. Such analysis comprises tokeniza-tion, sentence segmentation, part-of-speech tagging, and stop word removal.
- **Redundancy removal:** a Textual Entailment (TE) tool [6] is used to detect and remove repeated information. In this sense, two sentences containing a true entailment relationship are considered equivalent, and therefore, the one which is entailed is discarded.
- **Sentence relevance:** this stage computes a score for each sentence depend-ing on its importance, relying on two features: Term Frequency (TF) [13] and the Code Quantity Principle (CQP) [8]. On the one hand, TF allows us to determine the topic of a document[2], whereas on the other hand, the CQP states that the most important information within a text is expressed by

---

[2] Stop words are not taken into account.

a high number of units (for instance, words or noun phrases). In our TS approach, we select as units noun phrases because they are flexible coding units and can vary in the number of elements they contain depending on the information detail one wants to provide. Therefore, in order to generate the summary, sentences containing longer noun phrases of high frequent terms are considered more important, thus having more chances to appear in the final summary.

- **Summary generation:** finally, having computed the score for each sentence, in this last stage of the TS process sentences are ranked according to their relevance and the highest ones are selected and extracted in the same order as they appear in the original document, thus generating an extractive summary.

4.2   COMPENDIUM$_{E-A}$: Abstractive-Oriented Summarization

Our second TS approach, COMPENDIUM$_{E-A}$, combines extractive and abstractive techniques in the following manner: we take as a basis the COMPENDIUM$_E$ approach described in the previous section, and we integrate an ***information compression and fusion stage*** after the relevant sentences have been identified and before the final summary is generated, thus generating abstractive-oriented summaries. The goal of this stage is to generate new sentences in one of these forms: either a compressed version of a longer sentence, or a new sentence containing information from two individual ones. The main steps involved in this stage are:

- **Word graph generation:** for generating new sentences, we rely on word graphs adopting a similar approach to the one described in [7]. Specifically in our approach, we first generate an extractive summary in order to determine the most relevant content for being included in the summary. Then, a weighted directed word graph is built taking as input the generated extract, where the words represent the nodes of the graph, and the edges are adjacency relationships between two words. The weight of each edge is calculated based on the inverse frequency of co-occurrence of two words and taking also into account the importance of the nodes they link, through the Pagerank algorithm [3]. Once the extract is represented as a word graph, a pool of new sentences is created by identifying the shortest path between nodes (e.g. using Dijkstra's algorithm), starting with the first word of each sentence in the extract, in order to cover its whole content. The reason why we used the shortest path is twofold. On the one hand, it allows sentences to be compressed, and on the other hand, we can include more content in the summary, in the case several sentences are fused.
- **Incorrect paths filtering:** this stage is needed since not all of the sentences obtained by the shortest paths are valid. For instance, some of them may suffer from incompleteness (*"Therefore the immune system"*). Consequently, in order to reduce the number of incorrect generated sentences, we define a set of rules, so that sentences not accomplishing all the rules are not taken

into account. Three general rules are defined after analyzing manually the resulting sentences derived from the documents in the data set, which are: i) the minimal length for a sentence must be 3 words[3]; ii) every sentence must contain a verb; and iii) the sentence should not end in an article (e.g. a, the), a preposition (e.g. of), an interrogative word (e.g. who), nor a conjunction (e.g. and). In the future, we plan to extend the set of rules in order to detect a higher number of incorrect sentences, for instance, taking into account if the sentence ends with an adjective.

- **Given and new information combination:** the objective of the last step is to decide which of the new sentences are more appropriate to be included in the final summary. Since we want to develop a mixed approach, combining extractive and abstractive techniques, the sentences of the final summary will be selected following a strategy that maximizes the similarity between each of the new sentences and the ones that are already in the extract, given that the similarity between them is above a predefined threshold[4]. Finally, in the cases where a sentence in the extract has an equivalent in the set of new generated sentences, the former will be substituted for the latter; otherwise, we take the sentence in the extract.

5   Results and Discussion

In this Section, we explain the evaluation performed and we show the results obtained together with a discussion. We set up three different types of evaluation. In the first one, we use the human-written abstracts of the articles and we compare them with the ones generated by our approaches using a state-of-the-art evaluation tool. The second evaluation aims at determining to what extent our generated summaries contain the main topics of the research articles. In the third evaluation, we assess the summaries with regard to user satisfaction. Finally a discussion of the results is provided.

5.1   Comparison with Human Abstracts

The goal of this evaluation is to assess the informativeness of the summaries with respect to their content. Following the approaches described in Section 4, we generated summaries of approximately 162 words for the 50 documents pertaining to the data set (Section 3). We also use MS-Word Summarizer 2007[5] for producing summaries of the same length in order to allow us to compare our approaches with an state-of-the-art summarizer. We then use ROUGE-1 [11] for assessing how many common vocabulary there is between the generated summaries and the human-written abstract. Table 3 shows these results in percentages.

---

[3] We assume that three words (i.e., subject+verb+object) is the minimum length for a complete sentence.

[4] We use the cosine measure to compute the similarity between two sentences, and a threshold has been empirically set to 0.5.

[5] http://www.microsoft.com/education/autosummarize.aspx

Table 3. ROUGE-1 results for the different text summarization approaches

TS Approach	Recall	Precision	F-measure
COMPENDIUM$_E$	44.02	40.53	42.20
COMPENDIUM$_{E-A}$	38.66	41.81	40.20
MS-Word 2007	43.61	40.46	41.97

As can be seen, our both TS approaches are comparable with respect to the state of the art TS tool (i.e., MS-Word 2007 Summarizer). Regarding our abstractive-oriented approach (COMPENDIUM$_{E-A}$), it is worth mentioning that it obtains higher precision than the remaining approaches. However, its recall is lower, so in the end the final value of F-measure is negatively affected. This is due to the fact that for this TS approach we take as input the extracts previously generated, and we compress or merge some information within them. Therefore, the summaries produced using COMPENDIUM$_{E-A}$ will be shorter than the extracts, and since no extra information is added, the recall value will be never higher than it is for COMPENDIUM$_E$. In order to solve this problem, we could either generate the new sentences from the original article instead of using the extract as input; or include in the COMPENDIUM$_{E-A}$ summary the next highest ranked sentence in the document according to the relevance detection stage (Section 4.1) that we could not include it in the extractive summary, because the desired length had already been reached.

5.2 Topic Identification

The objective of this evaluation is to assess the generated summaries with respect to the topics they contain (i.e., how indicative they are). Together with the content of the article and the abstract, a number of keywords were also included (5 on average). These keywords usually reflect the most important topics dealt in the article. Consequently, we want to analyze to what extent such keywords appear in the summaries generated by our suggested TS approaches (COMPENDIUM$_E$ and COMPENDIUM$_{E-A}$). If our summaries are able to contain such keywords, it will mean that they are indicative of the content of the source document, and therefore, they will be appropriate to provide an idea of what the article is about. In order to compute the number of keywords a summary contains, we calculate how many of them a summary contains, and we divide this result by the total number of keywords the corresponding article has. Table 4 shows the results obtained for this evaluation.

Table 4. Percentage of topics that resulting summaries contain

	% Correct Topics			
	< 25%	< 50%	< 75%	75-100%
COMPENDIUM$_E$	5%	12.5%	47.5%	35%
COMPENDIUM$_{E-A}$	7.5%	17.5%	42.5%	32.5%

As it can be seen, a considerable percentage of summaries are able to reflect at least half of the topics of the articles (82.5% and 75%, for COMPENDIUM$_E$ and COMPENDIUM$_{E-A}$, respectively). It is worth stressing upon the fact that our approaches produce generic summaries and in none of the cases, the keywords provided in the article have been taken into account in the summarization process. Some of summaries generated employing the COMPENDIUM$_{E-A}$ approach do not contain as many topics as the ones for COMPENDIUM$_E$. This occurs because in the former approach the resulting summaries contain sentences that may have been compressed, so in some of these cases, there is a loss of information, although minimal.

5.3   User Satisfaction Study

In the last evaluation, we aim at assessing the user satisfaction with respect to the generated summaries. For this purpose, we performed a qualitative evaluation and we asked 10 humans to evaluate our summaries[6] according to a 5-level Likert scale (1= Strongly disagree...5=Strongly Agree) for three questions concerning the appropriateness of the summaries. Specifically, these questions were:

Q1: The summary reflects the most important issues of the document.
Q2: The summary allows the reader to know what the article is about.
Q3: After reading the original abstract provided with the article, the alternative summary is also valid.

The percentage of summaries for each question in a scale 1 to 5 is shown Table 5. As it can be seen from the results obtained, our abstractive-oriented approach (COMPENDIUM$_{E-A}$) obtains better results than the pure extractive one (COMPENDIUM$_E$). Although the results concerning the information contained in the summaries generated with COMPENDIUM$_{E-A}$ were slightly lower than the extractive approach, taking into consideration their quality from a human point of view, the abstractive-oriented summaries are much better than the extractive ones. When we have a look at the different percentages of summaries that have been rated in one of each categories, we observe that there is a higher percentage of abstractive-oriented summaries that humans agree with, compared to the extractive summaries for the same rating (e.g. fourth row of the table - *Agree* -). Moreover, it is worth stressing upon the fact that, analogously, the percentage of summaries with lower ratings for strongly disagree and disagree also decrease when COMPENDIUM$_{E-A}$ is employed.

Furthermore, regarding the average results, COMPENDIUM$_{E-A}$ achieves at most 3.37/5 for Q2 and 3.1/5 for Q1 and Q3, whereas the maximum average value for COMPENDIUM$_E$ is 2.83/5 for Q2, the remaining questions obtaining values lower than 2.60/5. In light of the results obtained, it has been proved that the combination of extractive and abstractive techniques is more appropriate and leads to better summaries than pure extractive summaries.

---

[6] The humans were also provided with the original articles and their abstracts.

Table 5. User satisfaction results for the different text summarization approaches

%	TS Approach	Q1	Q2	Q3
1. Strongly disagree	COMPENDIUM$_E$	9.76	19.51	19.51
	COMPENDIUM$_{E-A}$	2.44	0	2.44
2. Disagree	COMPENDIUM$_E$	41.46	19.51	34.15
	COMPENDIUM$_{E-A}$	31.37	21.95	31.71
3. Neither agree nor disagree	COMPENDIUM$_E$	24.39	29.27	26.83
	COMPENDIUM$_{E-A}$	21.95	29.27	26.83
4. Agree	COMPENDIUM$_E$	21.95	21.95	7.32
	COMPENDIUM$_{E-A}$	41.46	39.02	34.15
5. Strongly agree	COMPENDIUM$_E$	2.44	9.76	12.20
	COMPENDIUM$_{E-A}$	2.44	9.76	4.88

6   Conclusion and Future Work

In this paper we presented COMPENDIUM, a text summarization system applied to the generation of abstracts of research papers. In particular, two generic text summarisation approaches were analysed: one based on a pure extractive summarizer (COMPENDIUM$_E$) and a novel approach (COMPENDIUM$_{E-A}$) which combined extractive and abstractive techniques, by incorporating an information compression and fusion stage once the most important content is identified. We carried out an evaluation based on three criteria: i) the information contained in the summaries; ii) the topics identified; and iii) the users' satisfaction. From the results obtained, we can conclude that COMPENDIUM is useful for producing summaries of research papers automatically. Although extractive and abstractive-oriented summaries perform similarly as far as the information and topics they contain, abstractive-oriented summaries are more appropriate from a human perspective. However, there are some issues that have to be tackled in the short-term. We plan to analyze other variants of the proposed approach for building abstracts, such as taking the source document as a starting point. Moreover, we are interested in studying other graph-based algorithms.

References

1. Belz, A.: Automatic Generation of Weather Forecast Texts Using Comprehensive Probabilistic Generation-space Models. Natural Language Engineering 14(4), 431–455 (2008)
2. Bouras, C., Tsogkas, V.: Noun retrieval effect on text summarization and delivery of personalized news articles to the user's desktop. Data Knowledge Engineering 69, 664–677 (2010)
3. Brin, S., Page, L.: The anatomy of a large-scale hypertextual web search engine. Computer Networks ISDN Systems 30, 107–117 (1998)
4. Carenini, G., Cheung, J.C.K.: Extractive vs. NLG-based abstractive summarization of evaluative text: The effect of corpus controversiality. In: Proc. of the 5th International Natural Language Generation Conference, pp. 33–40 (2008)
5. Erkan, G., Radev, D.R.: Lexrank: Graph-based lexical centrality as salience in text summarization. Journal of Artificial Intelligence Research (JAIR) 22, 457–479 (2004)

6. Ferrández, Ó., Micol, D., Muñoz, R., Palomar, M.: A perspective-based approach for solving textual entailment recognition. In: Proc. of the ACL-PASCAL Workshop on Textual Entailment and Paraphrasing, pp. 66–71 (2007)
7. Filippova, K.: Multi-sentence compression: Finding shortest paths in word graphs. In: Proc. of the 23rd International Conference on Computational Linguistics, pp. 322–330 (2010)
8. Givón, T.: A functional-typological introduction, vol. II. John Benjamins, Amsterdam (1990)
9. Kumar, M., Das, D., Agarwal, S., Rudnicky, A.: Non-textual Event Summarization by Applying Machine Learning to Template-based Language Generation. In: Proc. of the 2009 Workshop on Language Generation and Summarisation, pp. 67–71 (2009)
10. Lal, P., Rüger, S.: Extract-based summarization with simplification. In: Workshop on Text Summarization in Conjunction with the ACL (2002)
11. Lin, C.Y.: ROUGE: a package for automatic evaluation of summaries. In: Proc. of ACL Text Summarization Workshop, pp. 74–81 (2004)
12. Liu, M., Li, W., Wu, M., Lu, Q.: Extractive summarization based on event term clustering. In: Proc. of the 45th ACL, pp. 185–188 (2007)
13. Luhn, H.P.: The automatic creation of literature abstracts. In: Mani, I., Maybury, M. (eds.) Advances in Automatic Text Summarization, pp. 15–22. MIT Press, Cambridge (1958)
14. Mohammad, S., Dorr, B., Egan, M., Hassan, A., Muthukrishan, P., Qazvinian, V., Radev, D., Zajic, D.: Using citations to generate surveys of scientific paradigms. In: Proc. of the North American Chapter of the ACL, pp. 584–592 (2009)
15. Pollock, J.J., Zamora, A.: Automatic abstracting research at chemical abstracts. In: Mani, I., Maybury, M. (eds.) Advances in Automatic Text Summarization, pp. 43–49. MIT Press, Cambridge (1999)
16. Saggion, H.: A classification algorithm for predicting the structure of summaries. In: Proc. of the Workshop on Language Generation and Summarisation, pp. 31–38 (2009)
17. Saggion, H., Lapalme, G.: Selective analysis for automatic abstracting: Evaluating indicativeness and acceptability. In: Proceedings of Content-Based Multimedia Information Access, pp. 747–764 (2000)
18. Sauper, C., Barzilay, R.: Automatically generating wikipedia articles: A structure-aware approach. In: Proc. of the 47th Association of Computational Linguistics, pp. 208–216 (2009)
19. Shen, D., Yang, Q., Chen, Z.: Noise Reduction through Summarization for Web-page Classification. Information Processing and Management 43(6), 1735–1747 (2007)
20. Spärck Jones, K.: Automatic summarising: The state of the art. Information Processing & Management 43(6), 1449–1481 (2007)
21. Wong, K.F., Wu, M., Li, W.: Extractive summarization using supervised and semi-supervised learning. In: Proc. of the 22nd International Conference on Computational Linguistics, pp. 985–992 (2008)
22. Yu, J., Reiter, E., Hunter, J., Mellish, C.: Choosing the Content of Textual Summaries of Large Time-series Data Sets. Natural Language Engineering 13(1), 25–49 (2007)
23. Zhou, L., Ticrea, M., Hovy, E.: Multi-document biography summarization. In: Proc. of the International Conference on Empirical Methods in NLP, pp. 434–441 (2004)

Automatic Generation of Semantic Features and Lexical Relations Using OWL Ontologies

Maha Al-Yahya[1], Hend Al-Khalifa[1], Alia Bahanshal[2], and Iman Al-Oudah[2]

[1] Information Technology Department, King Saud University, Riyadh, Saudi Arabia
{malyahya,hendk}@ksu.edu.sa
[2] Research Institute, King Abdulaziz City for Science and Technology (KACST),
Riyadh, Saudi Arabia
{abahanshal,ialoudah}@kacst.edu.sa

Abstract. Semantic features are theoretical units of meaning-holding components which are used for representing word meaning. These features play a vital role in determining the kind of lexical relation which exists between words in a language. Although such model of meaning representation has numerous applications in various fields, the manual derivation of semantic features is a cumbersome and time consuming task. We aim to elevate this process by developing an automated semantic feature extraction system based on ontological models. Such an approach will provide explicit word meaning representation, and enable the computation of lexical relations such as synonym and antonymy.

This paper describes the design and implementation of a prototype system used for automatically deriving componential formulae, and computing lexical relations between words from a given OWL ontology. The system has been tested on a number of ontologies, both English and Arabic. Results of the evaluation indicate that the system was able to provide necessary componential formulae for highly-axiomed ontologies. With regards to computing lexical relations, the system performs better when predicting antonyms, with an average precision of 40%, and an average recall of 75%. We have also found a strong relation between ontology expressivity and system performance.

Keywords: Semantic Feature Analysis, Ontology, OWL, Antonymy, Synonymy.

1 Introduction

The componential (semantic features) approach to meaning [1] [2] [3] views word meaning as a set of atomic meaning-holding components, called *features*. The methodology by which these features are identified and extracted is called Semantic Feature Analysis (SFA). SFA has been fruitfully applied in various disciplines such as linguistics (contrastive linguistic studies [4], and identifying lexical relations [2]), psycholinguistics (used as a model for understanding and studying conceptual representations of meaning in the human mind [6]), and language learning and teaching (for vocabulary acquisition, and reading comprehension [5], [6] [7]).

Despite its numerous applications, SFA is not without its problems, one being a cumbersome and time consuming task which requires intensive human effort and

R. Muñoz et al. (Eds.): NLDB 2011, LNCS 6716, pp. 15–26, 2011.
© Springer-Verlag Berlin Heidelberg 2011

expertise. In this paper, we approach this problem by utilizing ontologies to automatically perform SFA, and generate componential formulae. Once generated, these formulae can be used to extract various lexical relations such as synonymy and antonymy.

Ontologies are computational models for representing semantics for a specific domain. With advancements in web technologies and the increase in web based applications, we are seeing interest in developing ontologies whether general and upper level such as CYC [8] and SUMO [9], or domain specific such as OWL WordNet [10]. Since ontological models provide an explicit and formal semantic representation, using such ontologies as a resource for meaning derivation will simplify the task of SFA, and thus lexical relation extraction.

Work reported in the literature on extracting semantic features of words and identifying various lexical relations is mostly based on analysis of large sets of textual corpora, and identification of lexico-syntactic patterns for which a specific lexical relation holds. These patterns can either be observed manually, or derived computationally by using lists of words related with a specific lexical relation, and then analyzing text corpora to find patterns. One of the earliest reported methods for lexical relation identification is that reported in [11]. Hearst [11] presents a method for automatic acquisition of the hyponymy relation between words by using manually and automatically generated lexico-syntactic patterns. Similarly, the work described in [12] uses lexico-syntactic patterns generated both manually and automatically to extract synonyms and antonyms from textual corpora. The Espresso system presented in [13], uses lexico-syntactic patterns for extracting semantic relations, with emphasis on generality of patterns usable across different corpora genre. The semantic relations extracted by Espresso are "is-a, part-of, succession, reaction, and production". The method described in [14] uses a score function composed of feature weights to classify pairs of words as antonyms. The features are extracted from contexts of word pairs, and include; Inversely Proportional Distance between pair words, lexico-syntactic patterns, and significance determined by a lexical co-occurrence network built for each word. Our approach is different from that reported in the literature. It is a novel approach which does not depend on lexico-syntactic patterns or large textual corpora, instead it is based on rules and logic inherent in OWL ontologies for extracting word features and determining the lexical relation that exists.

This paper is organized as follows. Section 2 presents background on semantic feature analysis and how it is performed and represented. Section 3 provides a brief background on OWL ontologies. Section 4 presents our prototype system, and the methodology by which semantic features are extracted from OWL ontologies and used to compute lexical relations. Section 5 presents results of testing and evaluating the system using a number of OWL ontologies. Section 6 concludes the paper and provides outlook for future work.

2   Semantic Feature Analysis

Semantic feature analysis is based on the componential analysis theory of meaning representation [1]. Within this theory, a word's meaning is represented in terms of separable components (*features*), which if taken together constitute the *componential* definition for that word. The primary objective of this theory is to represent and explain a word sense, and not encyclopedic knowledge [2], hence the representation is compact, and focuses only on the major semantic primitives. Word features can be represented either using the *feature matrix*, or *the componential formula* (CF) [1].

A *feature matrix*, as often referred in vocabulary acquisition literature, is a matrix in which the top row represents features, and the left row represents the words. Each cell is filled with either a "+" if the column feature exist in the row word, and a "-" if the column feature does not exist in the row word. If the feature does not apply, or if the value is unknown, then the cell is left blank or filled with a "?" mark. Both matrix, and formula models for representing semantic features have been used, however, for language teaching the most widely used is the matrix form.

The *componential formula* is similar to a mathematical formula using the operands "+", and "-" to compose the meaning of a word. The word is positioned in the left-hand side of the formula with an "=" sign, and the features are listed as operands on the right-hand side of the formula. Mathematical operators "+", and "-" are used to organize the right-hand side of the formula. If a feature exists, it is prefixed with a "+" sign, otherwise it is prefixed with a "-" sign.

Using the SFA approach to meaning representation, one can discover the crucial distinguishing meaning between closely related (similar) words. For example (adapted from [15]), the words "car" , "train", "airplane", "boat", and "bike" all belong to the same semantic category of "Transportation". However, there is difference in meaning between them, and using SFA we can distinguish this difference and make it explicit. Table 1 shows the meaning of the words expressed in componential formula form.

Table 1. Componential Formulae for Trasportation Devices

| Car=+Engine+ Steering wheel +Brakes +Windshield |
| Train= +Engine+ Breaks+ Windshield |
| Airplane= +Engine +Steering wheel +Breaks + Windshield |
| Boat= +Steering wheel +Brakes |
| Bike= +Pedals +Breaks |

Moreover, this approach to meaning representation is useful for identifying semantically correct sentences. For example(adapted from [16]), the sentences "The apple ate the lady", and "The lady ate the apple", are both syntactically correct, however semantically, only the second one is correct. The reason is that the nouns "apple" and "lady" contain some meaning component which makes one acceptable to use as a subject for the verb "ate", while the other not. The meaning component which makes the noun suitable for use as a subject for the verb "ate", is that "lady" is an animate (living thing), while "apple" is not.

Semantic feature analysis can be used in inferring various lexical relations among the words in a vocabulary [2]. The relation between two words depend on the number of features shared between two given words. Using SFA, lexical relations such as synonymy and antonymy can be computed. Synonymy between words can be identified by examining the componential formulae, features for each word, if they are identical, then the two words can be considered synonymous. To check for antonymy between two words, we analyse the formula for each word, if they are all equal, except for one feature or more features, which has opposing signs, then we can say that these two words are antonyms along the dimension for which the features have opposing signs.

The process of SFA is usually carried out by a linguistic analyst. There are a number of methods to reach at the componential formula. One such model involves analyzing word meaning using a number of dictionaries. The analyst extracts only the basic/primary definition of a word (denotation) and does not consider other definitions (connotations). From all referenced resources, the analyst then finds the common definition for the target word among these resources. In the final stage, the analyst dissects this meaning into its minimal components using his/her general knowledge and intuition. This usually involves analyzing the definition, and putting the most general concepts in the beginning of the formula, next the less general, and leaving the most specific concept, which is the distinguishing feature at the tail of the formula.

Another approach usually carried out by language teachers is the matrix or grid approach. This method involves first, selecting a specific category or field or semantic domain to analyze. Next, a set of words related to the major category are chosen. These words are listed along the first column of a matrix. Based on the analyst experiences, as well as with the aid of dictionaries, the analyst tries to identify major features or attributes that might characterize these elements. The set of chosen characteristics are listed across the top row of the matrix. Finally, the cells in the matrix are completed with appropriate markers for each word.

3   OWL Ontologies

OWL (Web Ontology Language) is a Semantic Web modeling language designed to provide a set of vocabulary (metadata) to define entities in a certain domain, how they relate to other entities, and under what rules or restrictions. It is a highly expressive language which enables the creation of complex concept definitions based on simpler concepts. An OWL ontology provides precise descriptions of entities in a domain, called *terminology*. Besides *terminology*, an OWL ontology may include specification of concrete objects within a certain domain, these are called *individuals*. *Terminology* describe basic concepts and notions in a domain, while *individuals* represents a specific notion or concept. For example the concept of *Mother*, can be defined by stating that a *Mother* is a *Woman*, who is a *Parent*. *Mary* is an individual who is a member of the class *Mother*. Some OWL ontologies may contain terminology only, these ontologies are usually called *schema* ontologies. Others may only contain individuals and are called *individuals* or *instances* ontology. Yet others may contain both terminology and individuals. Our model is based on the first type, terminology or schema ontologies, and therefore heavily relies on class descriptions and class axioms. The recent OWL 2 recommendation provides powerful constructs for defining concepts (types of things which may exists in a domain of interest). The Relevant OWL constructs include *classes, class hierarchies, equivalent classes, class disjoints,* and *complex class* definitions.

Within our model for representing word meaning, *classes* represent a *word*, and deriving the class definition is a method for computing the componential formula of the target word. *Class hierarchies* allow the means to describe the subclass relation (concept generalization) between two given concepts. For example, if someone is a *Woman*, they must be a *Human*. The *equivalent class* construct provide means to indicate that two classes are equivalent or the same. For example, the words *Person* and *Human* can be used interchangeably, therefore we can say that they are equivalent classes. *Class disjointness* refers to the incompatibility between classes. For example, *Adult* and *Child* are incompatible because an individual cannot be an adult and a child at the same time, so membership to one means exclusion from the other. *Complex Classes* enables us to combine class descriptions and provide more expressive power for concepts in the domain. OWL provides logical class constructs using logical operators (*and, or, not*) in order to combine primitive (atomic) class definitions with class expressions. Logical *and* is defined using *ObjectIntersctionOf* construct, logical *or* is defined using *ObjectUnionOf*, and logical *not* is defined using *ObjectComplementOf*. Property *restrictions* use property constructs to constrain class definitions. There are three types of property restrictions, *existential quantification, universal quantification,* and *value restrictions*.

4   From OWL Axioms to Semantic Features

In this section we describe the general architecture of the system, and how the OWL constructs described in section 3 are used to derive semantic features for words. The general architecture of the system is composed of five main processes. Fig. 1 shows the architecture of the system. The system can either provide a definition for a single word, or compute the lexical relation which exists between two given words. Once a word is entered, the first phase of the system involves searching for an OWL ontology, either local or remote, which contains the description of the given word. A basic requirement for the system to yield meaningful results, is that the ontology should be a class-based ontology (as opposed to an instance-based), and the ontology should contain detailed axiom definitions for vocabulary. If the word is located within more than one context (has more than one sense), for example such as *chair* (single seat furniture), and *chair* (the person in authority), process (2) aims to resolve the context and identify the required sense. In the current algorithm, this is a supervised process with input from the user. Once the sense is disambiguated, and the system selects the relevant definitions, it starts extracting semantic features and building the componential formula for the given word(s). The processes implemented so far in our prototype system are processes (1), (3), (4), and (5). During process (3) the system computes the componential formulae by following predefined mapping rules. These rules convert an OWL axiom into the corresponding operand(s) in a semantic formula. Table 2 displays each OWL class axiom [17] using RDF functional syntax, and its corresponding SFA mapping.

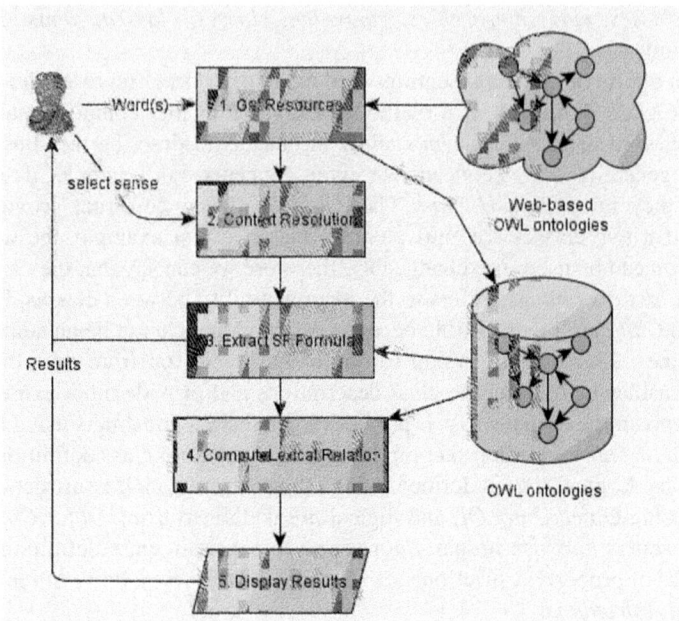

Fig. 1. System Architecture

Table 2. Mapping between OWL axioms and SFA formula

Class axiom	SFA formulation
EquivalentClasses(:Person :Human)	Person=+Human Human=+Person
DisjointClasses(:Woman :Man)	Woman= -Man Man= -Woman
SubClassOf(:Mother :Woman)	Mother= +Woman

EquivalentClasses(
  :Mother
  ObjectIntersectionOf(:Woman :Parent))
	Mother=+Woman + Parent

EquivalentClasses(
  :Parent
  ObjectUnionOf(:Mother :Father))
	Parent= +Mother
	Parent= +Father

EquivalentClasses(
  :ChildlessPerson
  ObjectIntersectionOf(
   :Person
    ObjectComplementOf(:Parent)))
	Childless Person= +Person –Parent

SubClassOf(
  :Grandfather
  ObjectIntersectionOf(:Man :Parent))
	Grandfather= +Man +Parent

EquivalentClasses(:Parent ObjectSomeValuesFrom(:hasChild :Person)	Parent= +has a Child which is a Person

4.1 Formula Extraction

For deriving the formula the system uses the asserted model of subclass hierarchies to generate the componential formulae. OWL language related axioms such as OWL:Resource, and OWL:Thing are excluded from the componential formula. All class axioms are considered semantic features, except for the word's (target class). Object and data type property axioms associated with the target are also considered semantic features.

4.2 Computing Lexical Relations

Depending on the user requirements, if only semantic features are required, the system displays the resulting formula. However, if the user would like to know the kind of lexical relation which exist between two words, it proceeds to process (4). Using the formulae composed in the previous process, the system computes the probability of a lexical relation existing. Since the system deals majorly with antonymy, and synonymy, the rules for computing these relations are described in the following algorithms, assuming w1 and w2 are two word formulae:

Algorithm for identifying antonyms

Compute $S = \{w1 \cap w2\}$
Remove elements of S from w1, and w2.

$$\acute{w}1 = \{w1 - S\}$$
$$\acute{w}2 = \{w2 - S\}$$

For remaining elements in $\acute{w}1$ and $\acute{w}2$

> *if there exists at least one operand which is common in both formulas, but with opposing signs, then w1, and w2 can be said to be antonyms (+Adult, -Adult).*
> *OR,*
> *if there exists at least one operand which is disjoint with one from the other formula, then w1, and w2 can be said to be antonyms (Male, Female), (Open, Close).*

Algorithm for identifying synonyms
Compute the similarity between word w1 and word w2, by dividing the cardinality of the intersection set by the cardinality of the union set:

$$sim_{(w1,w2)} = \frac{|w1 \cap w2|}{|w1 \cup w2|}$$

if (sim=1), then the w1 and w2 are synonyms,
else
> *if (sim>0), then w1 and w2 are synonymous with confidence value,*
> $$conf = sim \times 100$$
> *else*
>> *there is no synonymy relation between w1 and w2.*

5 Testing and Evaluation

The system developed was tested on three ontologies, the "Gender" ontology, part of the "SemQ' ontology [18], the "Generations" ontology [19], and the "NTNames" ontology [20]. The "Gender" ontology is an ontology developed for testing the prototype, it is based on gender definitions presented in [2]. The "SemQ" ontology is an ontology for representing Arabic vocabulary (Time nouns). The "Generations" ontology is an ontology developed to describe family relations. The "NTNames" ontology is an ontology which contains classes and properties for the new testament names created as part of the Semantic Bible project. These ontologies contain sufficient vocabulary and definitions to test the system in different langauges. Other ontologies are either instance ontologies, or do not contain detailed vocabulary descriptions, and therefore do not meet the system requirements. Ontology statistics are shown in Table 3. We define the *expressivity* metric of the ontology as the axiom to class ratio.

Table 3. Ontology Statistics

Ontology	No. of Classes	No. of subclass axioms	No. of equivalent class axioms	No. of disjoint class axioms	Expressivity Axiom/Class ratio
Gender	16	19	0	2	1.31
SemQ	53	62	0	8	1.32
Generations	18	0	17	0	1.13
NTNames	49	51	0	5	1.10

We tested the system on all ontologies. A sample set of the resulting formulae for the test ontologies is shown in Table 4. Statistics for the Componential formulae generated are shown in Table 5.

For computing lexical relations, each word was tested against all the other words in the same ontology. The system identified the type of relation which existed between

Table 4. Sample Output from Tested Ontologies

SemQ (*in Arabic*)
القائلة =الظهيرة + فيها نوم+ وسط النهار + النهار + زمن + محدد + فيها نور
العصر = + محدد +فيها نور +النهار + زمن +ميزها من_زوال_الشمس_الى_المغرب +آخر_النهار

Gender
Boy= +Male -Adult -Female +Human
Girl= +Female -Adult -Male +Human

Generations
Brother= +Person +hasSibling Person +hasSex has value MaleSex
Sister= +Person +hasSibling Person +hasSex has value FemaleSex

NTNames
SaltWaterArea= +WaterArea
FreshWaterArea= +WaterArea

Table 5. Componential Formulae Statistics

Ontology	No. of Componential Formulae generated	Componential formulae Average length
Gender	16	1
SemQ	53	4
Generations	18	2
NTNames	49	1

each pair, either antonymy or synonymy. We then evaluated the correctness of the identified synonyms and antonyms by comparing them to a to a gold-standard chosen to be the WordNet dictionary (for English). For the Arabic ontology, we relied on human evaluation of results. Three measures were computed; precision (P), recall (R), and the harmonic mean (F-measure) using the following formulae:

$T_{rp} = Total\ number\ of\ relations\ predicted$

$T_c = Total\ number\ of\ correctly\ predicated\ relations$

$T_r = Total\ number\ of\ relations\ in\ the\ set$

$P = \frac{T_c}{T_{rp}}, \ R = \frac{T_c}{T_r}, \ and\ F = 2 \cdot \frac{(P \cdot R)}{P+R}$

Table 6 shows results of the analysis for antonyms and Table 7 shows results for synonyms.

Table 6. Antonym Evaluation Results

Ontology	Antonyms Predicated	Antonyms in Test Set	Precision	Recall	F-measure $\beta = 1$
Gender	13	6	46%	100%	0.31
SemQ	83	76	91%	100%	0.95
Generations	0	5	0%	0%	0.0
NTNames	5	1	20%	100%	1

Table 7. Synonym Evaluation Results

Ontology	Synonyms Predicted	Synonyms in Test Set	Precision	Recall	F-measure $\beta = 1$
Gender	2	1	0%	0%	--
SemQ	127	1	0.78%	100%	0.013
Generations	18	1	0%	0%	--
NTNames	2	1	0%	0%	--

6 Results and Discussion

Results from testing the prototype system on a number of ontolgoies indicate that it was able to represent features for words in the ontologies using componential formulae. A total of 136 formulae were generated from the tested ontologies. Human evaluation of the formulae indicate that they were accurate and provide necessary meaning components for the target words. However, the problem of insufficient definitions (fewer arguments in the componential formulae) was observed. This suggests that there are not enough axioms associated with a certain word (class) in the given ontology. The problem of insufficient definitions can be resolved by using inference mechanisms in OWL ontologies. These mechanisms will provide longer definitions, as it will include not only asserted super-classes, but also inferred super-classes from the ontology.

With regards to computing lexical relations, a first glance at the tables shows that the system performed better at identifying antonyms than synonyms. A possible explanation is that the actual number of synonyms in the test set is low compared to the actual number of antonyms.

Further analysis of the phenomenon shows that precision and recall were extremely low despite the large number of synonyms predicted. A possible explanation is that some words correctly identified by the system as synonyms were not retrieved. The reason is that the gold standard chosen (WordNet) did not consider them to be synonyms. For example the pair (Woman, Lady) in the "Gender" ontology were not considered synonyms by WordNet, but other dictionaries such as the Roget's Thesaurus [21] consider them to be synonyms. Another issue with synonymy prediction is that some words considered as synonyms in WordNet, such as (Man, Human), are not returned by the system because the ontological definition of (Man) is different from that of (Human), (Man) comprises extra features which distinguish it from other words belonging to the (Human) category. Moreover, the large number of incorrectly predicated synonyms can be attributed to two other reasons, the first is the problem of short (insufficient) definitions described earlier, and the second is that some concepts in the ontology are not defined. For example the words "Concept", and "Lexical Relation" are both defined as being a "thing" in the ontology, in other words they are treated equal because there formulae are incomplete, and formulae equality is interpreted by the system as synonymy.

For computing antonymy relations, the system was able to predict antonyms for most ontologies, however it fails to do so for the "Generations" ontology, even though it contained five pairs of antonyms. Revisiting ontology statistics we can see a possible interpretation of the results. Since our system is heavily dependent on subclass, and disjoint axioms, which the "Generations" ontology lack, the system was not able to produce useful results for such an ontology. Moreover, some antonyms were not identified due to an important condition which should be included in the algorithm for identifying antonyms. We need to add a second condition which should check that each disjoint operand should be equivalent to the complement of the other (*Male, Female*). Class disjointness is not sufficient to deduce that two terms are antonymous, as if there are more than two siblings between disjoint classes antonymy might not hold. Therefore another necessary condition for antonymy to hold between two words is that a feature should not only be disjoint from the other, but also its complement as well.

At the level of ontologies, the best performance was shown using the SemQ ontology, and then the Gender ontology. Looking at ontology metrics, the expressivity value of these two ontologies is higher compared to the others. Although not conclusive, due to the small sample set, but this provides an indication that there is a strong relation between ontology expressiveness and system performance.

Finally, in order to improve the performance of the system, we can classify issues to be addressed as intrinsic issues and extrinsic issues. Intrinsic issues are those associated with the algorithms, and extrinsic issues are those associated with the type and structure of the input ontology. System algorithms need to be revised and enhanced to produce sufficient definitions for words. Moreover, when testing new versions of the system, we should include other external dictionaries and thesauri as gold standards in comparing results obtained by the system. With regards to extrinsic issues, it is important to consider the type of ontologies chosen. They should be heavily axiomed and contain deep sub-classing in order to produce meaningful formulae that posses enough primitives for clarifying meaning or making decisions with regards to lexical relations.

7 Conclusion and Future Work

In this paper we have described a system for automatic generation of semantic features for language vocabulary from OWL ontologies. This way of word meaning representation enables the computing of important lexical relations such as synonymy and antonymy. The prototype was tested and evaluated on a number of ontologies. Results of the evaluation indicate that the system was able to provide necessary componential formulae for highly axiomed ontologies. It has also been concluded that the system predicts antonyms with higher accuracy than synonyms. The average precision for predicting antonyms is 40%, and the average recall is 75%. We have also found a strong relation between ontology expressivity and system performance.

The evaluation highlighted a number of issues which need to be addressed in future versions of the system. In addition, future work include extending the type of lexical relations computed to include hyponymy, polysemy, and incompatibility. Although the system has its limitations, one of its major strengths is its multilingualism, it can identify antonyms and synonyms of any language as demonstrated with our tests.

Acknowledgements. This project is sponsored by KACST (King Abdulaziz City for Science and Technology). We would also like to thank Dr. Nawal Al-Helwa for providing insights into the process of Semantic Feature Analysis.

References

1. Nida, E.A.: Componential Analysis of Meaning: An Introduction to Semantic Structures. Mouton (1975)
2. Leech, G.N.: Semantics: The Study of Meaning. Penguin, UK (1974)
3. Loebner, S.: Understanding Semantics. Hodder Education (2002)
4. Wikberg, K.: Methods in Contrastive Lexicology. Applied Linguistics 4, 213–221 (1983)
5. Anders, P.L., Bos, C.S.: Semantic Feature Analysis; An Interactive Strategy for Vocabulary Development and Text Comprehension. Journal of Reading 29, 610–616 (1986)
6. Stieglitz, E.L.: A practical approach to vocabulary reinforcement. ELT Journal 37, 71–75 (1983)
7. Channell, J.: Applying Semantic Theory to Vocabulary Teaching. ELT Journal XXXV, 115–122 (1981)

8. Lenat, D.B.: CYC: a large-scale investment in knowledge infrastructure. Communications of the ACM 38, 33–38 (1995)
9. Pease, A., Niles, I., Li, J.: The Suggested Upper Merged Ontology: A Large Ontology for the Semantic Web and its Applications. In: Working Notes of The AAAI-2002 Workshop on Ontologies and The Semantic Web 2002 (2002)
10. Huang, X., Zhou, C.: An OWL-based WordNet lexical ontology. Journal of Zhejiang University - Science A 8, 864–870 (2007)
11. Hearst, M.A.: Automatic acquisition of hyponyms from large text corpora. In: Proceedings of the 14th Conference on Computational Linguistic, vol. 2, pp. 539–545. Association for Computational Linguistics, Stroudsburg (1992)
12. Lobanova, A., Spenader, J., van de Cruys, T., van der Kleij, T., Sang, E.T.K.: Automatic Relation Extraction - Can Synonym Extraction Benefit from Antonym Knowledge? In: NODALIDA 2009 Workshop WordNets and other Lexical Semantic Resources, Odense, Denmark (2009)
13. Pantel, P., Pennacchiotti, M.: Espresso: leveraging generic patterns for automatically harvesting semantic relations. In: Proceedings of the 21st International Conference on Computational Linguistics and the 44th Annual Meeting of the Association for Computational Linguistics, pp. 113–120. Association for Computational Linguistics, Stroudsburg (2006)
14. Lucerto, C., Pinto, D., Jiménez-Salazar, H.: An Automatic Method to Identify Antonymy Relations. In: Proceedings of Workshops on Artificial Intelligence, Puebla, Mexico, pp. 105–111 (2004)
15. Fisher, D., Frey, N.: Word Wise and Content Rich, Grades 7-12: Five Essential Steps to Teaching Academic Vocabulary. Heinemann (2008)
16. Yule, G.: The Study of Language. Cambridge University Press, Cambridge (1996)
17. OWL 2 Web Ontology Language Primer,
    http://www.w3.org/TR/owl2-primer/
18. Al-Yahya, M., Al_Khalifa, H., Bahanshal, A., Al-Odah, I., Al-Helwah, N.: An Ontological Model for Representing Semantic Lexicons: An Application on Time Nouns in the Holy Quran. The Arabian Journal for Science and Engineering (AJSE) 35, 21–35 (2010)
19. CO-ODE > Ontologies, http://www.co-ode.org/ontologies/
20. NTNames ontology,
    http://www.semanticbible.com/2006/11/NTNames.owl
21. Laird, C.: Webster's New World Roget's A-Z Thesaurus. Webster's New World (1999)

EmotiNet: A Knowledge Base for Emotion Detection in Text Built on the Appraisal Theories

Alexandra Balahur, Jesús M. Hermida, Andrés Montoyo, and Rafael Muñoz

Department of Software and Computing Systems, University of Alicante,
Apto. de correos 99, E-03080 Alicante, Spain
{abalahur,jhermida,montoyo,rafael}@dlsi.ua.es

Abstract. The automatic detection of emotions is a difficult task in Artificial Intelligence. In the field of Natural Language Processing, the challenge of automatically detecting emotion from text has been tackled from many perspectives. Nonetheless, the majority of the approaches contemplated only the word level. Due to the fact that emotion is most of the times not expressed through specific words, but by evoking situations that have a commonsense affective meaning, the performance of existing systems is low. This article presents the EmotiNet knowledge base – a resource for the detection of emotion from text based on commonsense knowledge on concepts, their interaction and their affective consequence. The core of the resource is built from a set of self-reported affective situations and extended with external sources of commonsense knowledge on emotion-triggering concepts. The results of the preliminary evaluations show that the approach is appropriate for capturing and storing the structure and the semantics of real situations and predict the emotional responses triggered by actions presented in text.

Keywords: EmotiNet, emotion detection, emotion ontology, knowledge base, appraisal theories, self-reported affect, action chain.

1 Introduction

The study of human affect-related phenomena has always been a challenge. Different scientific theories of emotion have been developed along the last century of research in philosophy, psychology, cognitive sciences or neuroscience. In Natural Language Processing (NLP), although different approaches to tackle the issue of emotion detection in text have been proposed, the complexity of the emotional phenomena led to a low performance of the systems implementing this task [1]. The main issue related to the present approaches is that they only contemplate the word level, while expressions of emotion are most of the times not present in text in specific words (e.g. "I am angry.") [2]. Most of the times, the affect expressed in text results from the interpretation of the situation presented therein [3,4]. Psychological theories of emotion give various explanations as to why certain episodes lead to a specific affective state [5]. Among them, the so-called "appraisal theories" [6] state that an emotion can only be experienced by a person if it is elicited by an appraisal of an object that directly affects them and that the result is based on the person's experience, goals and opportunities for action.

R. Muñoz et al. (Eds.): NLDB 2011, LNCS 6716, pp. 27–39, 2011.
© Springer-Verlag Berlin Heidelberg 2011

In the light of the appraisal theories, the aim of this research is to:

1) Propose a method for modelling affective reaction to real-life situations described in text, based on the psychological model of the appraisal theory.
2) Design and populate a knowledge base of action chains called EmotiNet, based on the proposed model. We subsequently extend the resource to include appraisal criteria, either by automatic extraction, extension with knowledge from other sources, such as ConceptNet [7] or VerbOcean [8].
3) Propose, validate and evaluate a method to detect emotion in text based on EmotiNet.

Results of the evaluations show that our approach to detecting emotion from texts based on EmotiNet outperforms existing methods, demonstrating its validity and the usefulness of the created resource for the emotion detection task.

2 State of the Art

In Artificial Intelligence (AI), the term *affective computing* (AC) was first introduced by Rosalind Picard [9]. Previous approaches to spot affect in text include the use of models simulating human reactions according to their needs and desires [10], fuzzy logic [11], lexical affinity based on similarity of contexts – the basis for the construction of WordNet Affect [12] or SentiWordNet [13], detection of affective keywords [14] and machine learning using term frequency [15], or term discrimination [16]. Other proposed methods include the creation of syntactic patterns and rules for cause-effect modelling [17]. Significantly different proposals for emotion detection in text are given in the work by [18] and the recently proposed framework of *sentic* computing [19], whose scope is to model affective reaction based on commonsense knowledge. For a survey on the affect models and their AC applications, see [5].

The set of models in psychology known as the appraisal theories claim that emotions are elicited and differentiated on the basis of the subjective evaluation of the personal significance of a situation, object or event [20, 21, 22]. These theories consider different elements in the appraisal process (see [6]), which are called appraisal criteria (e.g. familiarity, expectation). Scherer [24] later used the values of such criteria in self-reported affect-eliciting situations to construct the vectorial model in the expert system GENESIS. The appraisal models have also been studied and employed in systemic functional linguistics [25].

As far as knowledge bases are concerned, many NLP applications have been developed using manually created knowledge repositories such as WordNet [26], CYC (http://cyc.com/cyc/opencyc/overview), ConceptNet or SUMO (http://www.ontologyportal.org/index.html). Some authors tried to learn ontologies and relations automatically, using sources that evolve in time - e.g. Yago [27], which employs Wikipedia to extract concepts. Other approaches to knowledge base population were by Pantel and Ravichandran [28], and for relation learning [29]. DIPRE [30] and Snowball [31] create hand-crafted patterns to extract ontology concepts. Finally, Grassi [32] proposes a model of representing emotions using ontologies.

3 Motivation and Contribution

To illustrate the need to build a more robust model for emotion detection, we will start with series of examples.

Given a sentence such as (1) "I am happy", an automatic system should label it with "joy". Given this sentence, a system working at a lexical level would be able to detect the word "happy" (for example using WordNet Affect) and would correctly identify the emotion expressed as "joy". But already a slightly more complicated example – (2) "I am not happy" – would require the definition of "inverse" emotions and the approach would no longer be straightforward. In the second example, although emotion words are present in the text, additional rules have to be added in order to account for the negation. Let us consider another example: (3) "I'm going to a party", which should be labelled with "joy" as well. A system working at a lexical level would find in the text no word that is directly related to this emotion. A method to overcome this issue is proposed by Liu et al. [18] and Cambria et al. [19]. The main idea behind these approaches is to acquire knowledge on the emotional effect of different concepts. In this manner, the system would know that "going to a party" is something that produces "joy". These approaches solve the problem of indirectly mentioning an emotion by using the concepts that are related to it instead. However, they only spot the emotion contained in separated concepts and do not integrate their interaction or the context in which they appear. If the example we considered is extended as in (4) "I'm going to a party, although I should study for my exam", the emotion expressed is no longer "joy", but most probably "guilt". As it can be noticed, even if there are concepts that according to our general knowledge express a certain emotion, their presence in the text cannot be considered as a mark that the respective sentence directly contains that emotion. Finally, the same situations are associated with distinct emotion labels depending on properties of the actor, action or object (e.g. "The man killed the mosquito." versus "The man killed his wife"; or "The kitten climbed into my lap" versus "The pig climbed into my lap."; or "The dog started barking as I approached" versus "The dog started wagging his tail as I approached".) The properties of the actors, actions and objects, as we can see, are very important at the time of determining the emotional label of a situation. These properties actually translate into the different values of the appraisal criteria. Therefore, a resource aimed for emotion detection must include this information. The quantity of commonsense knowledge required is tremendous. However, most of it is already present in existing commonsense knowledge bases (e.g. CYC, SUMO or ConceptNet). Given a sufficiently flexible model of representation for action chains, the underlying ontology can be enriched with such knowledge from external sources, resulting in a deep semantic representation of the situations, from which emotion can be detected in a more precise way. In the light of these considerations, our contribution relies in proposing and implementing a framework for modelling affect based on the appraisal theories, which can support:

a) The automatic processing of texts to extract:

- The components of the situation presented (which we denote by "action chains") and their relation (temporal, causal etc.)
- The elements on which the appraisal is done in each action of the chain (agent, action, object);

- The appraisal criteria that can automatically be determined from the text (modifiers of the action, actor, object in each action chain);

  b) The inference on the value of the appraisal criteria, extracted from external knowledge sources (characteristics of the actor, action, object or their modifiers that are inferable from text based on common-sense knowledge);

  c) The manual input of appraisal criteria of a specific situation.

4   Building a Knowledge Base of Action Chains: EmotiNet

The general idea behind our approach is to model situations as chains of actions and their corresponding emotional effect using an ontological representation. According to the definition provided by Studer et al. [33], an ontology captures knowledge shared by a community that can be easily sharable with other communities. These two characteristics are especially relevant if we want the recall of our approach to be increased. Knowledge managed in our approach has to be shared by a large community and it also needs to be fed by heterogeneous sources of common knowledge to avoid uncertainties. However, specific assertions can be introduced to account for the specificities of individuals or contexts. In this manner, we can model the interaction of different events in the context in which they take place and add inference mechanisms to extract knowledge that is not explicitly present in the text. We can also include knowledge on the appraisal criteria relating to different concepts found in other ontologies and knowledge bases (to account for the different properties of the actor, action and object). At the same time, we can define the properties of emotions and how they combine. Such an approach can account for the differences in interpretation, as the specific knowledge on the individual beliefs or preferences can be easily added as action chains or properties of concepts.

Given these requirements, our approach defines a new knowledge base, called EmotiNet, to store action chains and their corresponding emotional labels from several situations in such a way that we will be able to extract general patterns of appraisal. From a more practical viewpoint, our approach defines an action chain as a sequence of action links, or simply actions that trigger an emotion on an actor. Each specific action link can be described with a tuple (actor, action type, patient, emotional reaction). Specifically, the EmotiNet KB was built by means of an iterative process that extracts the action chains from a document and adds them to the KB. This process is divided in a series of steps, explained in the subsequent sections.

4.1   ISEAR – Self-reported Affect

In the International Survey of Emotional Antecedents and Reactions (ISEAR) – [35], http://www.unige.ch/fapse/emotion/databanks/isear.html, the respondents were asked to report situations in which they had experienced all of 7 major emotions (*joy*, *fear*, *anger*, *sadness*, *disgust*, *shame*, and *guilt*). In each case, the questions covered the way they had appraised the situation and how they reacted. An example of entry in the ISEAR databank is: "I felt anger when I had been obviously unjustly treated and had no possibility to prove they were wrong." Each example is attached to one single emotion. In order to have a homogenous starting base, we selected from the 7667 examples in the ISEAR database only the 1081 cases that contained descriptions of situations involving kinship relationships. Subsequently, the examples were POS-tagged using TreeTagger. Within each emotion class, we then computed the similarity of the examples with one another, using the implementation of the Lesk distance in Ted Pedersen's Similarity Package. This score was used to split the examples in each emotion class into six clusters using the Simple K-Means implementation in Weka. The idea behind this approach, confirmed by the output of the clusters, was to group examples that are similar, in vocabulary and structure.

4.2 Modelling Situations with Semantic Roles

The next step was to extract, from each of the examples, the actions described. For this, we employed the semantic role labelling (SRL) system introduced by Moreda et al. [34]. In order to build the core of knowledge in the EmotiNet KB, we chose a subset of 175 examples (25 per emotion), which we denote by T. The criteria for choosing this subset were the simplicity of the sentences and the variety of actions described. In the case of these examples, we manually extracted the agent, the verb and the patient from the output of the SRL system (the remaining examples are used for testing). For example, if we use the situation "I borrowed my brother's car and I crashed it. My brother was very angry with me", we can extract three triples (or action links) with the main actors and objects of the sentences: (I, borrow, brother's car), (I, crash, brother's car) and (brother, feel, angry).

Further on, we resolved the anaphoric expressions automatically, using a heuristic selection of the family member mentioned in the text that is closest to the anaphoric reference and whose properties (gender, number) are compatible with the ones of the reference. The replacement of the references to the speaker, e.g. 'I', 'me', 'myself', is resolved by taking into consideration the entities mentioned in the sentence. Following the last example, the subject of the action is assigned to the daughter of the family and the triples are updated: (daughter, borrow, brother's car), (daughter, crash, brother's car) and (brother, feel, angry). Finally, the action links are grouped and sorted in action chains. This process of sorting is determined by the adverbial expressions that appear within the sentence, which actually specify the position of each action on a temporal line (e.g. "although", "because", "when"). We defined pattern rules according to which the actions introduced by these modifiers happen prior to or after the current context.

4.3 Models of Emotion

In order to describe the emotions and the way they relate and compose, we employ Robert Plutchik's wheel of emotion [36] and Parrot's tree-structured list of emotions [37]. These models are the ones that best overlap with the emotions comprised in the ISEAR databank. Moreover, they contain an explicit modelling of the relations between the different emotions. Plutchik's wheel of emotions contains 8 basic emotions and a set of advanced, composed emotions. The model described by Parrot comprises primary, secondary and tertiary emotions. Our approach combines both models by adding the primary emotions missing in the first model and adding the secondary and tertiary emotions as combinations of the basic ones. Using this combined model as a reference, we manually assigned one of the seven most basic emotions, i.e. *anger, fear, disgust, shame, sadness, joy or guilt*, or the *neutral* value to all the action links in the B_T set, thus generating 4-tuples *(subject, action, object, emotion)*, e.g. (daughter, borrow, brother's car, joy), (daughter, crash, brother's car, fear) and (brother, feel, angry, anger). This annotation was done in parallel by two annotators, obtaining a kappa value of 0.83. The cases where the annotators disagreed were discussed and a common decision was taken.

4.4 Designing the EmotiNet Ontology

The process of building the core of the EmotiNet knowledge base (KB) of action chains started with the design of the core of knowledge, in our case an ontology, whose design process was divided in three stages:

1) *Establishing the scope and purpose of the ontology.* The ontology we propose has to be capable of defining the concepts required in a general manner, which will allow it to be expanded and specialised by external knowledge sources. Specifically, the EmotiNet ontology needs to capture and manage knowledge from three domains: kinship membership, emotions (and their relations) and actions (characteristics and relations between them).

2) *Reusing knowledge from existing ontologies.* In a second stage, we searched for other ontologies on the Web that contained concepts related to the knowledge cores we needed. At the end of the process, we located two ontologies that would be the basis of our ontological representation: the ReiAction ontology (www.cs.umbc.edu/~lkagal1/rei/ontologies/ReiAction.owl), which represents actions between entities in a general manner, and the family relations ontology (www.dlsi.ua.es/~jesusmhc/emotinet/family.owl), which contains knowledge about family members and the relations between them.

3) *Building our own knowledge core from the ontologies imported.* This third stage involved the design of the last remaining core, i.e. emotion, and the combination of the different knowledge sources into a single ontology: EmotiNet. In this case, we designed a new knowledge core from scratch based on a combination of the models of emotion presented in Section 4.3 (see Fig.1). This knowledge core includes different types of relations between emotions and a collection of specific instances of emotion (e.g. anger, joy). In the last step, these three cores were combined using new classes and relations between the existing members of these ontologies (Fig. 1).

4.5 Extending and Populating EmotiNet with Real Examples

After designing the ontology core, we extended EmotiNet with new types of action and action chains (as instances of the ontology) using real examples from the ISEAR corpus. For this, we employed the T set. Once we carried out the processes described in sections 4.2 and 4.3 on the chosen documents, we obtained 175 action chains (ordered lists of tuples). In order to be included in the EmotiNet knowledge base, all their elements were mapped to existing concepts or instances within the KB. When these did not exist, they were added to it.

We would like to highlight that in EmotiNet, each "4-tuple" (actor, action, object, emotion) extracted from the process of action extraction and emotion assignation has its own representation as an instance of the subclasses of Action. Each instance of Action is directly related to an instance of the class Feel, which represents the emotion felt in this action. Subsequently, these instances (action links) were grouped in sequences of actions (class Sequence) ended by an instance of the class Feel, which, as mentioned before, determine the final emotion felt by the main actor(s) of the chain. In our example, we created two new classes Borrow and Crash (subclasses of DomainAction) and three new instances of them: instance "act1" ("Borrow", "daughter", "brother's car", "Joy"); instance "act2" ("Crash", "daughter", "brother's car", "Fear"). The last action link already existed within EmotiNet from another chain so we reused it: instance "act3" (Feel, "brother", "Anger").

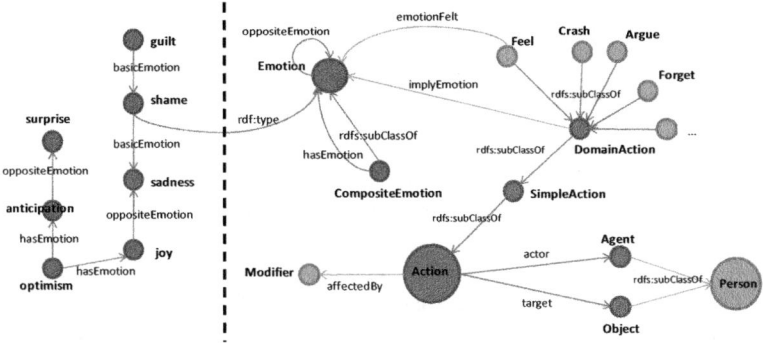

Fig. 1. Main concepts of EmotiNet and instances of the ontology of emotions

The next step consisted in grouping these instances into sequences using instances of the class Sequence, which is a subclass of Action that can establish the temporal order between two actions (which one occurred first). Fig. 2 shows an example of a RDF graph, previously simplified, with the action chain of our example. Following this strategy, we finally obtained a tight net of ontology instances that express different emotions and how actions triggered them. We used Jena (http://jena.sourceforge.net/) and MySQL for the management and storage of EmotiNet.

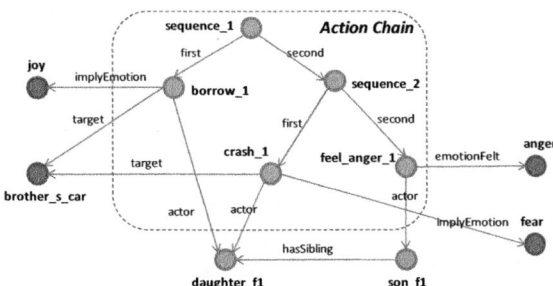

Fig. 2. RDF graph of an action chain

4.6 Expanding EmotiNet with Existing NLP Resources

In order to extend the coverage of the resource, we expanded the ontology with the actions and relations from VerbOcean [8]. In particular, 299 new actions were automatically included as subclasses of DomainAction, which were directly related to any of the actions of our ontology through three new relations: *can-result-in*, *happens-before* and similar. This process of expansion is essential for EmotiNet, since it adds new types of action and relations between actions, which might not have analysed before, thus reducing the degree of dependency between the resource and the initial set of examples. The more external sources of general knowledge added, the more flexible EmotiNet is, thus increasing the possibilities of processing unseen action chains.

5   Experiments and Evaluation

5.1   Experimental Setup

The evaluation of our approach consists in testing if by employing the model we built and the knowledge contained in the core of EmotiNet (which we denote by "training sets"), we are able to detect the emotion expressed in new examples pertaining to the categories in ISEAR.

The first training set (marked with A_T) contains a partial version of the EmotiNet core, with only 4 emotions (anger, disgust, guilt and fear) and 25 action chains per emotion. The second training set (marked with B_T) comprises the final version of the EmotiNet core knowledge base – containing all 7 emotions in ISEAR, and 25 action chains for each. These action chains are the ones obtained in Section 4.5.

The first test set (A) contains 487 examples from ISEAR (phrases corresponding to the anger, disgust, guilt and fear emotions, from which the examples used as core of EmotiNet were removed). The second test set (B) contains 895 examples (ISEAR phrases corresponding to the seven emotions modelled, from which core examples were removed).

In order to assess the system performance on the two test sets, we followed the same process we used for building the core of EmotiNet, with the exception that the manual modelling of examples into tuples was replaced with the automatic extraction of (actor, verb, patient) triples from the output given by the SRL system proposed by Moreda et al. [34]. Subsequently, we eliminated the stopwords in the phrases contained in these three roles and performed a simple coreference resolution (as presented in Section 4.2). Next, we ordered the actions presented in the phrase, using the adverbs that connect the sentences, through the use of patterns (temporal, causal etc.). The resulted action chains for each of the examples in the two test sets will be used in carrying different experiments:

(1). In the first approach, for each of the situations in the test sets (represented as action chains), we search the EmotiNet KB to encounter the sequences in which these actions in the chains are involved and their corresponding subjects. As a result of the search process, we obtain the emotion label corresponding to the new situation and the subject of the emotion based on a weighting function. This function takes into consideration the number of actions and the position in which they appear in the sequence contained in EmotiNet (normalising the number of actions from the new example found within an action chain, stored in the KB, by the number of actions in the reference action chain). The issue in this first approach is that many of the examples cannot be classified, as the knowledge they contain is not present in the ontology. The corresponding results using this approach on test sets A and B were marked with A1 and B1.

(2). A subsequent approach aimed at surpassing the issues raised by the missing knowledge in EmotiNet. In a first approximation, we aimed at introducing extra knowledge from VerbOcean, by adding the verbs that were similar to the ones in the core examples (represented in VerbOcean through the "similar" relation). Subsequently, each of the actions in the examples to be classified that was not already contained in EmotiNet, was sought in VerbOcean. In case one of the similar actions found in VerbOcean was already contained in the KB, the actions were considered equivalent. Further on, each action was associated with an emotion, using the ConceptNet relations and concepts (*EffectOf, CapableOf, MotivationOf, DesireOf*). Action chains were represented as chains of actions with their associated emotion. Finally, new examples were matched against chains of actions containing the same emotions, in the same order, this time, considering the additional knowledge obtained from the VerbOcen and ConceptNet knowledge bases While more complete than the first approximation, this approach was also affected by lack of knowledge about the emotional content of actions. To overcome this issue, we proposed two heuristics:

(2a) In the first one, actions on which no affect information was available, were sought within the examples already introduced in EmotiNet (extended with ConceptNet and VerbOcean) and were assigned the most frequent class of emotion labeling them. The corresponding results are marked with A2a and B2a, respectively.

(2b) In the second approximation, we used the most frequent emotion associated to the known links of a chain, whose individual emotions were obtained from ConceptNet. In this case, the core of action chains is not involved in the process. The corresponding results are marked with A2b and B2b.

5.2 Empirical Results

We performed the steps described in Section 4.1 on the two test sets, A and B. For the first approach, the queries to the test set A led to a result only in the case of 90 examples and for test set B only in the case of 571 examples. For the second approach, using the approximation (2a), 165 examples from test set A and 617 examples from test B were classified. In the case of approximation (2b), the queries obtained results in the case of 171 examples in test set A and 625 examples in test set B. For the remaining ones, the knowledge stored in the KB is not sufficient, so that the appropriate action chain can be extracted. Table 1 presents the results of the evaluations using as knowledge core the training set A_T Table 2 reports the results obtained using as training set the knowledge in B_T. The baselines are random, computed as average of 10 random generations of classes for all classified examples.

5.3 Discussion

From the results in Table 1 and 2, we can conclude that the approach is valid and improves the results of the emotion detection task. Nonetheless, much remains to be done to fully exploit the capabilities of EmotiNet. The model we proposed, based on appraisal theories, proved to be flexible, its level of performance improving – either by percentual increase, or by the fact that the results for different emotional categories become more balanced. We showed that the approach has a high degree of flexibility, i.e. new information can be easily introduced from existing common-sense knowledge bases, mainly due to its internal structure and degree of granularity.

Table 1. Results of the emotion detection using EmotiNet on test set A, using A_T as learning set

Emotion	Correct			Total			Accuracy			Recall		
	A1	A2a	A2b	A1	A 2a	A2b	A1	A2a	A2b	A1	A2a	A2b
disgust	10	28	29	41	52	67	24.39	53.85	43.28	16.95	47.46	49.15
anger	16	39	39	102	114	119	15.69	34.21	32.77	11.03	26.90	26.90
fear	37	43	44	55	74	76	67.27	58.11	57.89	43.53	50.59	51.76
guilt	27	55	59	146	157	165	18.49	35.03	35.76	13.64	27.78	29.80
Total	90	165	171	344	397	427	26.16	41.56	40.05	18.48	33.88	35.11
Recall Baseline	124	124	124	487	487	487	---	---	---	25.46	25.46	25.46

Table 2. Results of the emotion detection using EmotiNet on test set B, using B_T as learning set

Emotion	Correct			Total			Accuracy			Recall		
	B1	B2a	B2b	B1	B1	B1	B1	B2a	B2b	B1	B2a	B2b
disgust	16	16	21	26	59	63	36.36	38.09	52.50	27.11	27.11	35.59
shame	25	25	26	62	113	113	35.71	32.05	35.62	27.47	27.47	28.57
anger	31	47	57	29	71	73	29.52	40.86	47.11	21.37	32.41	39.31
fear	35	34	37	86	166	160	60.34	52.30	61.67	60.34	52.30	61.67
sadness	46	45	41	26	59	63	41.44	36.58	32.80	17.22	16.85	15.36
joy	13	16	18	62	113	113	52	55.17	51.43	26	32	36.00
guilt	59	68	64	29	71	73	37.34	41.21	37.43	29.79	34.34	32.32
Total	225	251	264	571	617	625	39.40	40.68	42.24	25.13	28.04	29.50
Recall Baseline	126	126	126	895	895	895	---	---	---	14.07	14.07	14.07

From the error analysis we performed, we could determine some of the causes of error in the system. The first important finding is that extracting only the action, verb and patient semantic roles is not sufficient. There are other roles, such as the modifiers, which change the overall emotion in the text (e.g. "I had a fight with my sister" – sadness, versus "I had a fight with my stupid sister" – anger). Therefore, such modifiers should be included as attributes of the concepts identified in the roles, and, additionally, added to the tuples, as they can account for other appraisal criteria. This can also be a method to account for negation. Given that just 3 roles were extracted and the accuracy of the SRL system, there were also many examples that did not make sense when input into the system (~20%). A further source of errors was that lack of knowledge on specific actions. As we have seen, VerbOcean extended the knowledge, in the sense that more examples could be classified. However, given the ambiguity of the resource and the fact that it is not perfectly accurate also introduced many errors. Thus, the results of our approach can be practically limited by the structure, expressivity and degree of granularity of the imported resources. Therefore, to obtain the final, extended version of EmotiNet we should analyse the interactions between the core and the imported resources and among these resources as well.

Finally, other errors were produced by NLP processes and propagated at various steps of the processing chain (e.g. SRL, coreference resolution). Some of these errors cannot be eliminated; others can be partially solved by using alternative NLP tools.

6   Conclusions and Future Work

This article presented our contribution concerning three major topics: the proposal of a method to model real-life situations described in text based on the appraisal theories, the design and population of EmotiNet, a knowledge base of action chains representing and storing affective reaction to real-life contexts and situations described in text and proposing and evaluating a method to detect emotion in text based on EmotiNet, using new texts. We conclude that our approach is appropriate for detecting emotion in text, although additional elements should be included in the model and extra knowledge is required. Moreover, we found that the process of automatic evaluation was influenced by the low performance of the NLP tools used. Thus, alternative tools must be tested in order to improve the output. We must also test our approach on corpora where more than one emotion is assigned per context.

Future work aims at extending the model by adding properties to the concepts included, so that more of the appraisal criteria can be introduced in the model, testing new methods to assign affective value to the concepts and adding new knowledge from sources such as CYC. We also plan to improve the extraction of action chains and adapt the emotion detection process to the persons describing the emotional experiences, by including special attributes to the concepts involved.

Acknowledgements. This paper has been partially supported by the Spanish Ministry of Science and Innovation (grant no. TIN2009-13391-C04-01), the Spanish Ministry of Education under the FPU Program (AP2007-03076) and Valencian Ministry of Education (grant no. PROMETEO/2009/119 and ACOMP/ 2010/288).

References

1. Strapparava, C., Mihalcea, R.: Semeval 2007 Task 14: Affective Text. In: Proceedings of the 4th International Workshop on Semantic Evaluations (SemEval-2007), Sattelite Workshop to ACL 2007, Prague, pp. 70–74 (June 2007)
2. Pennebaker, J.W., Mehl, M.R., Niederhoffer, K.: Psychological aspects of natural language use: Our words, our selves. Annual Review of Psychology 54, 547–577 (2003)
3. Balahur, A., Montoyo, A.: Applying a Culture Dependent Emotion Triggers Database for Text Valence and Emotion Classification. In: Proceedings of the AISB 2008 Convention "Communication, Interaction and Social Intelligence" (2008)
4. Balahur, A., Steinberger, R.: Rethinking Opinion Mining in Newspaper Articles: from Theory to Practice and Back. In: Proceedings of the First Workshop on Opinion Mining and Sentiment Analysis, WOMSA 2009 (2009)
5. Calvo, R.A., D'Mello, S.: Affect Detection: An Interdisciplinary Review of Models, Methods and Their Applications. IEEE Transactions on Affective Computing 1(1) (January-June 2010)

6. Scherer, K.: Appraisal Theory. Handbook of Cognition and Emotion. John Wiley & Sons Ltd., Chichester (1999)
7. Liu, H., Singh, P.: ConceptNet: A Practical Commonsense Reasoning Toolkit. BT Technology Journal 22 (2004)
8. Chklovski, T., Pantel, P.: VerbOcean: Mining the Web for Fine-Grained Semantic Verb Relations. In: Proceedings of EMNLP 2004 (2004)
9. Picard, R.: Affective computing, Technical report, MIT Media Laboratory (1995)
10. Dyer, M.: Emotions and their computations: three computer models. Cognition and Emotion 1, 323–347 (1987)
11. Subasic, P., Huettner, A.: Affect Analysis of text using fuzzy semantic typing. IEEE Trasactions on Fuzzy System 9, 483–496 (2000)
12. Strapparava, C., Valitutti, A.: Wordnet-affect: an affective extension of WordNet. In: Proceedings of the 4th International Conference on Language Resources and Evaluation, LREC 2004 (2004)
13. Esuli, A., Sebastiani, F.: Determining the semantic orientation of terms through gloss analysis. In: Proceedings of CIKM 2005 (2005)
14. Riloff, E., Wiebe, J.: Learning extraction patterns for subjective expressions. In: Proceedings of EMNLP 2003 (2003)
15. Pang, B., Lee, L., Vaithyanathan, S.: Thumbs up? Sentiment classification using machine learning techniques. In: Proceedings of EMNLP 2002 (2002)
16. Danisman, T., Alpkocak, A.: Feeler: Emotion Classification of Text Using Vector Space Model. In: Proceedings of the AISB Convention, "Communication, Interaction and Social Intelligence" (2008)
17. Mei Lee, S.Y., Chen, Y., Huang, C.-R.: Cause Event Representations of Happiness and Surprise. In: Proceedings of PACLIC 2009 (2009)
18. Liu, H., Lieberman, H., Selker, T.: A Model of Textual Affect Sensing Using Real-World Knowledge. In: Proceedings of IUI 2003 (2003)
19. Cambria, E., Hussain, A., Havasi, C., Eckl, C.: Affective Space: Blending Common Sense and Affective Knowledge to Perform Emotive Reasoning. In: Proceedings of the 1st Workshop on Opinion Mining and Sentiment Analysis, WOMSA (2009)
20. De Rivera, J.: A structural theory of the emotions. Psychological Issues 10(4) (1977); Monograph 40
21. Frijda, N.: The emotions. Cambridge University Press, Cambridge (1986)
22. Ortony, A., Clore, G.L., Collins, A.: The cognitive structure of emotions. Cambridge University Press, Cambridge (1988)
23. Johnson-Laird, P.N., Oatley, K.: The language of emotions: An analysis of a semantic field. Cognition and Emotion 3, 81–123 (1989)
24. Scherer, K.R.: Studying the Emotion-Antecedent Appraisal Process: An Expert System Approach. Cognition and Emotion 7(3/4) (1993)
25. Martin, J.R., White, P.R.: Language of Evaluation: Appraisal in English. Palgrave Macmillan, Basingstoke (2005)
26. Fellbaum, C.: WordNet: An Electronic Lexical Database. MIT Press, Cambridge (1998)
27. Suchanek, F., Kasnei, G., Weikum, G.: YAGO: A Core of Semantic Knowledge Unifying WordNet and Wikipedia. In: Proceedings of WWW (2007)
28. Pantel, P., Ravichandran, D.: Automatically Labeling Semantic Classes. In: Proceedings of HLT/NAACL 2004, Boston, MA, pp. 321–328 (2004)
29. Berland, M., Charniak, E.: Finding parts in very large corpora. In: Proceedings of ACL (1999)

30. Brin: Extracting patterns and relations from the World-Wide Web. In: Proceedings on the 1998 International Workshop on Web and Databases (1998)
31. Agichtein, E., Gravano, L.: Snowball: Extracting Relations from Large Plain-Text Collections. In: Proceedings of the 5th ACM International Conference on Digital Libraries, ACM DL (2000)
32. Grassi, M.: Developing HEO Human Emotions Ontology. In: Fierrez, J., Ortega-Garcia, J., Esposito, A., Drygajlo, A., Faundez-Zanuy, M. (eds.) BioID MultiComm2009. LNCS, vol. 5707, pp. 244–251. Springer, Heidelberg (2009)
33. Studer, R., Benjamins, R.V., Fensel, D.: Knowledge engineering: Principles and methods. Data & Knowledge Engineering 25(1-2), 161–197 (1998)
34. Moreda, P., Navarro, B., Palomar, M.: Corpus-based semantic role approach in information retrieval. Data Knowl. Eng. (DKE) 61(3), 467–483 (2007)
35. Scherer, K., Wallbott, H.: The ISEAR Questionnaire and Codebook. Geneva Emotion Research Group (1997)
36. Plutchik, R.: The Nature of Emotions. American Scientist 89, 344 (2001)
37. Parrott, W.: Emotions in Social Psychology. Psychology Press, Philadelphia (2001)

Querying Linked Data Using Semantic Relatedness: A Vocabulary Independent Approach

André Freitas[1], João Gabriel Oliveira[1,2], Seán O'Riain[1],
Edward Curry[1], and João Carlos Pereira da Silva[2]

[1] Digital Enterprise Research Institute (DERI)
National University of Ireland, Galway
[2] Computer Science Department
Universidade Federal do Rio de Janeiro

Abstract. Linked Data brings the promise of incorporating a new dimension to the Web where the availability of Web-scale data can determine a paradigmatic transformation of the Web and its applications. However, together with its opportunities, Linked Data brings inherent challenges in the way users and applications consume the available data. Users consuming Linked Data on the Web, or on corporate intranets, should be able to search and query data spread over potentially a large number of heterogeneous, complex and distributed datasets. Ideally, a query mechanism for Linked Data should abstract users from the representation of data. This work focuses on the investigation of a vocabulary independent natural language query mechanism for Linked Data, using an approach based on the combination of entity search, a Wikipedia-based semantic relatedness measure and spreading activation. The combination of these three elements in a query mechanism for Linked Data is a new contribution in the space. Wikipedia-based relatedness measures address existing limitations of existing works which are based on similarity measures/term expansion based on WordNet. Experimental results using the query mechanism to answer 50 natural language queries over DBPedia achieved a mean reciprocal rank of 61.4%, an average precision of 48.7% and average recall of 57.2%, answering 70% of the queries.

Keywords: Natural Language Queries, Linked Data.

1 Introduction

The last few years have seen Linked Data [1] emerge as a de-facto standard for publishing data on the Web, bringing the potential of a paradigmatic change in the scale which users and applications reuse, consume and repurpose data. However, together with its opportunities, Linked Data brings inherent challenges in the way users and applications consume existing Linked Data. Users accessing Linked Data should be able to search and query data spread over a potentially large number of different datasets. The freedom, simplicity and intuitiveness

R. Muñoz et al. (Eds.): NLDB 2011, LNCS 6716, pp. 40–51, 2011.
© Springer-Verlag Berlin Heidelberg 2011

provided by search engines in the Web of Documents were fundamental in the process of maximizing the value of the information available on the Web, approaching the Web to the casual user.

However the approaches used for searching the Web of Documents cannot be directly applied for searching/querying data. From the perspective of structured/ semi-structured data consumption, users are familiar with precise and expressive queries. Linked Data relies on the use of ontologies (also called vocabularies) to represent the semantics and the structure of datasets. In order to query existing data today, users need be aware of the structure and terms used in the data representation. In the Web scenario, where data is spread across multiple and highly heterogeneous datasets, the semantic gap between users and datasets (i.e. the difference between user queries terms and the representation of data) becomes one of the most important issues for Linked Data consumers. At Web scale it is not feasible to become aware of all the vocabularies that the data can be represented in order to formulate a query. From the users' perspective, they should be abstracted away from the data representation. In addition, from the perspective of user interaction, a query mechanism for Linked Data should be simple and intuitive for casual users. The suitability of natural language for search and query tasks was previously investigated in the literature (Kauffman [2]).

This work focuses on the investigation of a fundamental type of query mechanism for Linked Data: the provision of *vocabulary independent and expressive natural language queries for Linked Data*. This type of query fills an important gap in the spectrum of search/query services for the Linked Data Web, allowing users to expressively query the contents of distributed linked datasets without the need for a prior knowledge of the vocabularies behind the datasets.

This paper is structured as follows. Section 2 describes the proposed approach, detailing how the three elements (*entity search, spreading activation* and *semantic relatedness*) are used to build the query mechanism. Section 3 covers the evaluation of the approach, followed by section 4 which describes related work in the area. Finally, section 5 provides a conclusion and future work.

2   Query Approach

2.1   Introduction

The central motivation behind this work is to propose a Linked Data query mechanism for casual users providing flexibility (vocabulary independence), expressivity (ability to query complex relations in the data), usability (provided by natural language queries) and ability to query distributed data. In order to address these requirements, this paper proposes the construction of a query mechanism based on the combination of *entity search, spreading activation* and a *semantic relatedness measure*. Our contention is that the combination of these elements provides the support for the construction of a natural language and data model independent query mechanism for Linked Data. The approach represents a new contribution in the space of natural language queries over Linked Data. The remainder of this section describes the proposed approach and its main components.

2.2 Description of the Approach

The query mechanism proposed in this work receives as an input a natural language query and outputs a set of triple paths, which are the triples corresponding to answers merged into a connected path.

The query processing approach starts by determining the *key entities* present in the natural language query. Key entities are entities which can be potentially mapped to instances or classes in the Linked Data Web. After detected, key entities are sent to the entity search engine which determines the pivot entities in the Linked Data Web. A pivot entity is an URI which represents an entry point for the spreading activation search in the Linked Data Web (figure 1). The processes of key entity and pivot entity determination are covered in section 2.3.

After the key entities and pivots are determined, the user natural language query is analyzed in the query parsing module. The output of this module is a structure called *partial ordered dependency structure* (PODS), which is a reduced representation of the query targeted towards maximizing the matching probability between the structure of the terms present in the query and the *subject, predicate, object* structure of RDF. The partial ordered dependency structure is generated by applying Stanford dependency parsing [3] over the natural language query and by transforming the generated Stanford dependency structure into a PODS (section 2.4).

Taking as an input the list of URIs of the pivots and the partial ordered dependency structure, the algorithm follows a spreading activation search where nodes in the Linked Data Web are explored by using a measure of semantic relatedness to match the query terms present in the PODS to terms representing Linked Data entities (classes, properties and individuals). Starting from a pivot entity, the node exploration process in the spreading activation search is done by computing the relatedness measure between the query terms and terms corresponding to entities in the Linked Data Web. The semantic relatedness measure combined with a statistical threshold which represents the discrimination of the winning relatedness scores works as a spreading activation function which will determine the nodes which will be explored in the Linked Data Web.

Figure 1 depicts the core part of the spreading activation process for the example query *'From which university did the wife of Barack Obama graduate?'*. After parsing the natural language query into a partial ordered dependency structure (PODS) (light gray nodes), and after the pivot is determined (*dbpedia: Barack_Obama*) by the entity search, the algorithm follows computing the semantic relatedness between the next query term (*'wife'*) and all the properties, associated types and instance labels linked to the node dbpedia:Barack_Obama (*dbpedia-owl:spouse, dbpedia-owl:writer, dbpedia-owl:child, ...*). Nodes above a certain relatedness threshold are further explored (dereferenced). The matching process continues until all query terms are covered.

In the example, after the matching between *wife* and *dbpedia-owl: spouse* is defined (2), the object pointed by the matched property (*dbpedia: Michelle _Obama*) is dereferenced (3), and the RDF of the resource is retrieved. The next node in the PODS is *graduate*, which is mapped to both *dbpedia-owl:University* and *dbpedia-owl:Educational_Institution* (4) specified in the types. The algorithm then navigates to the last node of the PODS, *university*, dereferencing *dbpedia:Princeton_University* and *dbpedia:Harvard_Law_School* (5), matching for the second time with their type (6). Since the relatedness between the terms is high, the terms are matched and the algorithm stops, returning the subgraph containing the triples which maximize the relatedness between the query terms and the vocabulary terms. The proposed algorithm works as a best-effort query approach, where the semantic relatedness measure provides a semantic ranking of returned triples.

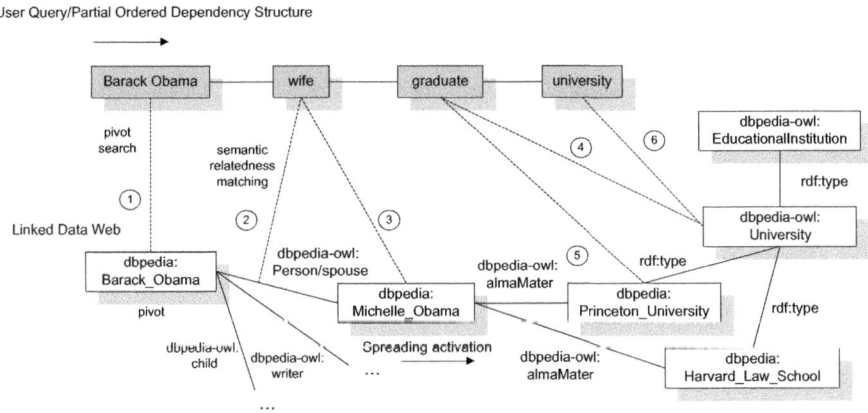

Fig. 1. The relatedness spreading activation for the question *'From which university did the wife of Barack Obama graduate?'*

The output of the algorithm is a list of ranked triple paths, triples following from the pivot entity to the final resource representing the answer, ranked by the average of the relatedness scores in the path. Answers are displayed to users using a list of triple paths and a graph which is built by merging the triple paths on a simple post-processing phase. This graph can be used by the user for further complementary exploration by navigation over the answer set. The query mechanism described above was implemented in a prototype named *Treo*, the word for *direction* in Irish, representing the direction that the algorithm takes in the node exploration process using semantic relatedness.

2.3 Entity Recognition and Entity Search

The query processing approach starts by determining the set of key entities (pivot candidates) that will be used in the generation of the partial ordered dependency structure and in the determination of the final pivot entity. The process of generating a pivot candidate starts by detecting named entities in the query. The named entity recognition (NER) approach used is based on Conditional Random Fields sequence models [4] trained in the CoNLL 2003 English training dataset [5], covering people, organizations and locations. Named entities are likely to be mapped to the URIs of individuals in the Linked Data Web. After the named entities are identified, the query is tagged by a part-of-speech (POS) tagger, which assigns grammatical tags to the query terms. The POS tags are used to determine pivot candidates which are not named entities (typically classes or individuals representing categories). The POS tagger used is a log-linear POS tagger [6].

The terms corresponding to the pivot candidates are sent to an entity centric-search engine that will resolve the terms into the final pivot URIs in the Linked Data Web. Entity-centric search engines for Linked Data are search engines where the search is targeted towards the retrieval of individual instances, classes and properties in the Linked Data Web. This work uses the entity search approach proposed by Delbru et al. [7], implemented in SIREn, the Semantic Information Retrieval Engine. The approach used in SIREn uses a variation of the TF-IDF weighting scheme (term frequency - inverse subject frequency: TF-ISF) to evaluate individuals by aggregating the values of partial scores for predicates and objects. The TF-ISF scheme gives a low weight to predicates or objects which occur in a large number of entities. SIREn's keyword based search is used for resolving entities in the Linked Data Web, where the recognized entities (pivot candidates) are sent to the search engine and a list of URIs is returned. The query mechanism prioritizes named entities as pivots. In case the query has more than one named entity, both candidate terms are sent to the entity search engine and the number of properties connected to the URI is used to determine the final pivot entity.

2.4 Query Parsing

The relatedness spreading activation algorithm takes as one of its inputs a partial ordered dependency structure (PODS) which is a directed acyclic graph connecting a subset of the original terms present in the query. The idea behind PODSs is to provide a representation of the natural language input which could be easily mapped to the *(subject, predicate, object)* structure of an RDF representation. Partial ordered dependency structures are derived from Stanford typed dependencies [3] which represents a set of bilexical relations between each term of a sentence, providing grammatical relation labels over the dependencies. Additional details covering Stanford dependencies can be found in [3].

The query parsing module builds PODSs by taking as inputs both Stanford dependencies and the detected named entities/pivots and by applying a set of operations over the original Stanford dependencies. These operations produce a reduced and ordered version of the original elements of the query. The pivots and named entities combined with the original dependency structure determine the ordering of the elements in the structure.

Definition I: Let T(V, E) be a typed Stanford dependency structure over the question Q. The partial ordered dependency structure D(V, E) of Q is defined by applying the following operations over T:

1. *merge* adjacent nodes V_K and $V_{K+1} \in T$ where $E_{K,K+1} \in \{$nn, advmod, amod$\}$.

2. *eliminate* the set of nodes V_K and edges $E_K \in T$ where $E_K \in \{$advcl, aux, auxpass, ccomp, complm, det$\}$.

3. *replicate* the triples where $E_K \in \{$cc, conj, preconj$\}$.

In the definition above the *merge* operation consists in collapsing adjacent nodes into a single node for the purpose of merging multi-word expressions, in complement with the NER output. The *eliminate* operation is defined by the pruning of a node-edge pair and eliminates concepts which are not semantically relevant or covered in the representation of data in RDF. The *replicate* operation consists in copying the remaining elements in the PODS for each coordination or conjunctive construction.

The transversal sequence should maximize the likelihood of the isomorphism between the partial ordered dependency structure and the subgraph of the Linked Data Web. The transversal sequence is defined by taking the pivot entity as the root of the partial ordered dependency structure and following the dependencies until the end of the structure is reached. In case there is a cycle involving one pivot entity and a secondary named entity, the path with the largest length is considered.

2.5   Semantic Relatedness

The problem of measuring the semantic relatedness and similarity of two concepts can be stated as follows: given two concepts A and B, determine a measure f(A,B) which expresses the semantic proximity between concepts A and B. The notion of semantic *similarity* is associated with taxonomic (is-a) relations between concepts, while semantic *relatedness* represents more general classes of relations. Since the problem of matching natural language terms to concepts present in Linked Data vocabularies can cross taxonomic boundaries, the generic concept of semantic relatedness is more suitable to the task of semantic matching for queries over the Linked Data Web. In the example query, the relation between *'graduate'* and *'University'* is non-taxonomic and a purely similarity analysis would not detect appropriately the semantic proximity between these two terms. In the context of semantic query by spreading activation, it is necessary to use a relatedness measure that: (i) can cope with terms crossing part-of-speech boundaries (e.g. verbs and nouns); (ii) measure relatedness among multi-word expressions; (iii) are based on a comprehensive knowledge base.

Distributional relatedness measures [11] meet the above requirements but demand the processing of large Web corpora. New approaches propose a better balance between the cost associated in the construction of the relatedness measure and the accuracy provided, by using the link structure of Wikipedia. Wikipedia Link-based Measure (WLM), proposed by Milne & Witten [12], achieved high correlation measurements with human assessments. This work uses WLM as the relatedness measure for the spreading activation process. The reader is directed to [12] for further details on the construction of the relatedness measure.

2.6   The Semantic Relatedness Spreading Activation Algorithm

The semantic relatedness spreading activation algorithm takes as an input a partial ordered dependency structure $D(V, E)$ and searches for paths in the Linked Data graph $W(V, E)$ which maximizes the semantic relatedness between D and W taking into account the ordering of both structures. The first element in the partial ordered dependency structure is the pivot entity which defines the first node to be dereferenced in the graph W. After the pivot element is dereferenced the algorithm computes the semantic relatedness measure between the next term in the PODS and the *properties, type terms* and *instance terms* in the Linked Data Web. Type terms represent the types associated to an instance through the *rdfs:type* relation. While properties and ranges are defined in the terminological level, type terms require an instance dereferenciation to collect the associated types. The relatedness computation process between the next query term k and a neighboring node n takes the maximum of the relatedness score between properties p, types c and instance terms i:

$$r_{k,n} = max(r(k, p), r(k, i), \max_{\forall c \in C}(r(k, c)))$$ (1)

Nodes above a relatedness score threshold determine the node URIs which will be activated (dereferenced). The activation function is given by an adaptive discriminative relatedness threshold which is defined based on the set of relatedness scores. The adaptive threshold has the objective of selecting the relatedness scores with higher discrimination and it is defined as a function of the standard deviation σ of the relatedness scores. The activation threshold of a node I is defined as:

$$a(I) = \mu(r) + \alpha \times \sigma(r)$$ (2)

where I is the node instance, $\mu(r)$ is the mean of the relatedness values for each node instance, $\sigma(r)$ is the standard deviation of the relatedness values and α is a empirically determined constant. The value of α was determined by calculating the difference in terms of $\sigma(r)$ of the relatedness value of the correct node instances and the average relatedness value for a random 50% sample of the nodes instances involved in the spreading activation process in the query dataset. The empirical value found for α is 2.185. In case no node is activated for the first value of α, the original value decays by an exponential factor of 0.9 until it finds a candidate node above *a(I)*.

In case the algorithm finds a node with high relatedness which has a literal value as an object (non dereferenceable), the value of the node can be re-submitted to entity search engine. In the case an URI is mapped, the search continues from the same point in the partial ordered dependency structure in a different pivot in the Linked Data Web (working as an entity reconciliation step).

From a practical perspective the use of type verification in the node exploration process can bring high latencies in the node exploration process. In order
to be effective, the algorithm should rely on mechanisms to reduce the number
unnecessary HTTP requests associated with the dereferenciation process, unec-
essary URI parsing or label checking and unnecessary relatedness computation.
The *Treo* prototype has three local caches implemented: one for RDF, the second
for relatedness pairs and the third for URI/label-term mapping. Another important practical aspect which constitutes one of the strengths of the approach is
the fact that it is both highly and easily parallelizable in the process of semantic
relatedness computation and on the de-referenciation of URIs.

3   Evaluation

The focus of the evaluation was to determine the quality of the results provided
by the query mechanism. With this objective in mind the query mechanism was
evaluated by measuring *average precision*, *average recall* and *mean reciprocal
rank* for natural language queries using DBPedia [8], a dataset in the Linked
Data Web. DBpedia 3.6 (February 2011) contains 3.5 million entities, where
1.67 million are classified in a consistent ontology. The use of DBPedia as a
dataset allows the evaluation of the system under a realistic scenario. The set
of natural language queries annotated with answers were provided by the training dataset released for the Question Answering for Linked Data (QALD 2011)
workshop [9] containing queries over DBPedia 3.6. From the original query set,
5 queries were highly dependent on comparative operations (e.g. *'What is the
highest mountain?'*). Since the queries present in the QALD Dataset did not
fully explore more challenging cases of query-vocabulary semantic gap match
ing, the removed queries were substituted with 5 additional queries. The reader
can find additional details on the data used in the evaluation and the associated
results in [10].

   In the scope of this evaluation an answer is a set of ranked triple paths. Different from a SPARQL query, the algorithm is a best effort approach where the
relatedness activation function works both as a ranking and a cut-off function and
the final result is a merged and collapsed subgraph containing the triple paths.
For the determination of *precision* we considered a correct answer a triple path
containing the URI of the answer. For the example query used in this article, the
triple path containing the answer *Barack Obama → spouse → Michelle Obama
→ alma mater → Princeton University* and *Harvard Law School* is the answer
provided by the algorithm instead of just *Princeton University* and *Harvard Law
School*. To determine both precision and recall, triple paths strongly supporting
semantically answers are also considered. For the query *'Is Natalie Portman an
actress?'*, the expected result is the set of nodes which highly supports the answer for this query, including the triples stating that she is an actress and that
she acted on different movies (this is used for both precision and recall). The
QALD dataset contains aggregate queries which were included in the evaluation. However, since the post-processing phase does not operate over aggregate
operators we considered correct answer triples supporting the answer.

Table 1 shows the quality metrics collected for the evaluation for each query. The final approach achieved an *mean reciprocal rank=***0.614**, *average precision=***0.487**, *average recall=***0.57** and % of *answered queries=* **70%**.

Table 1. Query dataset with the associated reciprocal rank, precision and recall

#	query	rr	precision	recall
1	From which university the wife of Barack Obama graduate?	0.25	0.333	0.5
2	Give me all actors starring in Batman Begins.	1	1	1
3	Give me all albums of Metallica.	1	0.611	0.611
4	Give me all European Capitals!	1	1	1
5	Give me all female German chancellors!	0	0	0
6	Give me all films produced by Hal Roach?	1	1	1
7	Give me all films with Tom Cruise.	1	0.865	1
8	Give me all soccer clubs in the Premier League.	1	0.956	1
9	How many 747 were built?	0.5	0.667	1
10	How many films did Leonardo DiCaprio star in?	0.5	0.733	0.956
11	In which films did Julia Roberts as well as Richard Gere play?	0	0	0
12	In which programming language is GIMP written?	0	0	0
13	Is Albert Einstein from Germany?	0.5	0.5	1
14	Is Christian Bale starring in Batman Begins?	0.125	0.071	1
15	Is Einstein a PHD?	1	1	1
16	Is Natalie Portman an actress?	1	0.818	0.273
17	Is there a video game called Battle Chess?	1	1	0.023
18	List all episodes of the first season of the HBO television series The Sopranos!	0.333	0.090	1
19	Name the presidents of Russia.	1	1	0.167
20	Since when is DBpedia online?	1	0.667	1
21	What is the band of Lennon and McCartney?	0	0	0
22	What is the capital of Turkey?	1	1	1
23	What is the official website of Tom Hanks?	1	0.333	1
24	What languages are spoken in Estonia?	1	1	0.875
25	Which actors were born in Germany?	1	0.033	0.017
26	Which American presidents were actors?	1	0.048	1
27	Which birds are there in the United States?	0	0	0
28	Which books did Barack Obama publish?	1	0.5	1
29	Which books were written by Danielle Steel?	1	1	1
30	Which capitals in Europe were host cities of the summer olympic games?	0	0	0
31	Which companies are located in California, USA?	0	0	0
32	Which companies work in the health area as well as in the insurances area?	0	0	0
33	Which country does the Airedale Terrier come from?	1	1	0.25
34	Which genre does the website DBpedia belong to?	1	0.333	1
35	Which music albums contain the song Last Christmas?	0	0	0
36	Which organizations were founded in 1950?	0	0	0
37	Which people have as their given name Jimmy?	0	0	0
38	Which people were born in Heraklion?	1	1	1
39	Which presidents were born in 1945?	0	0	0
40	Which software has been developed by organizations in California?	0	0	0
41	Who created English Wikipedia?	1	1	1
42	Who developed the video game World Warcraft?	1	0.8	1
43	Who has been the 5th president of the United states?	0	0	0
44	Who is called Dana?	0	0	0
45	Who is the wife of Barack Obama?	1	1	1
46	Who owns Aldi?	0.5	1	0.667
47	Who produced films starring Natalie Portman?	1	0.3	0.810
48	Who was the wife of President Lincoln?	1	1	1
49	Who was Tom Hanks married to ?	1	0.214	1
50	Who wrote the book The pillars of the Earth?	1	0.5	0.5

To analyze the results, queries with errors were classified according to 5 different categories, based on the components of the query approach. The first category, *PODS error*, contains errors which were determined by a difference between the structure of the PODS and the data representation which led the algorithm to an incorrect search path (Q35). In this case, the flexibility provided by semantic relatedness and spreading activation was unable to cope with this difference. The second error category, *Pivot Error*, includes errors in the determination of the correct pivot. This category includes queries with non-dereferenceable pivots (i.e. pivots which are based on literal resources) or errors in the pivot determination process (Q5, Q27, Q30, Q44). Some of the difficulty in the pivot determination process were related to overloading classes with complex types (e.g. for the query Q30 the associated pivot is a class *yago:HostCitiesOfTheSummerOlympicGames*). *Relatedness Error* include queries which were not addressed due to errors in the relatedness computation process, leading to an incorrect matching and the elimination of the correct answer (Q11, Q12). The fourth category, *Excessive Dereferenciation Timeout Error* covers queries which demanded a large number of dereferenciations to be answered (Q31, Q40). In the query Q40, the algorithm uses the entity *California* as a pivot and follows each associated *Organization* to find its associated type. This is the most challenging category to address, putting in evidence a limitation of the approach. The last categories cover small errors outside previous categories or combined errors in one query (Q32, Q39, Q43).

Table 2. Error types and distribution

Error Type	% of Queries
PODS Error	2%
Pivot Error	10%
Relatedness Error	4%
Excessive Dereferenciation Timeout Error	6%
Combined Error	8%

The approach was able to answer 70% of the queries. The relatedness measure was able to cope with non-taxonomic variations between query and vocabulary terms, showing high average discrimination in the node selection process (average difference between the relatedness value of answer nodes and the relatedness mean is 2.81 $\sigma(r)$). The removal of the queries with errors that are considered addressable in the short term (PODS Error, Pivot Error, Relatedness Error) leads to precision=0.64, recall=0.75 and mrr=0.81.

From the perspective of *query execution time* an experiment was run using an Intel Centrino 2 computer with 4 GB RAM. No parallelization or indexing mechanism outside the pivot determination process was implemented in the query mechanism. The average query execution time for the set of queries which were answered was 635s with no caching and 203s with active caches.

4   Related Work

Different natural language query approaches for Semantic Web/Linked Data datasets have been proposed in the literature. Most of the existing query approaches for semantic approximations are based on WordNet. PowerAqua [13] is a question answering system focused on natural language questions over Semantic Web/Linked Data datasets. PowerAqua uses PowerMap to match query terms to vocabulary terms. According to Lopez et al. [17], *PowerMap is a hybrid matching algorithm comprising terminological and structural schema matching techniques with the assistance of large scale ontological or lexical resources.* PowerMap uses WordNet based similarity approaches as a semantic approximation strategy. NLP-Reduce [15] approaches the problem from the perspective of a lightweight natural language approach, where the natural language input query is not analyzed at the syntax level. The matching process between the query terms and the ontology terms present in NLP-Reduce is based on a WordNet expansion of synonymic terms in the ontology and on matching at the morphological level. The matching process of another approach, Querix [16], is also based on the expansion of synonyms based on WordNet. Querix, however, uses syntax level analysis over the input natural language query, using this additional structure information to build the corresponding query skeleton of the query. Ginseng [14] follows a controlled vocabulary approach: the terms and the structure of the ontologies generate the lexicon and the grammar for the allowed queries in the system. Ginseng ontologies can be manually enriched with synonyms.

Compared to existing approaches, *Treo* provides a query mechanism which explores a more robust semantic approximation technique which can cope with the variability of the query-vocabulary matching on the heterogeneous environment of Linked Data on the Web. Additionally, its design supports querying dynamic and distributed Linked Data. The proposed approach also follows a different query strategy, by following sequences of dereferenciations and avoiding the construction of a SPARQL query and by focusing on a best-effort ranking approach.

5   Conclusion and Future Work

This paper proposes a vocabulary independent natural language query mechanism for Linked Data focusing on addressing the trade-off between expressivity and usability for queries over Linked Data. To address this problem, a novel combination for querying Linked Data is proposed, based on entity search, spreading activation and Wikipedia-based semantic relatedness. The approach was implemented in the *Treo* prototype and evaluated with an extended version of the QALD query dataset containing 50 natural language queries over the DBPedia dataset, achieving an overall *mean reciprocal rank* of 0.614, *average precision* of 0.487 and *average recall* of 0.572, answering 70% of the queries. Additionally, a set of short-term addressable limitations of the approach were identified. The result shows the robustness of the proposed query mechanism to provide a vocabulary independent natural language query mechanism for Linked Data. The proposed approach is designed for querying live distributed Linked Data. Directions for future investigations include addressing the set of limitations identified during the experiments, the incorporation of a more sophisticated post-processing mechanism and the investigation of performance optimizations for the approach.

Acknowledgments. The work presented in this paper has been funded by Science Foundation Ireland under Grant No. SFI/08/CE/I1380 (Lion-2).

References

1. Berners-Lee, T.: Linked Data Design Issues (2009),
   http://www.w3.org/DesignIssues/LinkedData.html
2. Kaufmann, E., Bernstein, A.: Evaluating the usability of natural language query languages and interfaces to Semantic Web knowledge bases. J. Web Semantics: Science, Services and Agents on the World Wide Web 8, 393–377 (2010)
3. Marneffe, M., MacCartney, B., Manning, C.D.: Generating Typed Dependency Parses from Phrase Structure Parses. In: LREC 2006 (2006)
4. Finkel, J.R., Grenager, T., Manning, C.D.: Incorporating Non-local Information into Information Extraction Systems by Gibbs Sampling. In: Proceedings of the 43rd Annual Meeting of the Association for Computational Linguistics (ACL 2005), pp. 363–370 (2005)
5. Sang, F., Meulder, F.: Introduction to the CoNLL-2003 shared task: language-independent named entity recognition. In: Proceedings of the Seventh Conference on Natural Language Learning at HLT-NAACL (2003)
6. Toutanova, K., Klein, D., Manning, C.D., Singer, Y.: Feature-Rich Part-of-Speech Tagging with a Cyclic Dependency Network. In: Proceedings of HLT-NAACL 2003, pp. 252–259 (2003)
7. Delbru, R., Toupikov, N., Catasta, M., Tummarello, G., Decker, S.: A Node Indexing Scheme for Web Entity Retrieval. In: Aroyo, L., Antoniou, G., Hyvönen, E., ten Teije, A., Stuckenschmidt, H., Cabral, L., Tudorache, T. (eds.) ESWC 2010. LNCS, vol. 6089, pp. 240–256. Springer, Heidelberg (2010)
8. Bizer, C., Lehmann, J., Kobilarov, G., Auer, S., Becker, C., Cyganiak, R., Hellmann, S.: DBpedia - A crystallization point for the Web of Data. J. Web Semantics: Science, Services and Agents on the World Wide Web (2009)
9. Hellmann, S.: 1st Workshop on Question Answering over Linked Data, QALD-1 (2011), http://www.sc.cit-ec.uni-bielefeld.de/qald-1
10. Evaluation Dataset (2011), http://troo.dori.io/rcoulto/nldb2011.htm
11. Gabrilovich, E., Markovitch, S.: Computing semantic relatedness using Wikipedia-based explicit semantic analysis. In: Proceedings of the International Joint Conference On Artificial Intelligence (2007)
12. Milne, D., Witten, I.H.: An effective, low-cost measure of semantic relatedness obtained from Wikipedia links. In: Proceedings of the First AAAI Workshop on Wikipedia and Artificial Intelligence (WIKIAI 2008), Chicago, I.L (2008)
13. Lopez, V., Motta, E., Uren, V.S.: PowerAqua: Fishing the semantic web. In: Sure, Y., Domingue, J. (eds.) ESWC 2006. LNCS, vol. 4011, pp. 393–410. Springer, Heidelberg (2006)
14. Bernstein, A., Kaufmann, E., Kaiser, C., Kiefer, C.: Ginseng A Guided Input Natural Language Search Engine for Querying Ontologies. In: Jena User Conference (2006)
15. Kaufmann, E., Bernstein, A., Fischer, L.: NLP-Reduce: A naive but Domain-independent Natural Language Interface for Querying Ontologies. In: Franconi, E., Kifer, M., May, W. (eds.) ESWC 2007. LNCS, vol. 4519, pp. 1–2. Springer, Heidelberg (2007)
16. Kaufmann, E., Bernstein, A., Zumstein, R.: Querix: A Natural Language Interface to Query Ontologies Based on Clarification Dialogs. In: Cruz, I., Decker, S., Allemang, D., Preist, C., Schwabe, D., Mika, P., Uschold, M., Aroyo, L.M. (eds.) ISWC 2006. LNCS, vol. 4273, pp. 980–981. Springer, Heidelberg (2006)
17. Lopez, V., Sabou, M., Motta, E.: PowerMap: Mapping the real semantic web on the fly. In: Cruz, I., Decker, S., Allemang, D., Preist, C., Schwabe, D., Mika, P., Uschold, M., Aroyo, L.M. (eds.) ISWC 2006. LNCS, vol. 4273, pp. 414–427. Springer, Heidelberg (2006)

Extracting Explicit and Implicit Causal Relations from Sparse, Domain-Specific Texts

Ashwin Ittoo and Gosse Bouma

University of Groningen, 9747 AE
Groningen, The Netherlands
{r.a.ittoo,g.bouma}@rug.nl

Abstract. Various supervised algorithms for mining causal relations from large corpora exist. These algorithms have focused on relations explicitly expressed with causal verbs, e.g. "to cause". However, the challenges of extracting causal relations from domain-specific texts have been overlooked. Domain-specific texts are rife with causal relations that are implicitly expressed using verbal and non-verbal patterns, e.g. "reduce", "drop in", "due to". Also, readily-available resources to support supervised algorithms are inexistent in most domains. To address these challenges, we present a novel approach for causal relation extraction. Our approach is minimally-supervised, alleviating the need for annotated data. Also, it identifies both explicit and implicit causal relations. Evaluation results revealed that our technique achieves state-of-the-art performance in extracting causal relations from domain-specific, sparse texts. The results also indicate that many of the domain-specific relations were unclassifiable in existing taxonomies of causality.

Keywords: Relation exaction, Causal relations, Information extraction.

1 Introduction

Causal relations, between causes and effects, are a complex phenomenon, pervading all aspects of life. Causal relations are fundamental in many disciplines, including philosophy, psychology and linguistics. In Natural Language Processing (NLP), algorithms have been developed for discovering causal relations from large general-purpose [2,4,7,8,14] and bio-medical corpora [9]. These algorithms rely extensively on hand-coded knowledge (e.g. annotated corpora), and only extract explicit causal relations. Explicit relations are realized by explicit causal patterns, predominantly assumed to be causal verbs [4,7,14]. Causal verbs (e.g. *"induce"*) are synonymous with the verb *"to cause"*. They establish a causal link between a distinct causal-agent (e.g. "rain"), and a distinct effect (e.g. "floods"), as in *"rain causes floods"*.

The discovery of causal relations from texts in other domains (e.g. business/corporate) has been largely overlooked despite numerous application opportunities. In Product Development/Customer Service, for instance, causal relations encode valuable operational knowledge that can be exploited for improving product quality. For example, in *"broken cable resulted in voltage loss"*, the causal relation between

R. Muñoz et al. (Eds.): NLDB 2011, LNCS 6716, pp. 52–63, 2011.
© Springer-Verlag Berlin Heidelberg 2011

the cause "broken cable" and the effect "voltage loss", established by the pattern "*resulted in*", helps engineers during product diagnosis. Similarly, in *"new analog processor causes system shutdown"*, the causal relation between "new analog processor" and "system shutdown", realized by the pattern *"causes"*, provides business organizations with insights on customer dissatisfaction.

However, extracting causal relations from domain-specific texts poses numerous challenges to extant algorithms. A major difficulty in many domains is the absence of knowledge resources (e.g. annotated data), upon which traditional algorithms rely. In addition, current techniques are unable to detect implicit causal relations, which are rife in the English language. Implicit relations are realized by implicit causal patterns. These patterns do not have any (explicit) causal connotation. But they subtly bias the reader into associating certain events in the texts with causal-agents or effects [10]. Thus, implicit patterns have a causal valence, even though they are not synonymous with *"to cause"*. We consider 3 main types of implicit causal relations. Relations of the first type, *T1*, are realized by resultative and instrumentative verbal patterns. These verbs, for e.g. *"increase"*, *"reduce"*, *"kill"*, inherently specify (part of) the resulting situation, as in *"the temperature increased"*. The second type of implicit causal relations, *T2*, involves patterns that make the causal-agents inseparable from the resulting situations [10]. Such patterns include *"mar (by)"*, *"plague(by)"*. For example, in *"white spots mar the x-ray image"*, the causal-agent "white spots" is an integral component of the result "marred x-ray image". The last type of implicit causal relations, *T3*, involves non-verbal patterns, for e.g. the preposition *"due to"*, as in *"replaced camera due to horizontal calibration problem"*. Besides the difficulties posed by implicit patterns, existing algorithms are also unable to disambiguate ambiguous causal relations that involve polysemous patterns (e.g. *"result in"*, *"lead to"*). These patterns express causality only in restricted contexts. For e.g., the pattern *"lead to"* establishes a causal relation in *"smoking leads to cancer"*, but not in *"path leads to garden"*.

To address these challenges, we develop and present a framework for automatically extracting high quality causal relations from domain-specific, sparse corpora. We implemented our methodology in a prototype as part of the DataFusion initiative[1], which aims at enhancing product quality and customer satisfaction using information extracted from corporate texts. The crux of our approach lies in acquiring a set of explicit and implicit causal patterns from Wikipedia, which we exploit as a knowledge-base. We then use these patterns to extract causal relations from domain-specific documents. Our strategy of applying the knowledge acquired from Wikipedia to specialized documents is based on domain-adaptation [3]. It circumvents the data sparsity issues posed by the domain-specific, corporate documents.

Our contributions are as follows. We present a minimally-supervised algorithm that extracts causal relations without relying on hand-coded knowledge. Also, our algorithm accurately disambiguates polysemous causal patterns, and discovers both explicit and implicit causal relations. In addition, we represent the extracted causal patterns as sophisticated syntactic structures, which overcome the shortcomings of traditional pattern representations based on surface-strings.

---

[1] DataFusion is a collaboration between academia and industry, sponsored by the Dutch Ministry of Economic Affairs.

Our experiments revealed that our approach achieved state-of-the-art performance in extracting causal relations from real-life, domain-specific and sparse documents. These results also demonstrate that Wikipedia can be effectively leveraged upon as a knowledge-base for extracting domain-specific causal relations. Furthermore, we found out that some of the identified domain-specific relations were not conclusively classifiable in Barriere's taxonomy of causality [1].

2 Related Work

Causality is a complex, non-primitive relation, which can be refined into more specialized sub-relations. A recent taxonomy of causal relations is that of Barriere [1], which distinguishes between *existential and influential causality*. *Existential causality* pertains to the creation, destruction, prevention and maintenance of events (or entities). *Influential causality* modifies features of events by increasing, decreasing or preserving their values.

Algorithms for extracting causal relations from texts adopt either a pattern-based or supervised-learning approach. The pattern-based algorithms in [8,9] extract text segments that match hand-crafted patterns, e.g. *"is the result of"*. These techniques detect explicit causal relations from Wall Street Journal (WSJ) articles and from Medline abstracts with precisions of 25% and 76.8% respectively. Their corresponding recall scores are respectively 68% and 75.9%. In the supervised approach of [7], sentences in the L.A. Times corpus that contain explicit causal verbs (e.g. *"to cause"*) are manually annotated as positive or negative examples of causal relations. They are used to train a decision tree classifier, which detects new relations with a precision of 73.9% and recall of 88.7%. In [2], a support vector machine (SVM) trained over the manually annotated SemEval 2007 Task 4 corpus achieved an accuracy of 77.5% in identifying cause-effect noun pairs. SVMs are also used in [4], where they are trained over manually annotated texts of the WSJ, and detect causal relations with a precision of 24.4% and recall of 79.7%. Another supervised-learning approach is that of [14]. An SVM, trained on sub-graphs from annotated WSJ sentences, identifies causal relations with a precision of 26% and recall of 78%.

Existing algorithms have been applied solely to large general-purpose texts or to bio-medical documents. The discovery of causal relations from other domains (e.g. corporate documents) poses new challenges yet to be addressed. Readily-available resources (e.g. hand-crafted patterns and annotated data), which are extensively used by traditional algorithms, are inexistent in many domains. Also, domain-specific texts are sparse, and negatively impact the performance of relation extraction techniques [6]. In addition, these texts are rife with implicit causal relations, which are realized by implicit verbal and non-verbal causal patterns. Implicit patterns and relations are more complex and difficult to detect than their explicit counterparts [7], traditionally extracted by current algorithms. Furthermore, extant algorithms are unable to precisely disambiguate ambiguous causal relations, which are realized by polysemous patterns.

We address these challenges by developing a framework for extracting explicit and implicit causal relations from domain-specific texts in a minimally-supervised fashion. Our proposed approach is described in the next section.

3   Methodology for Extracting Causal Relations

Our overall framework for mining causal relations from a domain-specific corpus is depicted in Figure 1 (dotted arrows correspond to inputs to the various phases, filled and solid arrows represent the outputs). To circumvent the data-sparsity issues of domain-specific texts, we first acquire a set of causal patterns from Wikipedia. We choose Wikipedia as a knowledge-base since its large size offers ample evidence for accurate statistical inferences. As it is a broad-coverage resource, it is also likely to contain the wide variety of explicit and implicit linguistic patterns that express causality. In addition, Wikipedia is readily-available, and has been successfully employed in NLP applications [12]. We transform Wikipedia's sentences into lexico-syntactic patterns in the Pattern Acquisition phase (Section 3.1). In Causal Pattern Extraction (Section 3.2), a minimally-supervised algorithm selects those patterns that encode causality. The harvested patterns are used during Causal Relation Extraction (Section 3.3) to discover causal relations from domain-specific, sparse documents. We refer to these documents as the *target texts.*

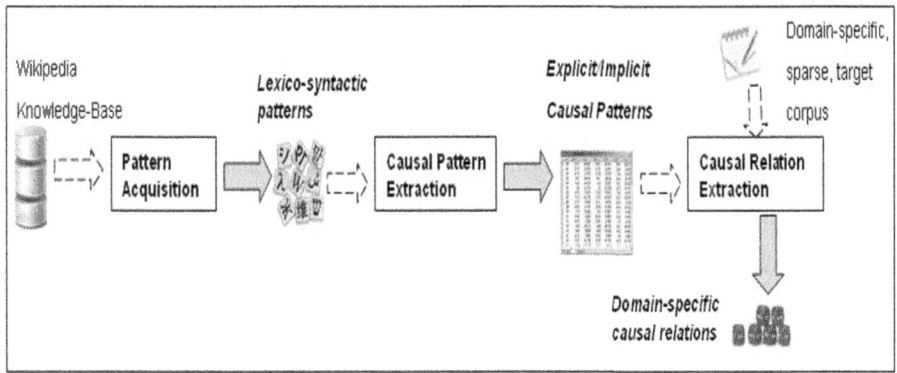

Fig. 1. Overall architecture of framework for extracting domain-specific causal relations

3.1   Pattern Acquisition

We syntactically parse Wikipedia's sentences, and represent the relation between each event-pair as the shortest path that connects the pair in the parse trees[2]. Such a path corresponds to a lexico-syntactic pattern. Sample (lexico-syntactic) patterns, the pairs they sub-categorize in Wikipedia, and the pair-pattern frequency are shown in the 4th, 3rd, 2nd and 1st columns of Figure 2. These patterns encode different semantic relations, as expressed in their corresponding Wikipedia sentences. The first 2 patterns are respectively inferred from sentences *S1 = "hurricanes are the most severe climatic disturbance in this area and have been known to cause extensive damage"* and *S2="hiv, the virus which causes aids, is transmitted through contact between blood"*. They express causal relations between the pairs "hurricane-damage" and "hiv-aids". The 3rd

---

[2] At this stage, we only consider nominal events, corresponding to noun phrases.

pattern, derived from *S3="the poem consists of five stanzas written in terza rima"*, expresses a part-whole relation between the pair "stanza-poem". ARG1 and ARG2 are generic placeholders, representing the pairs connected by the patterns.

```
14 | hurricane | damage | ARG1+nsubj < cause > dobj+ARG2
11 | hiv | aids | ARG1+nsubj < cause > dobj+ARG2
5  | stanza | poem | ARG2+nsubj < consist > prep+of+pobj+ARG1
```

Fig. 2. Lexico-syntactic patterns, pairs, and statistics from Wikipedia

Compared to the conventional surface-strings that existing algorithms employ to represent their patterns, our lexico-syntactic patterns neutralize word order and morphological variations. Thus, they alleviate the manual authoring of a large number of surface-patterns. For example, we derive a single general pattern to represent the relations in *S1* and *S2*. Our patterns also capture long range dependencies, regardless of the distance between related pairs and their positions in surface texts, for e.g. between "hurricane-damage" in *S1*.

3.2   Causal Pattern Extraction

We develop a minimally-supervised algorithm for determining which of the patterns are causal. Unlike traditional algorithms, it does not require annotated training data. It is initialized with cause-effect pairs (e.g. "hiv-aids"), called seeds. Our algorithm then starts by identifying patterns in Wikipedia that connect these pairs (seeds). The reliability, $r(p)$, of a pattern, p, is computed with equation (1) [13]. It measures the association strength between p and pairs, e, weighted by the pairs' reliability, $r(e)$. Initially, $r(e)=1$ for the seeds. In (1), E refers to the set of pairs, and $pmi(e,p)$ is the point-wise mutual information [5] between pattern p (e.g. "*cause*") and pair $e=x$-y (e.g. "hiv-aids")

$$r(p) = \frac{\sum_{e \in E}\left(\frac{pmi\ (e,p)}{max_{pmi}} \times r(e) \right)}{|E|} \quad (1)$$

Then, we select the top-k most reliable patterns, and identify other pairs that they connect in Wikipedia. The pairs' reliability is estimated using equation (2). In the next iteration, we select the top-m most reliable pairs, and extract new patterns that connect them.

$$r(e) = \frac{\sum_{p \in P}\left(\frac{pmi(e,p)}{max_{pmi}} \times r(p) \right)}{|P|} + purity(e) \quad (2)$$

The reliability, $r(e)$, of a pair, e, is defined by two components. In the first component, which is analogous to the pattern reliability, $|P|$ is a set of patterns. The second component determines the pair's *purity* in instantiating a causal relation. It stems from the Latent Relation hypothesis that pairs which co-occur with similar patterns instantiate similar semantic relations [15]. Thus, if a pair, $e=x$-y, is connected by a pattern, p_{ref}, which also connects our initial seeds, then e instantiates causality. As p_{ref}, we choose the pattern "*caused by*" since it explicitly and unambiguously expresses causality, and specifies the causal link between a distinct causal-agent and an effect. Also, "*caused by*" was found to co-occur with all our seeds. The *purity* of a pair $e=x$-y (e.g. "rain-flooding") is then calculated as its probability of being sub-categorized by p_{ref}. This is obtained by querying the Yahoo! search engine[3] with $q=$ "*y* p_{ref} *x*" (e.g. "*flooding caused by rain*"), and determining the fraction of the top-50 search results that contains phrase q in their summaries. In this way, invalid pairs, for e.g. "trail-summit", identified by ambiguous causal patterns, for e.g. "*lead to*" ("*trail leads to summit*"), are assigned lower purity values. This is because queries like $q=$"*summit caused by trail*" (formed by the pair "trail-summit" and p_{ref}) are incoherent, and are unlikely to return any search results. The overall reliability scores of these invalid pairs will then be smaller, and they will be discarded. Otherwise, the invalid pairs will be selected in the next iteration, and incorrect patterns connecting them, for e.g. "*pass through*" ("*trail passes though summit*"), will be extracted, degrading our performance. Conversely, valid pairs will be awarded much higher purity values, increasing their reliability scores. They are selected in the next iteration, enabling our algorithm to extract both explicit and implicit causal patterns that connect them.

Our recursive procedure of learning new patterns from pairs, and vice-versa is repeated until a suitable number, t, of causal patterns have been harvested. Parameter values for k, m, and t will be defined during Experimental Evaluation (Section 4.2). Figure 3 shows 2 example patterns extracted by our minimally-supervised algorithm from Wikipedia. ARG1 and ARG2 respectively denote the cause and effect events.

```
ARG1+nsubj < induce > dobj+ARG2
ARG2+nsubj < increase > prep+by+pobj+ARG1
```

Fig. 3. Example of causal patterns learnt from Wikipedia

The 1st pattern explicitly indicates causality with a causal verb, namely "*induce*", as in "*a non-zero current induces a magnetic field by Ampere's law*". The 2nd pattern implicitly expresses causality with a resultative/instrumentative verb, viz. "*increase (by)*", as in "*the population of the state of Nebraska was increased by positive birth-rates*". In this sentence, "*increase (by)*" biases the reader into ascribing the causal-agent role to "positive birthrates", as in "*positive birthrates caused an increase in the population*".

3.3   Causal Relation Extraction

We use the reliable causal patterns harvested from Wikipedia to extract relations from a domain-specific, sparse (target) corpus. A domain-specific causal relation is made up of a causal pattern and the events that it connects in the target corpus. We consider both nominal events like noun phrases (e.g. "loose connection"), and verbal events

---

[3] Using the Yahoo! Boss API, http://developer.yahoo.com/search/boss/boss_guide/

like verb phrases (e.g. "replacing the cables"). Figure 4 illustrates a domain-specific causal relation, extracted from the sentence *S4="replacing the cables generated intermittent x-rays"* in the target corpus.

```
<pattern = "generate" , cause="replace cable" , effect="intermittent x-ray">
```

Fig. 4. Domain-specific causal relation triples

4 Experimental Evaluation

This section describes experiments to evaluate the performance of our approach in extracting causal relations from domain-specific, sparse texts.

4.1 Pattern Acquisition

Using the Stanford parser [11], we syntactically parsed the sentences of the English Wikipedia collection [16] (August 2007 dump, around 500 million words). We identified 2,176,922 distinct lexico-syntactic patterns that connected 6,798,235 distinct event-pairs. Sample patterns, event-pairs they sub-categorized and the pair-pattern co-occurrence frequencies were shown in Figure 2.

4.2 Causal Pattern Extraction

To identify patterns that express causality, we implemented a minimally-supervised algorithm. It was initialized with 20 seeds. Seeds were cause-effect pairs that unambiguously instantiated causal relations (e.g. "bomb-explosion") and that occurred at least 3 times in Wikipedia. The 1st iteration of our algorithm extracted 10 most reliable patterns connecting the seeds. Then, 100 other pairs that were also connected by these patterns were extracted. In subsequent iterations, we identified 5 additional patterns (with the previously extracted pairs), and 20 additional pairs (with the previously extracted patterns). The values of parameters k and m (Section 3.2) were therefore $k=|P|+5$ and $m=|E|+20$, where $|P|$ and $|E|$ are respectively the number of previously collected patterns and pairs.

The recursive process of learning new pairs from patterns, and vice-versa was repeated until we observed a drop in the quality of the harvested patterns (e.g. when non-causal patterns were extracted). The performance peaked in the 16th iteration, where we harvested $t=81$ causal patterns[4] from Wikipedia. Examples are in Table 1. Fifteen of the patterns were causal verbs that explicitly indicated causality, for e.g. "*induce*" (column T0). Resultative/instrumentative verbs, for e.g. "*increase*", which implicitly expressed causality, accounted for 50 patterns (column T1). Fourteen non-verbal, implicit causal patterns were also found (column T3). They included nominalizations of resultative/instrumentative verbs, for e.g. "*increase in*"; adjectives, for e.g. "*responsible for*"; and prepositions, for e.g. "*due to*". Implicit causal patterns,

---

[4] Precision was roughly estimated as 79%.

Table 1. Explicit and implicit causal patterns extracted from Wikipedia

T0 (15/81=18.5%)	T1 (50/81 =61.7%)	T2 (2/81=2.5%)	T3 (14/81=17.3%)
cause (by, of)	affect (with, by)	mar (by)	drop (in)
induce	decrease (by, to)	plague (by)	due to
lead to	increase (by, to)		increase (in)
result in	inflict (by, on)		rise (in)
spark	limit (by, to)		source of
trigger	prevent (by, to)		responsible for

which made the causal-agent inseparable from the result (effect), were least frequent. Only 2 such patterns, viz. *"mar (by)"* and *"plague (by)"*, were detected (column T2). We can deduce from these results that causality in the English Wikipedia corpus is more commonly expressed by implicit causal patterns, particularly resultative/instrumentative verbs, than by explicit causal verbs.

4.3 Causal Relation Extraction

The domain-specific (target) corpus from which we extracted causal relations contained 32,545 English documents (1.1 million words). The documents described customer complaints and engineers' repair actions on professional medical equipment. They were linguistically pre-processed (e.g. to derive syntactic information) prior to their analysis for relation extraction.

Out of the 81 causal patterns from Wikipedia, 72 were found to connect nominal and verbal events in the target corpus, yielding a total of 9,550 domain-specific causal relations. Examples are presented in Table 2. The 3rd column shows the causal patterns that realized these relations. The 2nd column gives the percentage of the 9,550 relations that contained these patterns. The domain-specific causal relations are

Table 2. Domain-specific causal relations from target corpus

Id	Freq (%)	Causal Pattern	Linguistic realization
T1	55	*destroy*	*"short-circuit in brake wiring destroyed the power supply"*
		prevent	*"message box prevented viewer from starting"*
		exceed	*"breaker voltage exceeded allowable limit"*
		reduce	*"the radiation output was reduced"*
T0	23	*cause (by)*	*"gray lines caused by magnetic influence"*
		induce	*"bad cable extension might have induced the motion problem"*
T3	21	*due to*	*"replacement of geometry connection cable due to wear and tear"*
		drop in	*"there was a slight drop in the voltage"*
T2	1	*mar*	*"cluttered options mars console menu"*

depicted in the 4th column as their linguistic manifestations in the target corpus. The 1st column is an identifier for each pattern/relation group.

The most frequent patterns, participating in around 55% of the extracted relations, were resultative/instrumentative verbs (e.g. *"destroy"*, *"exceed"*), which implicitly expressed causality. The high frequency of these patterns in our target corpus could be attributed to their common use in describing product failures (e.g. *"short-circuit in brake wiring destroyed the power supply"*), and in reporting observations on product behavior (e.g. *"breaker voltage exceeded allowable limit"*). We observed that these relations had optional causal-agents, and that they could not be conclusively classified in Barriere's taxonomy of causality [1]. When the causal-agents were specified, as in *"[short-circuit in brake wiring$_{causal-agent}$] destroyed the power supply"*, the relations indicated the creation or destruction of events. Thus, according to Barriere's taxonomy, they established *existential causality*. However, when their causal-agents were unspecified, they expressed changes in magnitude, as in *"breaker voltage exceeded allowable limit"*. Then, these relations established *influential causality* based on Barriere's taxonomy. In the domain of Product Development/Customer Service (PD-CS), these relations provide useful information for product quality improvement. The next most frequent patterns, appearing in around 23% of the extracted relations, were explicit causal verbs (e.g. *"cause by"*, *"induce"*). These relations always specified a distinct causal-agent and its effect, as in *"[gray lines$_{effect}$] caused by [magnetic influence$_{causal-agent}$]"*. They established *existential causality* since they were realized by verbs synonymous with *"to cause (to exist)"*. In the PD-CS domain, these relations can be exploited to facilitate the diagnosis procedure of engineers. Around 21% of the relations were realized by non-verbal implicit causal patterns, such as noun phrases (e.g. *"drop in"*) and prepositions (e.g. *"due to"*). These relations were not conclusively classifiable in Barriere's taxonomy. When they were realized by noun phrases, they described unexplained phenomena with unknown causes. Hence, their causal-agents were often unspecified, as in *"there was a slight drop in the voltage"*. The relations then established *influential causality*. Conversely, when they were realized by prepositions, they established *existential causality* since they always specified a causal-agent that brought a resulting situation (effect) into existence, as in *"[replacement of geometry connection cable$_{effect}$] due to [wear and tear$_{causal-agent}$]"*. In the PD-CS domain, these relations provide pertinent information on customer dissatisfaction and on repair actions of engineers. The rarest patterns, appearing in less than 1% of the relations, were those that implicitly expressed causality by making the causal-agent inseparable from the result (e.g. *"mar"*, *"plague"*). For example, in *"cluttered options mars console menu"*, [cluttered options$_{causal-agent}$] is an integral part of the [marred console menu$_{effect}$]. These relations could not be unambiguously classified in Barriere's taxonomy. They established neither *existential* nor *influential causality*.[5]

Nine (out of 81) patterns from Wikipedia were not found in the target corpus. They included explicit causal verbs, for e.g. *"spark"*; implicit resultative/instrumentative verbs, for e.g. *"end"* and *"outstrip"*; and non-verbal expressions for e.g. *"growth"*. These patterns were unlikely to occur in our corpus of customer complaints and of engineers' repair actions.

---

[5] To some extent, they can be treated as influential causality.

To evaluate the performance of our approach, we randomly selected 3000 of the extracted causal relations, equally distributed across the groups T0, T1, and T3 of Table 2 (i.e. 1000 explicit relations with causal verbs, 1000 implicit relations involving resultative/instrumentative verbs, and 1000 implicit relations with non-verbal expressions). Relations in the group T2 were omitted as they were too sparse. The evaluation set of 3000 relations was manually inspected by human judges, and 2295 were deemed to correctly express causality (*true_positive*). The remaining 705 relations were incorrect (*false_positive*). They were realized by causal patterns, for e.g. *"due to"*, which did not connect valid events in the target corpus, as in *"screen due to arrive today"*. Using equation (3), we then estimated the precision of our approach as 76.5%. To determine the recall, a gold-standard of 500 valid causal relations was manually constructed from a subset of the target corpus. The sub-corpus was then automatically analyzed by our approach. We detected 410 of the gold-standard relations (*true_positive*), but failed to discover remaining 90 (*false_negative*). The recall was computed as 82% using equation (4).

$$precision = \frac{true\_positive}{true\_positive + false\_positive} \qquad (3)$$

$$recall = \frac{true\_positive}{true\_positive + false\_negative} \qquad (4)$$

Our scores of precision (76.5%) and recall (82%) compare favorably with those reported by other state-of-the-art algorithms [2,4,7,8,9,14]. These latter techniques, however, extensively relied on manually-crafted knowledge (e.g. annotated data). Also, they focused solely on detecting causal relations that were explicitly realized by causal verbs in large corpora. Our approach, on the other hand, extracts both explicit and implicit causal relations, expressed by verbal and non-verbal causal patterns, from sparser texts. In addition, it is minimally-supervised, and alleviates the need for hand-coded knowledge, which is expensive to generate in most domains.

We conducted another set of experiments to illustrate the significance of our event-pair *purity* measure (Section 3.2) in extracting the most reliable causal patterns and relations. We re-implemented our minimally-supervised algorithm such that the reliability, $r(e)$, of an event-pair, e, was estimated with only the 1st component of equation (2). That is, the *purity* estimation was omitted. The pattern reliability measure in equation (1) was unchanged. Our algorithm was initialized with seeds, and was ran over the Wikipedia collection as before. We observed that invalid pairs, for e.g. "street-exit", which were connected by many ambiguous patterns, for e.g. *"lead to"*, *"result in"* (*"street leads to exit"*, *"street results in exit"*), were awarded higher reliability scores than valid ones. Subsequently, these invalid pairs were selected, and other incorrect patterns that connected them, for e.g. *"at"* (*"street at exit"*), were in turn extracted. The quality of the harvested patterns deteriorated in much earlier iterations. Optimal performance was achieved in the 7th iteration, where 34 causal patterns were identified[6]. We also found out that the new implementation failed to discover many of the implicit causal patterns that had been detected by our original algorithm. We used

---

[6] Precision was roughly estimated as 65%.

the 34 causal patterns discovered by the new implementation from Wikipedia to extract domain-specific causal relations from the target corpus. The precision and recall were calculated by equations (3) and (4) as 65.7% and 68.3% respectively. These scores indicate a drop of performance compared to our original algorithm, and suggest that the new implementation is less accurate in extracting causal patterns and relations. The results reveal that our event-pair *purity* measure for disambiguating polysemous patterns is crucial to reliably harvest the wide range of patterns participating in both explicit and implicit causal relations.

5   Conclusion

In this paper, we have described an approach for mining causal relations from texts. The novelty in our technique lies in the use of Wikipedia as a knowledge-base. We proposed a minimally-supervised algorithm, which unlike previous techniques, alleviates the need for manually-annotated training data. Our algorithm employs sophisticated statistical analyses for disambiguating and extracting both explicit and implicit causal patterns from Wikipedia. The causal patterns from Wikipedia are then used to detect causal relations from a domain-specific corpus of corporate documents. This strategy, of applying the knowledge acquired from one domain (Wikipedia) to another domain (corporate documents), is based on previous studies in domain-adaptation. It overcomes the data-sparsity issues that domain-specific texts pose to relation extraction techniques. Evaluations results on real-life data reveal that our approach achieves state-of-the-art performance in discovering explicit and implicit causal relations from domain-specific, sparse texts. As future work, we are investigating whether the performance can be improved by using certain keywords that indicate causality, for e.g. modal verbs ("can", "will", "may") and subordinating conjunctions ("after", "as", "because"). We are also interested in question-answering systems based on causal relations for answering "why" questions.

Acknowledgment. This work is being carried out as part of the DataFusion project, sponsored by the Netherlands Ministry of Economic Affairs, Agriculture and Innovation, under the IOP-IPCR program.

References

1. Barriere, C.: Hierarchical refinement and representation of the causal relation. Terminology 8(1), 91–111 (2002)
2. Beamer, B., Bhat, S., Chee, B., Fister, A., Rozovskaya, A., Girju, R.: UIUC: A knowledge-rich approach to identifying semantic relations between nominals. In: 4th International Workshop on Semantic Evaluations, pp. 386–389 (2007)
3. Ben-David, S., Blitzer, J., Crammer, K., Kulesza, A., Pereira, F., Vaughan, J.W.: A theory of learning from different domains. Machine Learning 79(1), 151–175 (2010)
4. Bethard, S., Corvey, W., Klingenstein, S., Martin, J.H.: Building a corpus of temporal-causal structure. In: 6th International Language Resources and Evaluation Conference (2008)
5. Church, K.W., Hanks, P.: Word association norms, mutual information, and lexicography. Computational Linguistics 16(1), 22–29 (1990)

6. Cimiano, P., Pivk, A., Schmidt-Thieme, L., Staab, S.: Learning taxonomic relations from heterogeneous sources of evidence. In: Buitelaar, P., Cimiano, P., Magnini, B. (eds.) Ontology Learning From Text: Methods, Evaluation, and Applications. IOS Press, Amsterdam (2005)
7. Girju, R.: Automatic detection of causal relations for question answering. In: ACL 2003 Workshop on Multilingual Summarization and Question Answering, pp. 76–83 (2003)
8. Khoo, C.S.G., Kornfilt, J., Oddy, R.N., Myaeng, S.H.: Automatic extraction of cause-effect information from newspaper text without knowledge-based inferencing. Literary and Linguistic Computing 13(4), 177–186 (1998)
9. Khoo, C.S.G., Chan, S., Niu, Y.: Extracting causal knowledge from a medical database using graphical patterns. In: 38th Annual Meeting on Association for Computational Linguistics, pp. 336–343 (2000)
10. Khoo, C., Chan, S., Niu, Y.: The semantics of relationships: an interdisciplinary perspective. Kluwer Academic, Dordrecht (2002)
11. Klein, D., Manning, C.D.: Accurate unlexicalized Parsing. In: 41st Annual Meetingon Association for Computational Linguistics, pp. 423–430 (2003)
12. Medelyan, O., Milne, D., Legg, C., Witten, I.H.: Mining meaning from Wikipedia. International Journal of Human-Computer Studies 67(9), 716–754 (2009)
13. Pantel, P., Pennacchiotti, M.: Espresso: Leveraging Generic Patterns for Automatically Harvesting Semantic Relations. In: COLING/ACL 2006, pp. 113–120 (2006)
14. Rink, B., Bejan, C., Harabagiu, S.: Learning Textual Graph Patterns to Detect Causal Event Relations. In: 23rd Florida Artificial Intelligence Research Society International Conference (2010)
15. Turney, P.D.: The latent relation mapping engine: algorithm and experiments. Journal of Artificial Intelligence Research 33, 615–655 (2008)
16. Wikipedia, XML format (August 2007),
    http://ilps.science.uva.nl/WikiXML

Topics Inference by Weighted Mutual Information Measures Computed from Structured Corpus

Harry Chang

AT&T Labs – Research, Austin, TX USA
Harry_chang@labs.att.com

Abstract. This paper proposes a novel topic inference framework that is built on the scalability and adaptability of mutual information (MI) techniques. The framework is designed to systematically construct a more robust language model (LM) for topic-oriented search terms in the domain of *electronic programming guide* (EPG) for broadcast TV programs. The topic inference system identifies the most relevant topics implied from a search term, based on a simplified MI-based classifier trained from a highly structured XML-based text corpus, which is derived from continuously updated EPG data feeds. The proposed framework is evaluated against a set of EPG-specific queries from a large user population collected from a real world web-based IR system. The MI-base topic inference system is able to achieve 98 percent accuracy in recall measurement and 82 percent accuracy in precision measurement on the test set.

Keywords: Mutual Information, Latent Semantic Analysis, Natural Language Processing, TF-IDF, EPG.

1 Introduction

A unique challenge in building *efficient* language models (LMs) for information retrieval (IR) systems that support natural language (NL) based search interface is the trade-off between the simplicity of NL-based input language for the end user and the complexity of statistical *n*-gram based LMs used by the IR systems. Influenced by popular web search engines, average users learn to compress their complex and often ambiguous search expressions into only a few words and rely on the smart system to decipher their intention. This places a heavy burden on speech-driven search applications requiring a *high coverage* of search expressions intended for specific targets and at the same time a *high precision* inference engine to determine the underlying topics implied in the search term. This is because individual words often either provide unreliable evidence towards the intended conceptual topics or carry ambiguous associations with unrelated topics.

One well-known technique for topic inference is the use of *latent semantic analysis* (LSA) [1,2] and its derivatives, such as *probabilistic latent semantic analysis* (PLSA) [3,4]. The basic paradigm for LSA rests on the assumption that there is some underlying semantic core associated with a group of words, phrases, or paragraphs that differentiate one topic from others. However, the entity representing a conceptual

R. Muñoz et al. (Eds.): NLDB 2011, LNCS 6716, pp. 64–75, 2011.
© Springer-Verlag Berlin Heidelberg 2011

topic is often obscured or masked by many *noisy* words in the group. To overcome this problem, the best known LSA implementation uses a technique called singular value decomposition (SVD). Unfortunately, SVD is known to be computationally expensive because it requires the iterative computation of a very large similarity matrix from a training corpus [5]. Similar PLSA methods are arguably even more expensive. All these methods are known to be lack of adaptability to fast changing text content [6]. Recent studies show that a *pointwise mutual information* (PMI) based similarity measure is able to achieve a comparable performance with LSA when using a large amount of training data, and at a substantially lower cost [6,7].

We propose a new topic inference framework that is built upon the scalability and more importantly, the adaptability of the PMI method. The framework uses a weighted MI model trained on a domain-specific and highly structured text corpus without any supervision. The model can be updated automatically from the new data source in the same metadata structure. A simple inference process is developed to compute the likelihood that the topics selected at runtime are most relevant to a search term using the MI-based model augmented with the domain knowledge. To identify the topics associated with the words in a search term, we use a two-pass procedure. First, it determines the existence (or not) of a unique topic intended by the user. If this first pass fails, it would suggest the most relevant topics that are inferred implicitly from the *n*-gram words in the original search term. The framework can be easily extended to support longer ranging *n*-grams beyond our experiments, which only used the first two *n*-grams. It only utilizes the relationships between different entity classes that are directly computable from a highly structured text corpus, such as the one used in our experiment. From a NL processing perspective, this framework offers a very simple conceptual model since it does not explicitly differentiate between syntactic entities and semantic elements. It draws powers of inference entirely from the domain-specific corpus and has a very low computational requirement at run time.

2   Domain of Interests

For over two decades, statistical *n*-gram based LMs have been effective in modeling human language for IR systems. Researchers commonly face a well-known trade-off between the amount of training data chosen for a given domain and the precision of the final search results at run time. However, if every unigram, bigram, trigram or even longer ranging *n*-gram for the domain is kept in the underlying LMs to ensure a perfect coverage (recall measurement) the precision would drop to a level that makes it unusable for many IR systems requiring a spoken search interface. This factor creates a unique motivation for this study because the targeted LMs are ultimately intended for a spoken IR system requiring an extremely high precision. This is especially true for a TV-centric online system with a NL-based search interface in both mobile and home entertainment environments.

To average online users, the most common form of an Electronic Programming Guide (EPG) for TV networks is a two-dimensional grid as illustrated in Figure 1. Besides *program title* and *channel name*, there is *other info* visible in the EPG grid such as *subtitle, description, genres, cast members*, and *movie rating* (*PG-13, R*, etc), which is illustrated by zooming into an example cell where the textual information for a movie title *"The World is Not Enough"* is scheduled for a specific *channel:time* slot. The language models for TV viewers to express their search intent are complex and fluid due to ever expanding TV lexicons and multi-dimensional descriptors for TV shows or movies as shown in Figure 1.

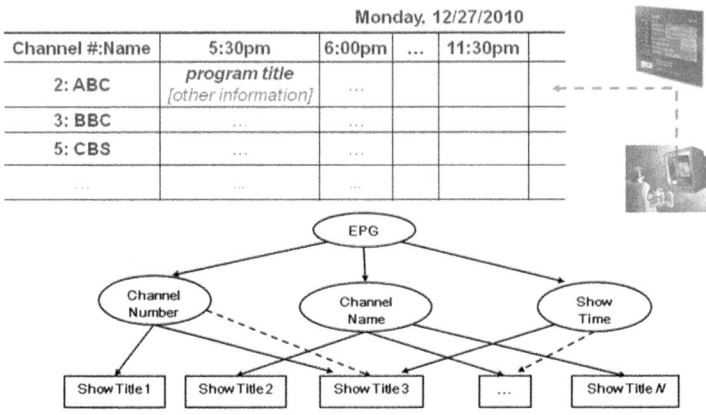

The textual information inside an example EPG cell

Title: "The World is Not Enough"		Genre: Action, Adventure, ...
Director: Michael Apted	Rating: PG-13	Date (Original Release): 1999
Actors: Pierce Brosnan, Robert Carlyle, Judi Dench …		
Description: James Bond goes to bat for queen and country by protecting an oil heiress from the terrorist who killed her father and retrieving M from his evil clutches.		

Fig. 1. A two-dimensional representation of a typical EPG and part of its taxonomy tree with a number of common conceptual nodes depicted

The titles of TV shows or movies showing on the TV networks in the U.S. have expanded considerably in the last 5 years. A recent study [8] shows that the EPG corpus created from the TV programming guides for ten major cities in the U.S. in 2008 alone contain over 45,000 unique titles with a vocabulary of approximately 26,400 words. The same study also suggests that power law weighted n-gram LMs can effectively cover over 75% of NL-based search queries by selecting the words occurring only in the title subset. However, our study shows that people do search EPG using NL-based expressions far beyond the title vocabulary, such as *"James Bond's movie,"* which can only be covered by the LMs created from the *description* field associated with a movie title in a typical EPG data feed as shown in Figure 1.

Using a highly structured metadata format, common EPG data sources contain not only the text descriptions for the programming content as illustrated above, but also the channel schedule associated with each program title. Thus, an intermediate goal in a NL-based search process for EPG-based TV content is to map the words expressed in a search term to the most relevant TV channels and subsequently, the most relevant programs. To follow this logic, the main objective for an NL-based topic inference system is to find a short list of TV channels carrying the programs most semantically relevant to the words in a search query not explicitly targeted for a specific TV show or movie, such as the 1-word expressions *Olympics* or *recipes,* or 2-word phrases like *Oklahoma football* and *nightly news.*

3 Methodology

The basic methodology used to develop the inference framework is to automatically acquire the computable knowledge from the structured corpus with very little or no manual supervision. This is because the framework has to deal with the structured corpus changed daily. The framework consists of four components: a domain-specific parser for extracting relevant entities from an EPG corpus with a well-defined taxonomy, a *n*-gram based classifier for program titles, an automatically trained (without any supervision) topic model from the entities extracted from the corpus, and a run-time topic inference engine.

3.1 TF-IDF Encoding with Domain-Based Weights

Techniques using Term Frequency Inverse Document Frequency (TF-IDF) to determine word relevance are well known to the IR research community [9,10]. For this study, a document in the EPG document space is automatically generated from a sequence of XML-based EPG data files using the text fields associated with a program title as illustrated in the movie example above. Each document has a title such as *"The World is Not Enough"* and many documents may share the same title such as *Seinfeld* which has over 100 different episodes. Each document starts with its title words followed by the words in *episodes* (if any) and/or *subtitles* (if any) data fields and then the *description* fields. In the movie example above, its corresponding document contains 33 words under this definition.

Term Frequency, *TF*, is defined in (1) where $n_{i,j}$ is the number of occurrences of the considered term (t_i) in document d_j, and the denominator is the sum of occurrences of all terms in document d_j, that is, the size of the document $\mid d_j \mid$. In the movie example *"The World is Not Enough"*, the word *Bond* has a single occurrence while the word *from* has two occurrences. Using (1) their corresponding *TF* values would be $TF_{Bond} = 0.0303$ and $TF_{from} = 0.0606$, respectively.

$$TF_{i,j} = \frac{n_{i,j}}{\sum_k n_{k,j}} \qquad (1)$$

For our domain, the relative importance of one document versus another is generally correlated with the popularity and schedule frequency of the underlying content. While the popularity of a TV show or movie is difficult to quantify, the schedule frequency can be directly computed from common EPG data sources, such as the one used in our experiment. Intuitively, if a TV show or a movie has been repeatedly airing on many channels over a recent time period, the viewers would be expected to register a higher semantic association for the words in the underlying document [which consists of program title and its various content descriptions, such as its cast members (e.g., *"Pierce Bronson"*) or the main characters (e.g., *"James Bond"*)], than for other words that only occur in a TV program or movie rarely shown during the same time period. To encode this "recency" factor in our overall weighting scheme, we also computed a weighted term frequency, $\widetilde{TF}$, based on the number of times a TV program has been showing on one or more channels covered by the EPG data source. For those programs airing during primetime, an extra weight is added in computing its schedule frequency.

The *inverse document frequency* (IDF) is a measure of the general importance of the term within a document space. For example, if the term t_i only occurs in a single document d within the document space D, it would be the most important semantically because it can be used to uniquely identify the document. Therefore, its IDF value would have the highest value as defined in (2) where $|D|$ is cardinality of D, or the total number of documents in the corpus.

$$IDF(t_i) = \log \frac{|D|}{|\{d: t_i \in d\}|} \tag{2}$$

The basic model for TF-IDF can be considered as a weighting function $\aleph$ incorporating both TF and IDF and can be formally defined in (3). This weight, $\aleph$, is a statistical measure used to evaluate how important a word is to one document relative to others in the same document space.

$$\aleph_{i,j} = \widetilde{TF}_{i,j} \times IDF_i \tag{3}$$

For the EPG domain, the most frequently used words in a program title are often the most common words, as evident in the names of many popular TV programs such as *24*, *Heroes*, *Two and a Half Men*, *Dancing with the Stars*, and etc. This factor makes the traditional unigram-based TF-IDF technique less effective in terms of distinguishing between terms. To overcome this problem unique to our domain, we expand the basic term definition from unigram to bigrams and compute the TF-IDF values for all bigrams in the corpus including their weighted term frequency $\widetilde{TF}$. Table 1 lists the unigrams with the 10 highest TF-IDF values and the first 10 bigrams with the TF-IDF values ranked from the 200th place. As shown in Table 1.b, when bigram words have a similar TF-IDF value, their weighted $\widetilde{TF}$ value can help further differentiate one term versus another.

Table 1. Selected unigram and bigrams with their TF-IDF values

unigram	$\widetilde{TF}$	TF-IDF	bigram	$\widetilde{TF}$	TF-IDF
the	64589	3.4846	the i	44409	0.0120
and	25246	2.0763	number 2	22460	0.0120
of	25355	2.0453	ice ice	47016	0.0120
in	15024	1.5841	fire fire	65193	0.0120
part	11749	1.2219	double double	33268	0.0120
to	10870	1.1915	the best	86322	0.0119
at	7126	0.8801	the last	72509	0.0117
for	6222	0.7748	a can	6849	0.0117
on	5730	0.7599	great the	61359	0.0116
with	5859	0.7381	a tale	15025	0.0116

a. The unigrams with the 10 highest TF-IDF values

b. The first 10 bigrams with the TF-IDF values ranked from the 200th places

3.2 Topic Clustering with Mutual Information Measures

In a traditional document-topic hierarchy, each TV program could be classified as a distinct topic by itself within the EPG document space, such as *Seinfeld*. This approach has two drawbacks. First, it would create too many topics (over 90,000 just over a 2-year time period). Second, the study [8] suggests that average TV viewers tend to associate different episodes of a TV series or movie sequels with the prime title. Many TV viewers also associate a group of similar programs with a broader topic known as genres. However, TV or movie genres are often ambiguous and overlapping. For this reason, we decided to use channels as the basic classification unit for the topics in the EPG document space. This scheme has two advantages. First, average TV viewers are very familiar with the relationship between a channel and a TV program as illustrated in the 2-dimensional EPG grid shown in Figure 1. Second, the relationship between each TV program and the channels is quantifiable by parsing the schedule elements in the XML-based metadata file for each EPG time period. In the remainder of this section, we describe a corpus-based topic classification method using *mutual information* technique where each topic corresponds to an EPG channel.

Mathematically, mutual information (MI) measures the information that two discrete random variables X and Y share. From a perspective of NL processing, MI provides a simple and quantitative measure on how much information the presence/absence of a term t contributes to making the correct identification of a class c where t is a member. The technique has been used in a number of NL research fields such as detection of constituent boundaries in parsing algorithm [11], resolving ambiguity in query terms [12], and semantic text clustering [13,14]. For our domain, a term t would be n-gram words such as "*James Bond*" or a longer ranging n-gram based short phrase such as "*The World is Not Enough*". The class c would be an EPG channel defined by the programming documents for the shows on the channel.

In this paper, we measure MI with a function $I(U,C)$ that is formally defined by Manning et al. [15] in equation (4), where U is a random variable that takes values $e_t = 1$ (the document contains term t) and $e_t = 0$ (the document does not contain term t), and C is another random variable that takes values $e_c = 1$ (the document is in class c) and $e_c = 0$ (the document is not in the class c).

$$I(U,C) = \sum_{e_t \in \{0,1\}} \sum_{e_c \in \{0,1\}} P(U = e_t, C = e_c) \log_2 \frac{P(U=e_t,C=e_c)}{P(U=e_t)P(C=e_c)} \qquad (4)$$

Based on the principle of *maximum likelihood estimation*, the probability distribution function P can be estimated from a document space where U and C can be observed. This leads the authors [15] to assert that equation (4) is equivalent to equation (5).

$$I(U,C) = \frac{N_{11}}{N} \log_2 \frac{N \cdot N_{11}}{N_{1x}N_{x1}} + \frac{N_{01}}{N} \log_2 \frac{N \cdot N_{01}}{N_{0x}N_{x1}} + \frac{N_{10}}{N} \log_2 \frac{N \cdot N_{10}}{N_{1x}N_{x0}} + \frac{N_{00}}{N} \log_2 \frac{N \cdot N_{00}}{N_{0x}N_{x0}} \qquad (5)$$

The two subscripts for N have the values of e_t and e_c. Thus, N_{10} is the number of documents that contain t ($e_t = 1$) and are not in c ($e_c = 1$). The character "x" in the subscripts serves as a wildcard so that $N_{1x} = N_{10} + N_{11}$ is the number of documents that contain t ($e_t = 1$) independent of class membership ($e_c \in \{0,1\}$). Therefore the total number of documents in the corpus is $N = N_{00} + N_{01} + N_{10} + N_{11}$.

The EPG corpus used for this study contains 663,354 documents, each representing a unique program that has been shown at least once on one of 446 channels during the lifetime of the corpus. Thus, the total number of topics inducible for any given search term is 446. With this topic scheme, equation (5) can be explained more intuitively using an example. Consider the channel *Bravo* and the term *gossip* in the corpus. There are 2,223 unique programs showing on the channel *Bravo*. The term *gossip* appears at least once in 18 of those program documents. For all other channels, there are 1,817 program documents that also contain the term *gossip*. The counts of the number of documents with the four possible combinations of MI indicator values are listed in Table 2. Inserting the four indicators in (4) produces the final MI score with a value of 0.0000163985. To improve readability, all MI scores are multiplied by a factor of 1000.

Table 2. MI worked examples for the channel *Bravo* and the term *gossip*

	$e_c = e_{Bravo} = 1$	$e_c = e_{Bravo} = 0$
$e_t = e_{gossip} = 1$	18	1,817
$e_t = e_{gossip} = 0$	2,205	659,314

$$I(U,C) = \frac{18}{663354}\log_2\frac{663354 \cdot 18}{(18+1817)(18+2205)} + \frac{2205}{663354}\log_2\frac{663354 \cdot 2205}{(2205+659314)(18+2205)}$$
$$+ \frac{1817}{663354N}\log_2\frac{663354 \cdot 1817}{(18+1817)(1817+659314)}$$
$$+ \frac{659314}{663354}\log_2\frac{663354 \cdot 659314}{(2205+659314)(1817+659314)} = 0.0000163985$$

To further illustrate the relationship between words with high MI scores and their corresponding channels, Table 3 shows 10 words with the highest MI scores on three selected TV channels computed from the corpus used for this study.

Table 3. The words of the highest MI scores on three selected channels

ESPU		FOOD		USA	
college	13.43502	dishes	6.02581	law	2.38374
basketball	6.08835	meals	5.72422	unit	1.92896
football	2.60865	delicious	5.36825	jag	1.88490
ncaa	2.00504	salad	5.30091	victims	1.78642
state	1.77148	food	4.80851	order	1.69596
stadium	1.67090	rachael	4.22995	ncis	1.43711
center	1.40430	cooking	4.20850	walker	1.14715
lacrosse	1.17815	recipes	4.00144	ranger	1.12078
women's	0.98827	chicken	3.99080	stabler	1.08099
wildcats	0.89582	sauce	3.30139	intent	1.05222

For each of 446 channels, a MI score table is generated using (4) after filtering out most common terms that are not program titles, such as prepositions and pronouns, which have a very weak semantic implication. These 446 MI score tables form the basis for the topic inference procedure described in the next section.

3.3 Topic Inference Procedure

For each query string, an n-gram based program title classifier is used to screen if the query is intended for a specific program title using a similar algorithm as described in our previous study [8]. For example, if a 1-word query string such as *recipes* is rejected for *not* being part of a program title or channel name, it is then used as a simple index to loop through all *MI* score tables where it appears. If the search term contains more than one word, the procedure is repeated for each unigram in the text string. The words of the highest *TF* values are excluded from the indexing because those words such as '*the*' or '*and*' produces too many unrelated topics. The three tables with the highest accumulative *MI* scores for the term are selected to represent the channel-based topics. Using Table 3 as an example, for the 1-word query *recipes* if there are no other *MI* tables containing the word with a value greater than 4.00144, the channel *FOOD* will be ranked as the best topic for the query.

4 Evaluations and Results

4.1 Training Corpus

The EPG training corpus used for this study was constructed from the EPG metadata source for 59 TV markets in the U.S. over a 32-month period, from January 2008 to July 2010. The corpus consists of the following five data sets which collectively define the search space for the domain.

PT (Program Titles): ~91,000 unique sentences
PD (Program Descriptions): ~400,000 unique sentences (including subtitle/episode fields)
CS (Channel Sign): ~1500 phrases
CN (Channel Number): ~750 phrases
PN (Persons' Names): 20,000 entries

The corpus has a vocabulary of approximately 210,000 words and contains over 1.8 billion word tokens. For this paper, the training corpus is referred to as EPG59. The parser described in Section 3 constructs 661,537 documents from the corpus. On average, each document contains 22 words. Let V_S be the vocabulary for a subset s in EPG59 where $s = PD, CS, CN, PN$. The language model used by the *MI*-based topic inference engine is constructed automatically from the corpus in the following steps:

Step 1: Let w_n be an n-gram ($n \leq 2$) text string that is a true substring of a sentence in the *PT* subset. Compute $\aleph(w_n)$ and discard those w_n with their values in the bottom 10 percent. The remaining unigram words form a vocabulary denoted as V_{PT}.

Step 2: Create a baseline vocabulary $V = V_{PT} \cup V_{CS} \cup V_{CN} \cup V_{PN}$

Step 3: Let w be a unigram word in V_{PD}. Compute the weighted $\widetilde{TF}(w)$ as described in Section 3.1. Add w to the baseline vocabulary if its $\widetilde{TF}$ value is greater than an empirically-determined threshold (λ), which was created to reduce the overall vocabulary size. The result is the domain vocabulary, $V_{EPG} = V \cup V_{PD} \{w | \widetilde{TF}(w) > \lambda\}$, which contains approximately 114,000 unique words.

Step 4: The documents associated with each EPG channel (446 to be exact) form a topic class, c. On average, each topic class contains approximately 5,200 documents. For each unigram word w in the topic class c, if $w \in V_{EPG}$, compute $MI(w)$ using (4).

At the end of the unsupervised training process, the language models used for the topic inference at the runtime, M, can be formally defined in (5) in the form of the following set of feature vectors. Each vector contains a finite number of n-gram text strings with a relative weight factor. Unless none of words in a given query string is in the domain vocabulary, V_{EPG}, it will always be mapped to one of the five possible content categories: PT, CS, CN, PN, and MI-based channel topics.

$$M = PT\{w \in \aleph\}, CS, CN, PN, MI\{c_w | w \in V_{EPG}\} \tag{5}$$

4.2   Testing Data

Approximately 1.5 million *typed* search texts were collected at an EPG website from an estimated user population of over 20,000. The raw text data was filtered with a domain-specific automatic spell check dictionary and then compared with the full program titles in the *PT* data set. All search queries mapped to an exact program title (e.g., *24*, *Seinfeld*, *Project Runway*, etc) or a substring belonging to a unique program title (e.g., *"Two and a Half"* ∈ *"Two and a Half Men"*) were considered too simplistic and excluded from this study. The remaining set containing over 560,000 search texts are considered as a *harder* test set because they reflect *worse-than-average* scenarios in terms of low-frequency words and query complexity. From this set, the queries with four words or less are selected to form our final test set, which contains 544,425 text strings (representing approximately 97 percent of the harder test set).

The main reason for excluding the longer test queries with 5 or more words is twofold. First, the n-gram based *PT* classifier is trained from the title substrings with string length between 1 and 4. Consequently, the classifier cannot produce a perfect match for any test query with more than 4 words. Second, our previous study indicates that the true intention of EPG search phrases with 5 words or more is difficult to model. Therefore, to determine if the topics identified for those longer test queries at the end of the inference procedure is correct or not, the scoring process described in the next section would have to include many ad-hoc rules.

4.3   Performance Evaluation of Topic Inference Engine and Results

Prior to the topic inference test each query string in the test set is pre-screened to determine if it is in the domain. A query string s is considered out of the domain if half of the words in the string are not in the domain vocabulary, V_{EPG}. As shown in Table 5, out of the whole test set, approximately 8.8 percent of the text strings (48,252) in the harder test set used for this study are deemed to be out of domain, most of due to spelling errors that were not detected by the spell check dictionary. Of the remaining in-domain subset (496,173), approximately 34 percent is matched to a *PT*-specific topic using the n-gram based *PT* classifier, 1.7 percent is matched to a channel number, and 39 queries are matched by a specific *PN* entry such as *"Tom Hanks"*. The remaining (318,994) are input to the topic inference engine. As shown in

Table 4. Testing Results of the topic inference engine

Out-domain	48,252	8.8%	Topic Inference Test (N=318,994)			
In-domain	496,173	91.2%	Accepted		Rejected	
Identified as PT	168,699	34%	314,120	98.5%	4,874	1.5%
Identified as CN	8,441	1.7%	Examples of rejected test queries: *pti,*			
Identified as PN	39	0.01%	*cltv, mhd, alms, 239, insanitarium, dci,*			
Topic Inference	318,994	64.3%	*rnc, csa, 850*			

Table 4, among the subset, 4,873 queries are rejected because none of the words in a test string can be found in any *MI* score table. The ten most frequently occurring test queries rejected by the topic inference engine are listed in Table 4.

The precision is computed as follows: based on the top 3 channel topics selected by the inference engine, c_i, whose MI score tables are best matched to a test query, q, the topics are considered to be correct if q occurs in at least once in one of the documents $d \in c_i$. This scoring method is best explained using the following example. Consider a test query *Queen Elizabeth*, among 446 channels there are 84 channels carrying one or more TV programs where both words appear in their corresponding programming documents. From these 84 channels, the inference engine selected three channels whose *MI* score tables produce the highest accumulative scores. The channel, *KVUE*, one of local TV stations affiliated with *ABC*, has the highest mutual information score among the top three. One of the matching documents associated with the channel *KVUE* is for a TV program called *The Royal Family*, which is a special episode of the TV series *20/20* with the following program description:

> *The Royal Family. Drawing on hundreds of hours of footage and unprecedented access to **Queen Elizabeth** II and her family, including Princes William and Harry, Barbara Walters looks into the private and public life of the world's most famous royal family.*

For this test case, the channel *KVUE* represents the most relevant topic for the test query, *Queen Elizabeth*, and the topic selection is deemed correct. From the 314,120 test queries accepted by the topic inference engine, three subsets are selected to compute the precision rate: 200 from the test set with the highest frequency counts, 200 from the middle bracket with their frequency counts closest to the overall median, and 200 from the test cases with the lowest frequency counts. From the 600 test results, 494 are verified as correct, producing a precision rate of 82.4 percent.

4.4 Discussions and Future Work

The MI-based topic inference framework described in this paper has two unique merits for our domain of interest. First, it takes advantage of the highly structured XML-based EPG data source for training and updating the title classifier and the *MI* score tables without any manual supervision. Secondly, the topic inference algorithm is highly efficient for channel-based topic selections at runtime. For the EPG domain, average query strings contain only 2.6 words. For an EPG with 500 channels, the *MI*-based topic inference procedure simply loops through a table of 500 hashes once for each word in a query string. This step only takes a few milliseconds of CPU time on a common computer server.

One future improvement to the two pass process used for this topic inference is to increase the coverage for the title classifier. If a test query intended for a program title were not identified by the title classifier during the first pass, it would be submitted to the *MI*-based topic inference engine. For those queries containing common words, it can cause the topic inference algorithm to select incorrect topics. This problem is best explained by analyzing three test queries that were mismatched to the incorrect topics as shown in Table 5.

Using the true substring method to build the *n*-gram based title classifier has one obvious drawback, that is, unable to counter deletion often found in the user queries, as reflected in the last example in Table 5. Similarly, the multi-category queries as shown in the second example below have to be segmented first. Otherwise, the words intended for a channel number (or a specific date) would lead to too many unrelated topics. Both problems deserve a future study in order to further reduce the probability of sending those *n*-gram words to the topic inference engine that would most likely produce too many unrelated topics.

Table 5. Selected error cases produced by the topic inference engine

Test Queries	Intended Titles	Channels identified
1000 bc	10,000 B.C.	OUT
Comment: The first word selected by the user in the search term was a mistake in terms of expressing the user's true intention. The word "*1000*" has a high mutual information score with the programs often showing on the *OUT* channel such as "*1000 Places to See Before You Die*".		
1006 house	House	HGTV
Comment: The user was looking for the TV show *House* on channel 1006. However, the title classifier failed to recognize the structure of the search term (*<channel number>* + *<title>*). Therefore the two-word search term was treated as a topic expression where the second word leads to an incorrect channel carrying programs related to *Home/Garden*.		
17 and counting	17 Kids and Counting	FUEL
Comment: Omitting the key word '*Kids*' in the user's query string causes it to escape from the title classifier. As a result, the two common words '*17*' and '*counting*' led to high mutual information score with programs showing on the *FUEL* channel, which is not correct.		

5   Conclusions

This paper describes a novel topic inference framework that is built upon the scalability and more importantly, adaptability of the PMI method. The framework uses a weighted *MI* model trained on a domain-specific and highly structured text corpus without supervision. The model can be refreshed automatically from the continuously updated EPG data source. By using an *n*-gram based title classifier to screen out those search terms targeted for a specific TV program title such as "*House*" or "*Dancing with the Stars*", a *MI*-based topic inference system is able to identify the most relevant channels for search terms not intended for a specific TV program or movie such as *"Queen Elizabeth"* or *"scariest movies"*, at a precision accuracy of 82.4 percent on a large test set collected from a real world web-based IR system.

References

1. Deerwester, S., Dumais, S.T., Furnas, G.W., Landauer, T.K., Harshman, R.: Indexing by Latent Semantic Analysis. J. Am. Soc. Inform. Science 41(6), 391–407 (1990)
2. Bellegarda, J.: Latent Semantic Mapping. IEEE Signal Processing Magazine 22, 70–80 (2005)
3. Papadimitriou, C.H., Raghavan, P., Tamaki, H., Vempala, S.: Latent Semantic Indexing: A Probabilistic Analysis. In: Proc. 17th ACM Symp. Princeples Database Systems, pp. 159–168 (1998)
4. Hofmann, T.: Probabilistic latent Semantic Analysis. Uncertainty in Artificial Intelligence (1999)
5. Landauer, T.K., Dumais, S.: A Solution to Plato's Problem: The Latent Semantic Analysis Theory of Acquisition, Induction, and Representation of Knowledge. Psychological Review 104, 211–214 (1997)
6. Recchia, G., Jones, M.N.: More Data Trumps Smarter Algorithms: Comparing Pointwise Mutual Information with Latent Semantic Analysis. Behavior Research Methods 41(3), 647–656 (2009)
7. Budiu, R., Royer, C., Pirolli, P.L.: Modeling Information Scent: A Comparison of LSA, PMI, and GLAS Similarity Measure on Common Tests and Corpora. In: Proc. of the 8th Annual Conference of the Recherche d'Information Assistee Par Ordinateur (2005)
8. Chang, H.M.: Conceptual Modeling of Online Entertainment Programming Guide for Natural Language Interface. In: Hopfe, C.J., Rezgui, Y., Métais, E., Preece, A., Li, H. (eds.) NLDB 2010. LNCS, vol. 6177, pp. 188–195. Springer, Heidelberg (2010)
9. Berger, A., Lafferty, J.: Information Retrieval as Statistical Translation. In: Proc. of the 22nd ACM Conference on Research and Development in Information Retrieval, pp. 222–229 (1999)
10. Ramos, J.: Using TF-IDF to Determine Word Relevance in Document Queries. In: Proc. of the First Instructional Conference on Machine Learning (2003)
11. Magerman, D.M., Marcus, M.P.: Parsing a Natural Language Using Mutual Information Statistics. In: Proc. of the 8th National Conference on Artificial Intelligence, pp. 984–989 (1990)
12. Jang, M.-G., Myaeng, S.H., Park, S.Y.: Using Mutual Information to Resolve Query Translation Ambiguities and Query Term Weighting. In: Proc. of the 37th Annual Meeting of Association of Computational Linguistics, pp. 223–228 (1999)
13. Peters, J.: Semantic Text Clusters and Word Classes – The Dualism of Mutual Information and Maximum Likelihood. In: Proc. of the Workshop on Language Modeling and Information Retrieval, pp. 55–59 (2001)
14. Dhillon, I.S., Mallela, S., Modha, D.S.: Information-Theoretic Co-Clustering. In: Proc. of the 9th ACM SIGKDD International Conference on Knowledge Discovery and Data Mining, pp. 89–98 (2003)
15. Manning, C.D., Raghavan, P., Schutze, H.: Introduction to Information Retrieval. Cambridge University Press, Cambridge (2008)

Improving Subtree-Based Question Classification Classifiers with Word-Cluster Models

Le Minh Nguyen and Akira Shimazu

Japan Advanced Institute of Science and Technology
School of Information Science
{nguyenml,shimazu}@jaist.ac.jp

Abstract. Question classification has been recognized as a very important step for many natural language applications (i.e question answering). Subtree mining has been indicated that [10] it is helpful for question classification problem. The authors empirically showed that subtree features obtained by subtree mining, were able to improve the performance of Question Classification for boosting and maximum entropy models. In this paper, our first goal is to investigate that whether or not subtree mining features are useful for structured support vector machines. Secondly, to make the proposed models more robust, we incorporate subtree features with word-cluster models gained from a large collection of text documents. Experimental results show that the uses of word-cluster models with subtree mining can significantly improve the performance of the proposed question classification models.

1 Introduction

Question classification is the task of mapping a question in natural language to a pre-defined category. This task has been attractive many studies in recently years [8][10][20][19].

There are two main approaches for question classification. The first approach does question classification using handcrafted rules which are reported in[18]. Although it is possible to manually write some heuristic rules for the task, it usually requires tremendous amount of tedious work.

The second approach applies machine learning to question classification problems such as in [16], the machine learning tool Ripper has been utilized for this problem. The reported accuracy of their system on TREC dataset is 70%, which is not high enough for question classification. Li [8] presented the use of SNoW method in which the good performance of their system depends on the feature called "RelWords" (related word) which are constructed semi-automatically but not automatically.

Zhang and Lee[13] employed tree kernel with a SVM classifier and indicated that the syntactic information as well as the tree kernel is suitable for this problem. Ray et al [19] recently showed that their method can improve the question classification by incorporating semantic features from WordNet and Wikipedia. Huang et al [20] proposed the use of hyponyms and head words for question classification. All previous methods are required elaborate investigations on feature sets in term of improving the accuracy of question classification. On the other hand, the connection of data mining

R. Muñoz et al. (Eds.): NLDB 2011, LNCS 6716, pp. 76–87, 2011.
© Springer-Verlag Berlin Heidelberg 2011

techniques with machine learning problem recently shows that data mining can help improve the performance of classifier. Morhishita [11] showed that the use of association rule mining can be helpful for enhancing the accuracy of the boosting algorithm.

Subtree features are successfully applied for question classification [10] in two machine learning method including boosting and maximum entropy models. In this paper, we would like to explore appropriate machine learning models for question classification based subtree mining. We see that the number of class labels within the question classification is large (i.e 50 labels), so a suitable machine learning model that can deal well with the large number of labels is needed. In order to do that, we propose the use of structure support vector machines [12] using subtree features obtained by using subtree mining for question classification. We empirically compare the proposed method with other subtree-based question classification methods (i.e subtree maximum entropy and boosting).

On the other hand, the limitation of question classification using supervised methods is that it takes time consuming for constructing such annotated corpora. To deal with this limitation, we select a simple semi-supervised learning approach, which relies on the uses of unsupervised data for improving the performance of question classification. Our motivation is generally based on an observation that the main situation caused the question classification fair when it copes with a test sentence from other domains than the training data. One solution is to use a large un-annotated corpus to enrich feature-set in the discriminative models in order to make it to be correlated with features of test data. The simple method is to use the brown-clustering method [2] to cluster words and map them as features for improving the performance of discriminative learning models.

The original idea of combining word clusters with discriminative learning has been previously explored by Miller et al.(2004), which is mainly applied for Named Entity recognition. For subtree-based question classification, the challenge of incorporating word-cluster models and subtree spaces is novel and worth investigating. The contribution of our works is to introduce a novel framework for question classification which combines subtree features and word-cluster information via a structured classification method. It is also indicated that we can use a large amount of unlabeled data to improve the performance of question classification.

The remainder of this paper is organized as follows: Section 2 presents the subtree mining for question classification. Section 3 shows a framework of using word-cluster model for subtree-based question classification methods. Section 4 shows experimental results and Section 5 gives some conclusions and plans for future work.

2   Subtree Mining for Question classification

2.1   Tree Classification

In [10], the question classification problem is considered as classifier for trees in which given questions are parsed into the trees, then a subtree mining process is performed on the trees, to obtain appropriate features for the question classification task. The authors investigated two machine learning methods for question classification using subtree mining. In this section, we would like to investigate other classification methods that can deal well with the large number of labels in question classification.

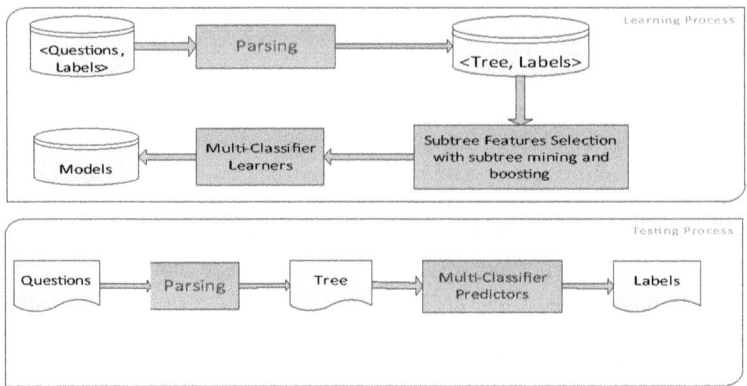

Fig. 1. A framework of subtree mining for question classification

Figure 1 shows a framework of subtree-based for question classification. All questions are parsed to trees using a parsing method. Note that a tree can be a sequence of words, a dependency tree, and a syntactic tree. All trees are mined to obtain a set of subtrees then a machine learning model is used to learn the model for question classification.

The problem of question classification is equivalent to classifying a syntactic tree into a set of given categories. We reduce the problem of tree classification in multi-class problem to the binary classification problem by using the one-vs.-all strategy.

Definition. The tree classification problem is to induce a mapping $f(x) : X \rightarrow \{+1, -1\}$, from given training examples $T = \{\langle x_i, y_i \rangle\}_{i=1}^{L}$, where $x_i \in X$ is a labeled ordered tree and $y_i \in \{+1, -1\}$ is a class label associated with each training data. Let t and x be labeled ordered trees, and y be a class label ($y_i \in \{+1, -1\}$), a decision stump classifier for trees is given by:

$$h_{<t,y>}(x) = \begin{cases} +y \ t \subseteq x \\ -y \ \text{otherwise} \end{cases}$$

Decision stump functions are observed on the training data (the set of trees and their labels), then these functions are incorporated to a maximum entropy model and a boosting model.

2.2   Subtree Features Selection with Boosting

Assume that we are given a set of decision stump functions and we want to apply it for classification problem. The boosting framework [15] by repeatedly calling a given weak learner to finally produce hypothesis f, which is a linear combination of K hypotheses produced by the prior weak learners (K is the number of iterations), i.e.:
$f(x) = \text{sgn}(\sum_{k=1}^{K} \alpha_k h_{<t_k,y_k>}(x))$

A weak learner is built at each iteration k with different distributions or weights $d^{(k)} = (d_i^{(k)}, ..., d_L^{(k)})$, (where $\sum_{i=1}^{N} d_i^{(k)} = 1, d_i^{(k)} \geq 0$). The weights are calculated in such a way that hard examples are focused on more than easier examples.

$$gain(< t, y >) = \sum_{i=1}^{L} y_i d_i h_{<t.y>}(x_i)$$

There exist many Boosting algorithm variants, however the original and the best known algorithm is AdaBoost[15]. For this reason we used the Adaboost algorithm with the decision stumps serving as weak functions.

As described in [10], we applied the subtree mining method [14] which enumerates all subtrees from a given tree. First, the algorithm starts with a set of trees consisting of single nodes, and then expands a given tree of size $(l..1)$ by attaching a new node to this tree to obtain trees of size l. However, it would be inefficient to expand nodes at arbitrary positions of the tree, as duplicated enumeration is inevitable. The algorithm, rightmost extension, avoids such duplicated enumerations by restricting the position of attachment. To mine subtrees we used the parameters bellow:

- minsup: the minimum frequency of a subtree in the data
- maxpt: the maximum depth of a subtree
- minpt: the minimum depth of a subtree

Table 1 shows an example of mining results.

Table 1. Subtrees mined from the corpus

Frequency	Subtree
4	(VP(VBZfeatures)(PP))
7	(VP(VP(VBNconsidered)))
2	(VP(VP(VBNcredited)))
2	(VP(VP(VBNdealt)))
5	(VP(VP(VBNinvented)))

After mining the subtrees, we used a boosting algorithm to select appropriate subtrees for question classification problems. Note that the subtree selection algorithm is based on the decision stump function.

Formally, let $T = \{< x_1, y_1, d_1 >, ..., < x_L, y_L, d_L >\}$ be training data, where x_i is a tree and y_i is a labeled associated with x_i and d_i is a normalized weight assigned to x_i. Given T find the optimal rule $< t^0, y^0 >$ that maximizes the gain value.

The most naive and exhaustive method, in which we first enumerate all subtrees F and then calculate the gains for all subtrees, is usually impractical, since the number of subtrees is exponential to its size. In order to define a canonical search space, we applied the efficient mining subtree method as described in [14]. After that, we used the branch-and-bound algorithm in which we defined the upper bound for each gain function in the process of adding a new subtree. To define a bound for each subtree, we based our calculation on the following theorem [11]

For any $t' \supseteq t$ and $y \in \{+1, -1\}$, the gain of $< t', y >$ is bounded by $\mu(t)$

$$\mu(t) = \max \left\{ 2 \sum_{i|y_i=+1, t \in x_i} d_i - \sum_{i=1}^{L} y_i d_i, 2 \sum_{i|y_i=-1, t \in x_i} d_i + \sum_{i=1}^{L} y_i d_i \right\} \quad (1)$$

We can efficiently prune the search space spanned by right most extension using the upper bound of gain $\mu(t)$. During the traverse of the subtree lattice built by the recursive process of rightmost extension, we always maintain the temporally suboptimal gain δ among all gains calculated previously. If $\mu(t) < delta$, the gain of any super-tree $t' \in t$ is no greater than $delta$, and therefore we can safely prune the search space spanned from the subtree t. Otherwise, we can not prune this branch. The detail of this algorithm is described in [10].

2.3 Structured SVMs with Subtree Features

We used subtree features mined from the training data with the subtree mining technique. We then apply the boosting method to obtain appropriate subtree features for structured SVMS learner. After that, all subtree features are indexed to convert instance in boosting to the format of SVM. We got totally 5500 instances with the number of labels is 50 and 6 for fine labels and coarse, respectively. Before describing the method in detail, let us briefly summary the structured SVMs as following subsections.

Structured Support Vector Models. Structured classification is the problem of predicting y from x in the case where y has a meaningful internal structure. Elements $y \in Y$ may be, for instance, sequences, strings, labelled trees, lattices, or graphs.

The approach we pursue is to learn a discriminant function $F : X \times Y \rightarrow R$ over $< input, output >$ pairs from which we can derive a prediction by maximizing F over the response variable for a specific given input x. Hence, the general form of our hypotheses f is

$$f(x; w) = \arg \max_{y \in Y} F(x; y; w)$$

where w denotes a parameter vector.

For convenience, we define

$$\delta \psi_i(y) \equiv \psi(x_i, y_i) - \psi(x_i, y)$$

where (x_i, y_i) is the training data and $\psi(x_i, y_i)$ is a feature vector relating input x and output y.

The hard-margin optimization problem is:

$$\text{SVM}_0 : min_w \frac{1}{2} \|w\|^2 \quad (2)$$

$$\forall i, \forall y \in Y \backslash y_i : \langle w, \delta \psi_i(y) \rangle > 0 \quad (3)$$

where $\langle w, \delta\psi_i(y)\rangle$ is the linear combination of feature representation for input and output.

in order to allow errors in the training set, by introducing slack variables.

$$\text{SVM}_1 : \min \frac{1}{2}\|w\|^2 + \frac{C}{n}\sum_{i=1}^{n}\xi_i \tag{4}$$

$$\text{s.t.}\forall i, \xi_i \geq 0, \forall y \in Y\backslash y_i : \langle w, \delta\psi_i(y)\rangle \geq 1 - \xi_i \tag{5}$$

Alternatively, using a quadratic term $\frac{C}{2n}\sum_i \xi_i^2$ to penalize margin violations, we obtained SVM_2. Here $C > 0$ is a constant that control the tradeoff between training error minimization and margin maximization.

$$\text{SVM}_2 : \min \frac{1}{2}\|w\|^2 + \frac{C}{2n}\sum_{i=1}^{n}\xi_i * \xi_i \tag{6}$$

$$\text{s.t.}\forall i, \xi_i \geq 0, \forall y \in Y\backslash y_i : \langle w, \delta\psi_i(y)\rangle \geq 1 - \xi_i \tag{7}$$

To deal with problems in which $|Y|$ is large, such as question classification [1], we used the framework which is proposed by [12] that generalize the formulation SVM_0 and SVM_1 to the cases of arbitrary loss function. The first approach is to re-scale the slack variables according to the loss incurred in each of the linear constraints.

$$\text{SVM}^{\Delta_s} : \underbrace{\min}_{w,\xi} \frac{1}{2}\|w\|^2 + \frac{C}{n}\sum_{i=1}^{n}\xi_i, \text{s.t.}\forall i, \xi_i \geq 0 \tag{8}$$

$$\forall i, \forall y \in Y\backslash y_i : \langle w, \delta\psi_i(y)\rangle \geq \frac{1 - \xi_i}{\Delta(y_i, y)} \tag{9}$$

The second approach to include loss function is to re-scale the margin as a special case of the Hamming loss. The margin constraints in this setting take the following form:

$$\forall i, \forall y \in Y\backslash y_i : \langle w, \delta\psi_i(y)\rangle \geq \Delta(y_i, y) - \xi_i \tag{10}$$

This set of constraints yields an optimization problem, namely $\text{SVM}_1^{\Delta_m}$.

Structured SVMs for Question Classification. The support vector learning algorithm aims at finding a small set of active constraints that ensures a sufficiently accurate solution. The detailed algorithm, as presented in [12] can be applied to all SVM formulations mentioned above. For the sake of simplicity, we assume that the margin re-scaling ($\text{SVM}_2^{\Delta_m}$) is mainly used for our work.

Typically, the way for appling structured SVMs is to implement a feature mapping function $\psi(x, y)$, a loss function $\Delta(y_i, y)$, as well as a maximization algorithm. Since the question classification is a multi-classification problem, so we can easily consider the feature mapping as the feature space using subtree mining. We applied the subtree

---

[1] Number of labels is 50.

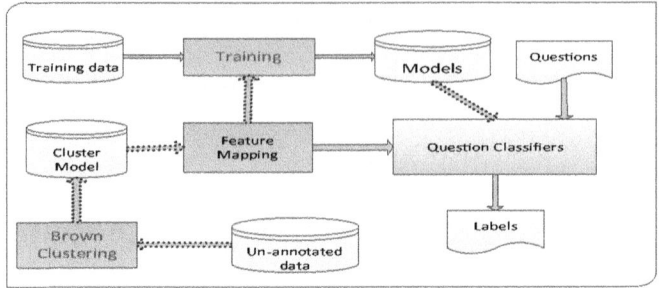

Fig. 2. A Word-Cluster Model for Question Classification

feature selection method to obtain a set of subtrees. As mentioned earlier, we used all subtrees to generate examples for structured SVMs. In other aspects, the maximization algorithm is simple returning the label that has the highest score. The loss function is simply applied the hinge loss and it can be described as follows.

Let a fine category label be $X : y$, in which X is a course label and y is a subcategory among $X : y$. For example, in the label "NUM:count", X is "NUM" and y is "count", respectively. The loss function of two categories $X_i : y_i$ and $X_j : y_j$ is defined

$$\text{loss}(X_i : y_i, X_j : y_j) = 0 \text{ if } X_i = X_j \text{ and } y_i = y_j$$
$$\text{loss}(X_i : y_i, X_j : y_j) = a \text{ if } X_i = X_j \text{ and } y_i! = y_j$$
$$\text{loss}(X_i : y_i, X_j : y_j) = b \text{ if } X_i! = X_j \text{ and } y_i! = y_j$$

The parameters a and b are assigned by experiments. In our experiment we set a and b to 5 which give highest results.

3   Word-Cluster Model for Question Classification

In this section we present a framework of question classification using word-cluster model. Figure 2 shows a general framework for incorporating word-cluster modes to discriminative learning models such as Maximum Entropy, Boosting, and structured SVMs. Figure 2 also can be viewed as a semi-supervised learning framework for discriminative learning models. Larger un-annotated text documents are clustered using the brown-clustering method to obtain word-cluster models. A word-cluster model is then used to enrich features space for discriminate learning models in both the training and classification (testing) processes. For convenience, we briefly present a summary of the Brown Algorithm as follows.

3.1   The Brown Algorithm

The Brown algorithm is a hierarchical agglomerative word clustering algorithm [2]. The input of this algorithm is a large sequence of words $w_1, w_2, \ldots, w_n$, which are extracted from raw texts. The output of this algorithm is a hierarchical clustering of words—a binary tree—wherein a leaf represents a word, and an internal node represents a cluster containing the words in the sub-tree, whose root is that internal node.

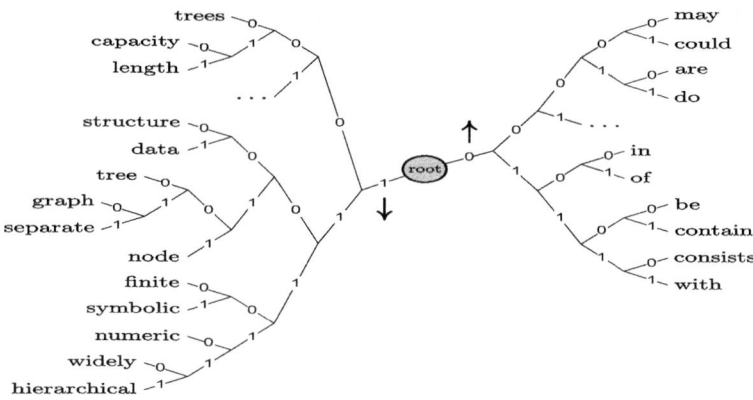

Fig. 3. An example of a hierarchical clustering. Each word at a leaf is encoded by a bit string with respect to the path from the root, where 0 indicates an *"up"* branch and 1 indicates a *"down"* branch.

This algorithm uses contextual information—the next word information—to represent properties of a word. More formally, (w) denotes the vector of properties of w (or w's context). We can think of our vector for w_i as counts, for each word w_j, of how often w_j followed w_i in the corpus:

$$(w_i) = (|w_1|, |w_2|, \ldots, |w_n|)$$

(w_i) is normalized by the count of w_i, and then we would have a vector of conditional properties $P(w_j|w_i)$. The clustering algorithm used here is HAC-based, therefore, at each iteration, it must determine which two clusters are combined into one cluster. The metric used for that purpose is the minimal loss of average mutual information.

Figure 3 shows a portion of a hierarchical clustering, which is derived from a small portion of text, which contains 12 sentences and 182 distinct words. This portion of text is about politics situation in Thailand.

From this tree, we can freely get a cluster of words by collecting all words at the leaves of the sub-tree, whose root is a chosen internal node. For instance, some clusters are shown in Figure 2.

To use word cluster information in our model at several levels of abstraction, we encode each word cluster by a bit string that describes the path from the root to the chosen internal node. The path is encoded as follows: we start from the root of the hierarchical clustering, "0" is appended to the binary string if we go up, and "1" is appended if we go down. For instance, to encode the above three clusters, we use the following bit strings "100", "110", "010", and "1111", respectively. If we want to use a higher level of abstraction, we can simply combine the clusters that have the same prefix.

3.2   Word-Cluster Based Features

In addition to the baseline features presented in the previous section, we used the word-cluster model to enrich the feature space by the simple method as follows.

– (1) Each word in the training data and testing data is mapped to a bit string. Each cluster of words is represented by a bit string as described in Sect. 3.1. For example the word "*structure*" and "*data*" are represented as "1100" and "1101", respectively. The cluster information of a word is also represented by the bit string of the cluster containing that word.

– (2) For each subtree mined from the corpus, we replace each word within the subtree with its bit representation. For internal nodes (syntactic nodes: NP, VP, PP, etc), we keep the same as the original tree.

– (3) We then create an indicator function for each cluster, and use it as a selection feature:

$$f_{110101}(w) = \begin{cases} 1 \text{ if } w \text{ has bit string } 110101, \\ 0 \text{ otherwise.} \end{cases}$$

4 Experimental Results

First, to build a hierarchical cluster of words, we use an external corpus, the BLLIP corpus [4], which is a collection of raw text with approximately 30 million words. The Brown algorithm implementation of Liang [9] ran on that corpus to produce 1,000 word clusters.

Second, We compared the proposed methods using word-cluster models with the previous works [10]. We investigate the use of subtree features with three machine learning models, including maximum entropy models, boosting models, and structured support vector machines.

To achieve this goal, we tested the proposed models on the standard data similarly experiments in previous works [8][13][10] which includes 5,500 questions for training and 500 questions for testing. The labels of each question is followed the two-layered question taxonomy proposed by [8], which contains 6 coarse grained category and 50 fine grained categories, as shown bellow. Each coarse grained category contains a non-overlapping set of fine grained categories.

In order to obtain subtrees for our proposed methods, we initially used the Chaniak parser [3] to parse all sentences within the training and testing data. We obtained a new training data consisting of a set of trees along with their labels. We then mined subtrees using the right most extension algorithm [14]. Table 2 shows a subtree features obtained by using subtree mining.

We conducted the following experiments to confirm our advantage of using subtree mining for question classification with the maximum entropy model, structured SVMs, and boosting models. The experiments were done using a significant test with 95% confident interval. Table 3 shows the accuracy of our subtree maximum entropy model (ST-Mem), subtree boosting model (ST-Boost), the tree kernel SVM (SVM-TK), and SVM-struct for the coarse grained category, respectively. It shows that the boosting model achieved competitive results to that of SVM-TK. The SVM-struct outperforms SVM-TK and achieves the best accuracy. The reason why SVM-struct outperforms

Table 2. A running example of using subtree mining subtrees [10]

Features	Weight
ENTY (ADJP(ADVP))	0.074701
ENTY (ADJP(INof))	0.433091
ENTY (ADJP(JJcelebrated))	0.084209
HUM (ADJP(JJcommon))	0.003351
HUM (ADJP(JJdead))	0.117682
ENTY:substance (NP(ADJP(RBSmost)(JJcommon)))	0.354677
NUM:count (ADJP(PP(INfor)))	0.157310
NUM:count (NP(DTthe)(NNnumber))	0.810490

other models is that it enjoys both the advantage of sub-tree features and the ability of large margin method estimating on the hierarchical structure of question labels[2].

Table 3 clearly indicated that the use of word-clusters is significantly contributed to the performance of question classification. The improvement of three subtree-based question classification models: subtree-MEM, subtree-Boosting, and subtree-SVMs are 3.4%, 2.2%, and 5.3%, respectively.

Table 3. Question classification accuracy using subtrees, under the coarse grained category definition (total 6 labels)[3]

With cluster	ST-Mem	ST-Boost	SVM-TK	SVM-Struct
No	88.4	90.6	90.0	86.73
Yes	91.8	92.8	-	92.04

Table 4. Question classification accuracy using subtrees, under the fine grained category definition (total 50 labels).

With cluster	ST-Mem	ST-Boost	SVM-struct
No	80.5	82.6	79.42
Yes	82.3	83.0	85.40

Table 4 shows the performance of subtree-mem [10], subtree-boost[10], and subtree-SVMs for question classification. Since there is no report on using SVM tree kernel method for question classification on fine grained categories. In [13], the author reported the performance of SVM using tree kernel on the coarse grained category only. Now on our paper we reported the performance of fine-grained category using subtrees under the maximum entropy, structured SVM, and the boosting model. We see that without using semantic features such as relation words we can obtain a better result

---

[2] We also tested SVM-struct with different parameters. The margin re-scaling ($SVM_2^{\Delta m}$) showed the highest results.

[3] With cluster means we enriched the feature set with cluster feature.

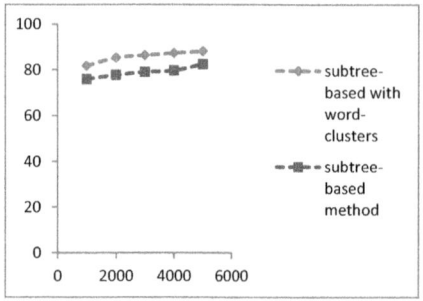 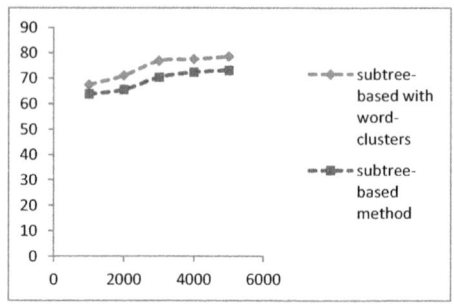

Fig. 4. Contribution of word-clusters for various sizes of training data from 1000 to 5000 samples. Note that the figure in the left and the right are for coarse labels and gain labels, respectively.

in comparison with previous works. It is similar to the case of question classification for coarse-grained categories, we also see that the uses of word-cluster features significantly contribute to the performance of question classification. Table 4 indicated that word-cluster features were able to significantly improve three sub-tree based machine learning models for question classification. The improvements of three subtree-based question classification models: subtree-MEM, subtree-Boosting, and subtree-SVMs are 1.8%, 0.4%, and 5.98%, respectively.

We also reported experiments on various sizes of training data while keeping the same test data using 500 questions. The experiment was done by comparing structured SVMs with subtree mining and word-cluster models and the subtree-based structured SVMs. Figure 4 shows that word-clusters is significantly improved the results of subtree-based question classification methods in various sizes of training data.

Since the subtree information and word cluster features are not overlapped with other kinds of semantic features such as the named entity types, so mixing these features together would improve the performance of question classification task. In future works, we plan to use both subtree features and those features using WordNet and headword as described in [19] and [20].

5   Conclusions

This paper proposes a method, which allows incorporating subtrees mined from training data to the question classification problem. We then investigate the use of three learning models for multi-class including MEMs, structured SVM, and Booting with subtrees as feature function and weak functions by considering subtree mining as subtree feature selection process. Experimental results show that the uses of subtree mining features are useful for the question classification task. The use of word-cluster features are significantly improved the subtree-based question classification models. It is also indicated that word-clusters for subtree-based structured SVMs attain the best result.

Acknowledgments. The work on this paper was partly supported by the grants for Grants-in-Aid for Young Scientific Research 22700139 .

References

1. Berger, A., Pietra, S.D., Pietra, V.D.: A maximum entropy approach to natural language processing. Computational Linguistics 22(1) (1996)
2. Brown, P.F., Della Pietra, V.J., de Souza, P.V., Lai, J.C., Mercer, R.L.: Class-Based n-gram Models of Natural Language. Computational Linguistics 18(4), 467–479 (1992)
3. Charniak, E.: A Maximum-Entropy Inspired Parser. In: Proc. ACL (2001)
4. Charniak, E., Blaheta, D., Ge, N., Hall, K., Hale, J., Johnson, M.: BLLIP 1987-1989 WSJ Corpus Release 1. Linguistic Data Consortium (2000)
5. Carlson, A., Cumby, C., Roth, D.: The SNoW learning architecture, Technical report UIUC-DCS-R-99-2101, UIUC Computer Science Department (1999)
6. Kadri, H., Wayne, W.: Question classification with Support vector machines and error correcting codes. In: Proceedings of NAACL-HLT 2003, pp. 28–30 (2003)
7. Kudo, T., Maeda, E., Matsumoto, Y.: An Application of Boosting to Graph Classification. In: Proceedings NIPS (2004)
8. Li, X., Roth, D.: Learning question classifiers. In: Proceedings of the 19th International Conference on Computational Linguistics, pp. 556–562 (2002)
9. Liang, P., Collins, M.: Semi-supervised learning for natural language. Master thesis, MIT (2005)
10. Nguyen, M.L., Shimazu, A., Nguyen, T.T.: Subtree mining for question classification problem. In: Proceedings IJCAI 2007, pp. 1695–1700 (2007)
11. Morishita, S.: Computing optimal hypotheses efficiently for boosting. In: Arikawa, S., Shinohara, A. (eds.) Progress in Discovery Science. LNCS (LNAI), vol. 2281, pp. 471–481. Springer, Heidelberg (2002)
12. Tsochantaridis, I., Hofmann, T., Joachims, T., Altun, Y.: Support Vector Machine Learning for Interdependent and Structured Output Spaces. In: Proceedings ICML 2004 (2004)
13. Zhang, D., Lee, W.S.: Question classification using Support vector machine. In: Proceedings of ACM SIGIR-2033, pp. 26–33 (2033)
14. Zaki, M.J.: Efficiently Mining Frequent Trees in a Forest. In: Proceedings 8th ACM SIGKDD 2002 (2002)
15. Schapire: A brief introduction to boosting. In: Proceedings of IJCAI 1999 (1999)
16. Radev, D.R., Fan, W., Qi, H., Wu, H., Grewal, A.: Probabilistic Question Answering from the Web. In: Proceedings of WWW (2002)
17. Vapnik, V.: The Nature of Statistical Learning Theory. Springer, N.Y (1995)
18. Voorhees, E.: Overview of the TREC 2001 Question Answering Track. In: Proceedings of TREC 2010, pp. 157–165. NIST, Gaithersburg (2001)
19. Ray, S.K., Singh, S., Joshi, B.P.: A semantic approach for question classification using WordNet and Wikipedia. Pattern Recognition Letters 31(13), 1935–1943 (2010)
20. Huang, Z., Thint, M., Kin, Z.: Question classification using head words and their hypernyms. In: Proceedings EMNLP 2008, pp. 927–936 (2008)

Data-Driven Approach Based on Semantic Roles for Recognizing Temporal Expressions and Events in Chinese

Hector Llorens[1], Estela Saquete[1], Borja Navarro[1], Liu Li[2], and Zhongshi He[2]

[1] University of Alicante, Alicante, Spain
[2] Chongqing University, Chongqing, China
{hllorens,stela,borja}@dlsi.ua.es, {liuli,zshe}@cqu.edu.cn

Abstract. This paper addresses the automatic recognition of temporal expressions and events in Chinese. For this language, these tasks are still in an exploratory stage and high-performance approaches are needed. Recently, in TempEval-2 evaluation exercise, corpora annotated in TimeML were released for different languages including Chinese. However, no systems were evaluated in this language. We present a data-driven approach for addressing these tasks in Chinese, TIRSemZH. This uses semantic roles, in addition to morphosyntactic information, as feature. The performance achieved by TIRSemZH over the TempEval-2 Chinese data (85% F1) is comparable to the state of the art for other languages. Therefore, the method can be used to develop high-performance temporal processing systems, which are currently not available for Chinese. Furthermore, the results obtained verify that when semantic roles are applied, the performance of a baseline based only on morphosyntax is improved. This supports and extends the conclusions reached by related works for other languages.

1 Introduction

The temporal information processing task implies the automatic annotation of temporal expressions (timexes), events, and their relations in natural language text. The output of such information extraction process is valuable to other fields such as information retrieval, question answering, or text summarization, provided that the performance is high-enough. Recently, the growing interest of the scientific community in this area has been reflected in specialized evaluation exercises such as TempEval-2 [14], whose objective was measuring the performance of the current approaches for different languages (i.e., English, Spanish, Italian, French, Korean and Chinese). However, there were only participants in two languages (English and Spanish).

Only one of the systems in TempEval-2 addressed the tasks in more than one language, TIPSem [9]. Furthermore, this data-driven approach showed a high-performance in both tasks. TIPSem uses semantic role-based features in addition to morphosyntactic features, and this demonstrated that semantic roles, as additional feature, are useful for recognizing timexes and events in English and Spanish.

For Chinese, these tasks are still in an exploratory stage and high-performance approaches are needed. In fact, no event and only few timex-recognition approaches are available. However, the available TempEval-2 data supposes a shared framework to evaluate and compare approaches for Chinese.

R. Muñoz et al. (Eds.): NLDB 2011, LNCS 6716, pp. 88–99, 2011.
© Springer-Verlag Berlin Heidelberg 2011

Taking this statements as starting point, the objectives of this paper are: (i) to develop a high-performance approach for Chinese based on TIPSem, and (ii) to analyze whether semantic roles are also useful in this language. In particular, this paper focuses on *timex recognition* and *event recognition*, which imply the exact bounding of such elements in text. Taking TIPSem architecture as basis, we present and evaluate a data-driven approach TIRSemZH (Temporal Information Recognition in Chinese) which includes semantic roles and morphosyntax-based features. In addition, to measure the influence of semantic roles, we evaluate a baseline, TIRSemZH-B, which only uses morphosyntax.

The next section includes the background and Section 3 describes our proposal. Section 4 includes the evaluation and a comparative analysis of the results. Finally, in Section 5, conclusions are drawn.

2   Background

This section includes (i) the definition of temporal expressions and events, (ii) the review of the related work on the computational recognition of such entities, and (iii) a discussed motivation for using semantic roles.

2.1   Temporal Expressions and Events

The aim of studying temporal information in natural language led the scientific community to define different temporal entities. In this paper, we address the following ones:

Temporal expressions (timex). A timex is a linguistic representation of a time point or period. It is normally denoted by a noun, adjective, adverb, or a noun phrase, adjective phrase, or adverb phrase (see Example 1).

(1) 1. 约翰星期一来的。
       EN: John came on Monday.

    2. 他要在这待两个星期。
       EN: She will be here for two weeks.

– **Events.** An "event" is defined as something that *happens, occurs, takes place, or holds*. They are generally expressed by verbs, nominalizations, adjectives, predicative clauses or prepositional phrases (see Example 2).

(2) 1. 约翰星期一来的。
       EN: John came on Monday.

    2. 他将出席 会议。
       EN: He will attend the meeting.

Concerning the linguistic expression of time, one of the peculiarities of Mandarin Chinese is that it lacks tense morphemes. Time is expressed by aspectual markers and also non-grammaticalized expressions of location in time like temporal adverbials (e.g., tomorrow) or explicit calendar references (e.g., 1991).

Several efforts have been made to define standard ways to represent temporal information in texts. The main objective of this representation is to make temporal information explicit through standard annotation schemes. This paper is focused on the

TimeML annotation scheme [10], which has been recently adopted as a standard by a large number of researchers due to the completeness of the comprehensive improvements it adds to the previous schemes. Example (3) shows a sentence annotated with TimeML timexes (TIMEX3) and events (EVENT).

(3) 约翰<TIMEX3>星期一</TIMEX3>
    <EVENT>来的</EVENT>。

    EN: John <EVENT>came</EVENT> on
    <TIMEX3>Monday</TIMEX3>

2.2   Related Work

Different computational approaches addressing the automatic recognition of temporal expressions and events for English were presented in the TempEval-2 evaluation exercise. Table 1, summarizes the best systems there evaluated for English which addressed both tasks and the linguistic knowledge they use.

Table 1. TempEval-2 best approaches for English (ms: morphosyntax)

System	Linguistic Info
HeidelTime [12]	ms
Edinburgh [4]	ms
TIPSem [9]	ms, semantic roles
TRIOS [13]	ms, semantic-logic

In Table 1, we considered *morphosyntactic (ms)* as the use of the following linguistic information: sentence segmentation, tokenization, word lemmatization, part-of-speech (PoS) tagging, syntactic parsing, word-triggers[2], and lexical semantics extracted from WordNet [3] such as synonymy. Finally, *semantic-logical form* and *semantic roles* refer to compositional semantics at sentence level and will be discussed later.

Table 2 summarizes the scores obtained by previously cited systems in the tasks they participated.

Table 2. TempEval-2 results for English (best systems)

Task	System	precision	recall	$F_{\beta=1}$
Timex	HeidelTime	0.90	0.82	0.86
	TRIOS	0.85	0.85	0.85
	TIPSem	0.92	0.80	0.85
Event	TIPSem	0.81	0.86	0.83
	Edinburgh	0.75	0.85	0.80
	(TRIOS)	0.80	0.74	0.77

These systems can be divided into those based on hand-crafted rules (Edinburgh) and (HeidelTime), those which are data-driven or corpus based (TIPSem), and those

---

[2] A predefined list of keywords that are likely to appear within a temporal expression or event (e.g., "year", "war").

which implement an hybrid strategy (TRIOS). From among the data-driven and hybrid approaches, the most popular machine learning technique was conditional random fields (CRF) [6]. Only one system, TIPSem, participated also in Spanish obtaining 0.91 F1 timex, and 0.88 F1 for event recognition.

Although, a Chinese TimeML dataset was developed for TempEval-2 [16], there were not participants for such language. To the best of our knowledge, the evaluation that is presented in Section 4 is the first one using this dataset.

Prior to TempEval-2, some authors tackled the automatic processing of temporal information in Chinese. An early proposal was presented by Li et al. [8]. Afterwards, the same authors addressed the temporal relation categorization task using machine learning techniques – decision tree and bayesian classifiers [7]. Focusing on the tasks tackled in this paper, only the temporal expression recognition task has been addressed [5,18,11], but never using the current TimeML standard. These approaches noted that "the main error causes for timex recognition were morphosyntactic ambiguities". Regarding events, there are no approaches addressing event recognition in Chinese.

This motivates us to adapt to Chinese the best state of the art approach for event recognition. Our approach differs from those cited for Chinese in the following respects: (i) we use a standard annotation scheme (TimeML), (ii) our approach has been evaluated over an available dataset (TempEval-2), and (iii) the approach uses semantic roles which may avoid the difficulties found in related works.

2.3 Semantic Roles

Only two of the reviewed systems for English went further than lexical semantics (TRIOS and TIPSem). This points out that the application of compositional semantic-knowledge is fairly novel.

Semantic roles capture the meaning of a sentence regarding how their arguments are related. Semantic role labeling (SRL) consists in grouping the sequences of words and then classify them according to the semantic roles they play, which is an important task of shallow semantic analysis and embodies the basic meaning of the sentence (see Example 4).

(4) 1. [他$_{agent}$] [一周后$_{tmp}$] 要参加 [比赛$_{patient}$]。
    EN: [He $_{agent}$] will attend [the competition $_{patient}$]
    [next week $_{tmp}$].

2. [一周后$_{tmp}$] [他$_{agent}$] 要参加 [比赛$_{patient}$]。
    EN: [Next week $_{tmp}$], [he $_{agent}$] will attend
    [the competition $_{patient}$].

From the SRL of these sentences, we can extract the information about agent, time, and location. As shown, the SRL is independent from the presentation of the arguments (e.g., argument order).

The most widely used semantic role sets for Chinese, are Chinese-FrameNet (CFN) [17] and Chinese-PropBank (CPB) [15]. Each proposal focuses on a different role granularity. CFN defines a detailed representation of situations including a set of very specific roles (frames), while CPB consists of a limited set of abstract roles. CFN avoids the problem of defining a small set of abstract roles defining as many roles as necessary with minimal information loss. However, the CPB set offers a wider lexical coverage than CFN, which only covers most general English verbs and nouns.

Regarding the application of roles to timex and event processing, they have never been applied in Chinese. In English, on the one hand, Uzzaman et al. [13] used a logical form representation of compositional semantics at sentence level based on a reduction of the FrameNet roleset. On the other hand, TIPSem used a standard role set, PropBank since it avoids the FrameNet's coverage and generalization problems.

The effect of applying of semantic roles for these tasks has not been analyzed for Chinese, being one of the main goals of this paper.

3   Our Proposal: TIRSemZH

As introduced previously, the tasks addressed by our approach are timex and event recognition (bounding). Example (5) illustrates a raw text input and the expected output after the described processing.

(5) **Input**: 约翰星期一来的。– EN：John came on Monday.
Output: 约翰<TIMEX3>星期一</TIMEX3>
<EVENT>来的</EVENT>。
EN: John <EVENT>came</EVENT> on <TIMEX3>Monday</TIMEX3>

The focus of our approach is the application of semantic roles in addition to morphosyntax for reaching a high-performance in these tasks for Chinese, and analyzing the advantages that semantic roles may offer.

Taking TIPSem [9] as reference, we developed a data-driven approach, TIRSemZH, which includes semantic roles and morphosyntactic features. Furthermore, in order to measure the influence of semantic roles on how temporal elements are dealt with, a baseline, TIRSemZH-B, was developed using only morphosyntactic features.

Apart from the language it tackles, the main difference between TIRSemZH and TIPSem is that TIRSemZH only uses semantic roles as semantic information. That is to say, TIRSemZH does not include features based on semantic networks.

The features concerning morphosyntax and semantic roles were obtained using the LTP tool [2]. This is a language platform which contains split sentence, word segment, PoS tagging, named entity recognition, word sense disambiguation, dependency parsing and semantic role labeling modules. The platform stated free sharing from 2006 and has been applied to research and international companies for the real business projects. The semantic role labeling module uses statistical methods based on maximum entropy, which got the first place in CoNLL 2009 shared task [1]. It obtains an F1 of 77.2% with an efficiency of about 1.3KB/s.

3.1   Learning Features

Given a set of annotated examples, this training process builds two models for each one of the approaches: timex recognition and event recognition. Once the models have been generated, the approaches use them to annotate any input raw text. Fig. 1 summarizes the architecture that the proposal follows.

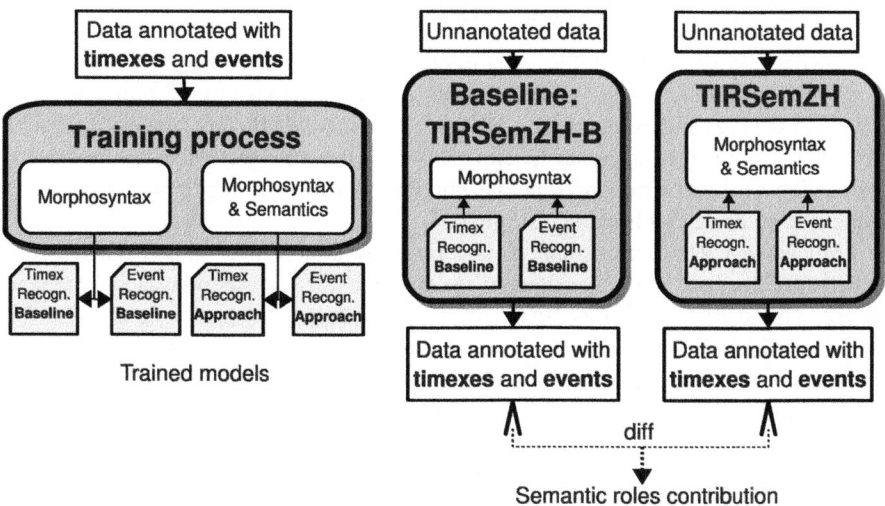

Fig. 1. Proposal architecture

The **morphosyntactic features** are shared by for both timex and event recognition: these are word and PoS context, in a 5-window (-2,+2).

The features based on **semantic roles** which enhance the training framework are:

- **Semantic Roles for Timex**: Semantic roles provide structural semantic relations of the predicates in which TimeML elements may participate. The temporal role (AM-TMP) represents the temporal adjunct and many times contains a time expression like in (6). For each token, we considered the $role$ played as a feature. Furthermore, the combination $role + word$ is included as feature.

  (6) [约翰$_{A0}$] [星期一 $_{AM-TMP}$] 来的。

      EN: [John $_{A0}$] came on [Monday $_{AM-TMP}$].

- **Semantic Roles for Event**: The SRL tool we used in this proposal annotates the main verbs of sentences as heads of the predicates that govern arguments playing different roles. Although, the tool may miss some of the events due to the fact that its performance is not perfect, it is still acceptable for Chinese language. Mainly, events in Chinese are identified by main verbs, and by nouns appearing under specific roles for particular verbs. For example in (7), the A1 role of "attend" contains an event. Three role-based features are used:

  • $role$: semantic role played by a token.
  • $role + word$: the combination of the token and the semantic role.
  • $role + verb$: the combination of the semantic role and the verb the token depends on. This distinguishes roles depending on specific verbs which is particularly important in the case of numbered roles (A0, A1, etc.) meaning different things when depending on different verbs.

  (7) [他$_{A0}$] 将 出席 [会议 $_{A1}$]。

      EN: [He $_{A0}$] will attend [the meeting $_{A1}$].

3.2 Machine Learning Technique: CRF

Following TIPSem proposal, the recognition task is defined as a sequence labeling problem. TIRSemZH employs conditional random fields (CRF) [6] to infer models for recognizing timexes and events in text. CRF is a popular and efficient supervised machine learning technique for sequence labeling. Not only the word sequence, but morphological, and semantic structure, present in the previously described features of our approach, benefit from using this learning technique.

For using CRFs, we assume X is a random variable over data sequences to be labeled (sentences), and Y is a random variable over the corresponding label sequences IOB2[2] labels (also known as BIO). All Y components (Y_i) are members of a finite alphabet of IOB2 labels $\gamma = \{$B-timex, I-timex, B-event, I-event, O$\}$. Given an input sentence, each token must be classified as being the beginning of a timex (B-timex), inside a timex (I-timex), the beginning of an event (B-event), inside an event (I-event), or outside any element (O). Example (8) shows a sentence labeled using these IOB2 tags.

(8) [John   came   last   week   .]
     [O   B-event   B-timex   I-timex   O]

The random variables X and Y are jointly distributed, and CRFs construct a conditional model from paired observation and label sequences: $p(Y|X)$. Once the models are constructed given a set of annotated examples and the previously described features (training), these can be applied to automatically solve the task given unannotated data, raw text plus the features described for the approach. For implementing CRF, the CRF++ toolkit[3] was used.

4   Evaluation: TempEval-2 Test

The objective of this evaluation is to measure the performance of TIRSemZH, and analyze the effect of applying semantic roles to timex and event recognition in Chinese.

4.1 Evaluation Framework

This consists of the description of the data used, as well as the evaluation criterion and measures to score the performance of the presented approaches.

Corpora. The approaches have been trained and tested using the datasets made available for TempEval-2[4], that way other evaluations can be compared. The Chinese dataset description is shown in Table 3.

The Chinese data comes from different newspaper articles of *Xinhua News Agency*, except the files chtb_0592-0596 that come from an article of *Sinorama*, a monthly magazine published in Taiwan.

---

[2] IOB2 format: (B) begin, (I) inside, and (O) outside.

[3] http://crfpp.sourceforge.net/

[4] http://semeval2.fbk.eu/semeval2.php?location=data.

Table 3. Chinese data (TempEval-2)

Set	Documents	Words	Element
Training	44	23K	TIMEX(766)
			EVENT(3744)
Test	8	5K	TIMEX(129)
			EVENT(1039)

Criterion and measures. The criterion applied for the evaluation was the same as in TempEval-2 (SemEval-2010). We used the following metrics:

$$precision = \frac{true\_positives}{true\_positives + false\_positives}$$

$$recall = \frac{true\_positives}{true\_positives + false\_negatives}$$

$$F_{\beta=1} = \frac{2 * precision * recall}{precision + recall}$$

where $true\_positives$ are the number of tokens that are part of an extent in the key file and the approach output, $false\_positives$ are the number of tokens that are part of an extent in the approach output but not in the key, and $false\_negatives$ are the number of tokens that are part of an extent in the key but not in the approach output.

4.2   TIRSemZH Results

The following subsections report on the results obtained for timex en event recognition.

Temporal Expression (timex) recognition:

Table 4 shows the results obtained in timex recognition for Chinese.

Table 4. Timex recognition results

Approach	precision	recall	$F_{\beta=1}$
TIRSemZH-B	0.94	0.74	0.83
TIRSemZH	0.97	0.76	**0.85**

As shown in the table, TIRSemZH offers a high performance 0.85 $F_{\beta=1}$. The application of semantic roles improved the precision and recall results of the baseline increasing the $F_{\beta=1}$ from 0.83 to 0.85. The relative error reduction over the baseline in $F_{\beta=1}$ is of 12%. This supports the hypothesis that semantic roles enable the proposal to learn more general models, which are also more precise. This is specially beneficial when the size of the annotated dataset is small because the approach is not only focused on words and PoS tags but also in the role the words play in the predicates.

The temporal expressions in the training data are contained the most of the times by an argument labeled with the temporal semantic role (AM-TMP). This role introduced two advantages in temporal expression recognition: (i) increasing the probability of any token appearing within this role, and (ii) aiding the approach in the recognition of multi-token temporal expressions such as 一月底 (EN: the end of January) and 去年六

月底 (EN: the end of June last year) represented by a sequence of tokens belonging to the AM-TMP role. Example (9) illustrates a timex that was missed by the baseline and correctly recognized using semantic roles.

(9) TempEval-2, file: chtb_0628 - sent.: 12
一季度末, 国家外汇储备余额达一千一百二十点六亿美元。
EN: At the end of the first quarter, the balance of foreign exchange reserves reached 112.06 billion US dollars.

In this sentence, all the tokens of the argument "一季度末 (EN: the end of the first quarter)" hold the AM-TMP role. The baseline missed this expression since this specific sequence of tokens is not very common among the training examples. However, with the use of semantic roles, the approach takes into account that this holds a AM-TMP role which is normally obtained by the temporal expressions annotated in the training and thus they are likely to be a temporal expression.

However, in some cases like in (10), the application of semantic roles decreased the performance.

(10) TempEval-2, file: chtb_0593 - sent.: 8
南韩目前利率已飙至百分之二十。
EN: South Korea's current interest rate has soared to twenty percent.

In this case, "current interest" does not hold a AM-TMP role and thus current was not tagged by the approach as timex. In this expression, "current" is a modifier of interest so this is not strictly a timex modifying the sentence, as it could be "currently" in "Currently, the interest has soared to twenty percent", but it is marked as a timex in the annotated data.

Event recognition:

Table 5 shows the results obtained in event recognition for Chinese.

Table 5. Event recognition results

Approach	precision	recall	$F_{\beta=1}$
TIRSemZH-B	0.86	0.80	0.83
TIRSemZH	0.90	0.80	**0.85**

As shown in the table, TIRSemZH obtains satisfactory results for event recognition. The scores obtained show that the application of semantic roles increases the $F_{\beta=1}$ from 0.83 to 0.85. In event recognition, the major relative error reduction over the baseline was obtained for precision (29%).

Primarily, semantic roles aided on increasing the precision, that is to say, reducing the false positives that the baseline presented. There are many words that have different meanings in different semantic surroundings. One word W can indicate an event in sentence A, or just a normal word in sentence B. While the baseline often obtains false positives, the semantics-aware approach, TIRSemZH, lead to the correct recognition.

Example (11) shows two sentences in which the word "出" appears. In the first sentence, it is a main-verb governing two arguments: "Annan" (A0) and "Money" (A1). In this case, "出 (EN: pay)" is an event. In the second sentence, "显示 (EN: showing)" is the main verb and "出 (EN: out)" is not an event but a verbal modifier. Hence, "出" has different meanings in different semantic surroundings in Chinese language and thus semantic roles aid on determining in which cases it is an event.

(11) TempEval-2, file: chtb_0595 - sent.: 29
    安南不出钱。
    EN: Annan does not <u>pay</u> money.

    TempEval-2, file: chtb_0604 - sent.: 11
    显示出成功率高、经营状况良好的特点。
    EN: Showing out a highly success rate and the characteristics of good business.

In Example (12), the baseline recognized "投资 (EN: investment)" as an event. This is because this word represents an event in many examples of the training data. This is because sometimes "investment" is referred to as an happening. However, it does not always indicate an event. In the sentence of the example, it appears within the argument "从资结构 (EN: from the investment structure)" holding an AM-DIS role of the verb "看 (EN: see)" and it does not represent an event because it, together with "structure", represents an entity from which something can be seen. Semantic roles aided on distinguishing which instances of "investment" are real events.

(12) TempEval-2, file: chtb_0618 - sent.: 10
    从投资结构看，科技含量高的大项目明显增多。
    EN: We can see from the investment structure, that large projects with high technological content significantly increased.

Although there were few cases and the recall was not significantly affected, some events that were missed by the baseline were correctly recognized by TIRSemZH. Example 13 shows how semantic roles aid on the recognition of stative-events. In this sentence, "充满 (EN: has full of)" is a main-verb that heads a predicate in the sentence and indicates an event. It was missed by the baseline because in the training it does not appear in this lemma-PoS context or a similar one. However, when using semantic roles, the approach is aware of the fact that "充满" is a main-verb governing two roles "社会公众 (EN: the public)" A0, and "信心 (EN: confidence)" A1. In the training, many events are found in this semantic role setting and hence the probability of the model for recognizing them as an event increases.

(13) TempEval-2, file: chtb_0628 - sent.: 8
    社会公众对宏观调控和经济发展充满信心。
    EN: The public has <u>full of</u> confidence to the macro-control and economic development.

Most of the errors of the semantic role-based approach, TIRSemZH, were due to SRL tool annotation errors. For example, the verb "减低 (EN: reduce)", which represents an event in a test sentence, was not labeled as main-verb by the SRL tool, and the event was missed by the approach. In the opposite case, the SRL tool labeled "贸易 (EN: trade)" as a main-verb in a sentence in which it represented just a noun and then the approach using roles produced a false positive.

5   Conclusions and Future Work

This paper addresses the automatic recognition of temporal expressions and events in Chinese text. The lack of systems recognizing TimeML timexes and especially events, together with the recent release of TimeML Chinese data, motivated us for developing a data-driven approach for this language, TIRSemZH. This is based on the best state of the art system for event recognition in English, TIPSem [9]. The paper reports on the performance of TIRSemZH and on the contribution of semantic roles when used in addition to morphosyntax for timex and event recognition in Chinese.

Unlike related works for Chinese, we use a standard annotation scheme (TimeML), and our approach is evaluated over a publicly available dataset (TempEval-2). Our approach uses semantic roles which have never been used in Chinese for these tasks. Furthermore, we approached the automatic TimeML-event recognition that, to the best of our knowledge, has never been approached before in Chinese.

From the results obtained it may be concluded that TIRSemZH is a good and effective approach for TimeML timex and event recognition (85% $F_{\beta=1}$). The use of semantic roles lead to the learning of more general models which improve the ones learned using only morphosyntax (baseline: TIRSemZH-B). In particular, role-based features contribute to the performance of the temporal expression recognition ($F_{\beta=1}$ error reduction: 12%) and, to the performance of event recognition, especially to the precision (precision error reduction: 29%).

For timex recognition, the roles improved the precision and recall in the identification and correct bounding of expressions that are ambiguous at lower language analysis levels and also the multi-token expressions (e.g., the end of the first quarter).

For event recognition, semantic roles increased the precision in the detection of main-verbs as events, as well as some nominal events when they play particular roles (e.g., A1) in some predicates (e.g., attend).

These conclusions match with the ones achieved by Llorens et al [9] for English and Spanish languages. This paper extends such the conclusions by proving that semantic role labeling is useful for timex and event recognition, not only in English and Spanish but also in Chinese, which is not an Indo-European language.

Unlike English, for example, Chinese sentences have not variation in tense and morphology; the sentence structure and phrase structure are normally identical in Chinese (i.e., realization relation between phrase and sentence); there is no a simple relationship in a one-one correspondence between PoS and syntax components; and Chinese lexical information is mainly based on conceptual and semantic attributes. If only simple methods based on morphology and syntax analysis were used, it would certainly affect the recognition results.

Semantic role labeling reflects the basic meanings of a sentence which is useful for the addressed tasks in Chinese. It should be taken into account when considering the presented results that not all the temporal entities in the dataset are ambiguous and therefore the advantages of semantics can only improve part of the cases.

As further work we propose to extend the application of semantic roles to temporal relation categorization. The semantic relations between the different arguments of sentences in which more than one event may participate, could be useful in the categorization of event-event intra-sentential temporal relations.

Acknowledgments. Paper supported by the Spanish Government, in projects: TIN-2006-15265-C06-01, TIN-2009-13391-C04-01 and PROMETEO/2009/119 and ACOMP/2011/001, where H.Llorens is funded (BES-2007-16256).

References

1. Che, W., Li, Z., Li, Y., Guo, Y., Qin, B., Liu, T.: Multilingual dependency-based syntactic and semantic parsing. In: Proceedings of the 13th Conference on Computational Natural Language Learning: Shared Task, CoNLL 2009, pp. 49–54. ACL (2009)
2. Che, W., Li, Z., Liu, T.: Ltp: A chinese language technology platform. In: Coling 2010: Demonstrations, Beijing, China, pp. 13–16 (2010)
3. Fellbaum, C.: WordNet: An Electronic Lexical Database (Language, Speech, and Communication). MIT Press, Cambridge (1998)
4. Grover, C., Tobin, R., Alex, B., Byrne, K.: Edinburgh-ltg: Tempeval-2 system description. In: Proceedings of SemEval-5, pp. 333–336. ACL (2010)
5. Hacioglu, K., Chen, Y., Douglas, B.: Automatic time expression labeling for english and chinese text. In: Gelbukh, A. (ed.) CICLing 2005. LNCS, vol. 3406, pp. 548–559. Springer, Heidelberg (2005)
6. Lafferty, J.D., McCallum, A., Pereira, F.C.N.: Conditional random fields: Probabilistic models for segmenting and labeling sequence data. In: Proceedings of the 18th ICML, pp. 282–289. Morgan Kaufmann, San Francisco (2001)
7. Li, W., Wong, K.-F., Cao, G., Yuan, C.: Applying machine learning to chinese temporal relation resolution. In: Proceedings of the 42nd Meeting of the ACL (ACL 2004), Barcelona, Spain, Main Volume, pp. 582–588 (2004)
8. Li, W., Wong, K.F., Yuan, C.: A model for processing temporal references in chinese. In: Workshop on Temporal and Spatial Information Processing, pp. 33–40. ACL (2001)
9. Llorens, H., Saquete, E., Navarro, B.: TIPSem (English and Spanish): Evaluating CRFs and Semantic Roles in TempEval-2. In: Proceedings of SemEval-5, pp. 284–291. ACL (2010)
10. Pustejovsky, J., Castaño, J.M., Ingria, R., Saurí, R., Gaizauskas, R., Setzer, A., Katz, G.: TimeML: Robust Specification of Event and Timexes in Text. In: IWCS-5 (2003)
11. Silberztein, M.: An alternative approach to tagging. In: Kedad, Z., Lammari, N., Métais, E., Meziane, F., Rezgui, Y. (eds.) NLDB 2007. LNCS, vol. 4592, pp. 1–11. Springer, Heidelberg (2007)
12. Strötgen, J., Gertz, M.: Heideltime: High quality rule-based extraction and normalization of temporal expressions. In: Proceedings of SemEval-5, pp. 321–324. ACL (2010)
13. UzZaman, N., Allen, J.F.: Trips and trios system for tempeval-2: Extracting temporal information from text. In: Proceedings of SemEval-5, pp. 276–283. ACL (2010)
14. Verhagen, M., Saurí, R., Caselli, T., Pustejovsky, J.: Semeval-2010 task 13: Tempeval-2. In: Proceedings of SemEval-5, pp. 57–62. ACL (2010)
15. Xue, N.: Labeling chinese predicates with semantic roles. Computational Linguistics 34(2), 225–255 (2008)
16. Xue, N., Zhou, Y.: Applying syntactic, semantic and discourse constraints in chinese temporal annotation. In: Coling 2010: Posters, Beijing, China, pp. 1363–1372 (2010)
17. You, L., Liu, K.: Building chinese framenet database. In: Proceedings of IEEE International Conference on Natural Language Processing and Knowledge Engineering (NLP-KE 2005), pp. 301–306 (2005)
18. Zhang, C., Cao, C.G., Niu, Z., Yang, Q.: A transformation-based error-driven learning approach for chinese temporal information extraction. In: Li, H., Liu, T., Ma, W.-Y., Sakai, T., Wong, K.-F., Zhou, G. (eds.) AIRS 2008. LNCS, vol. 4993, pp. 663–669. Springer, Heidelberg (2008)

Information Retrieval Techniques for Corpus Filtering Applied to External Plagiarism Detection

Daniel Micol[1], Óscar Ferrández[2], and Rafael Muñoz[1]

[1] Research Group on Natural Language Processing and Information Systems
Department of Software and Computing Systems
University of Alicante
San Vicente del Raspeig, Alicante, Spain
[2] Department of Biomedical Informatics
University of Utah
Salt Lake City, Utah, United States of America
dmicol@dlsi.ua.es, oscar.ferrandez@utah.edu, rafael@dlsi.ua.es

Abstract. We present a set of approaches for corpus filtering in the context of document external plagiarism detection. Producing filtered sets, and hence limiting the problem's search space, can be a performance improvement and is used today in many real-world applications such as web search engines. With regards to document plagiarism detection, the database of documents to match the suspicious candidate against is potentially fairly large, and hence it becomes very recommendable to apply filtered set generation techniques. The approaches that we have implemented include information retrieval methods and a document similarity measure based on a variant of *tf-idf*. Furthermore, we perform textual comparisons, as well as a semantic similarity analysis in order to capture higher levels of obfuscation.

1 Introduction

Within Natural Language Processing, external plagiarism detection is an increasingly popular task that attempts to determine if a suspicious document contains one or more appropriations of another text that belongs to a set of source candidates. It can be a very expensive task if the number of documents to compare against is large, which is often the case. The corpora used as input for external plagiarism detection systems are composed of two sets of documents, namely suspicious and source. The first of them includes those documents that may contain one or more plagiarisms extracted from documents belonging to the second of these sets.

To carry out the plagiarism detection task in an efficient way, it is recommended to perform a corpus filtering of the mentioned candidate source documents, in order to be able to later on apply over this subset a more complex and costly function that will detect the corresponding plagiarized document, if

R. Muñoz et al. (Eds.): NLDB 2011, LNCS 6716, pp. 100–111, 2011.
© Springer-Verlag Berlin Heidelberg 2011

any [Stein et al., 2007]. The resulting source documents not removed by the filtering process, and compared against the corresponding suspicious text in order to extract the exact position of the plagiarisms, are referred to as filtered set.

To develop and measure the approaches described in this paper we have used the *1st International Competition on Plagiarism Detection* [Potthast et al., 2009] as framework. The committee of this competition provides annotated corpora as well as a definition of the measures to evaluate our methods. Concretely, it provides a source documents corpus, containing those that may have been plagiarized, and a suspicious documents corpus, containing those that might include the plagiarism itself. The documents included are mostly written in English, but some of them are also in Spanish and German.

The system that we are developing to detect document plagiarism is composed of two major modules, similar to what is described in [Stein et al., 2007]: document corpus filtering and passage matching. The first reduces the candidate document search space by identifying those texts that are likely to be the source of the plagiarized document, and the second analyzes in more detail the documents resulting from the previous step in order to identify the exact fragments that have been plagiarized. In this paper we will describe the first of these components. To accomplish the corpus filtering, we apply different techniques that are detailed throughout this document.

The remainder of this paper is structured as follows. In the second section we will describe the state of the art in corpus filtering techniques in the context of external plagiarism detection. The third one contains the methods that we have developed, and the fourth our experimental results. Finally, the last section presents our conclusions and proposes future work based on our research.

2   State of the Art

In recent years, the generation and usage of filtered sets has become increasingly popular as the corpora sizes to process have increased due to computers' capabilities to handle larger amounts of information. Therefore, many recent research approaches in the field of external plagiarism detection contain a simple and efficient heuristic retrieval to reduce the number of source documents to compare every suspicious text against, and a more complex and costly detailed analysis that attempts to extract the exact position of the plagiarized fragment.

Some authors have proposed measures that analyze the document as a whole, trying to determine if it is roughly the same as another one, or if it is contained in the other. This is the case of [Broder, 1997], given that the measures presented in this work are based on word occurrences and positions, allowing word position shifting within a given window. In addition, other authors create a representation of the document's structure and compare those of two given texts, trying to determine from these structures if they are plagiarized. This is the case presented in [Si et al., 1997], where the authors generate a tree data structure based on the sections of a document, and assign the words found in every part to its corresponding tree node. Then, they calculate normalized weight vectors based on the appearance of every keyword in different sections. Later on, a comparison of tree nodes is done for every suspicious-source pair by using the dot product.

[Kasprzak et al., 2009] create an inverted index of the corpus documents' contents in order to be able to retrieve efficiently a set of texts that contain a given n-gram. This is very similar to using a full-text search engine, although they preferred to implement their own approach integrated into their software.

[Basile et al., 2008, Grozea et al., 2009] decided to apply document similarity measures that would be used as heuristic to determine if a given suspicious and source documents are similar enough to hold a plagiarism relation, being these measures based on information retrieval functions like *tf-idf* and the cosine similarity. On the other hand, [Barrón-Cedeño et al., 2009] decided to apply the Kullback-Leibler distance for this same purpose.

[Grozea et al., 2009] implement a character-level n-gram comparison and apply a cosine similarity based on term frequency weights. They extract the 51 most similar source documents to the suspicious one being analyzed. Other authors, such as [Basile et al., 2009, Kasprzak et al., 2009], decided to implement a word-level n-gram comparison. [Basile et al., 2009] keep the 10 most similar using a custom distance function based on frequency weights, while [Kasprzak et al., 2009] keep those source documents that share at last 20 word-grams of length 5.

3 Methods

We have developed two methods for corpus filtering. The first of them uses a full-text search engine to index and retrieve the similar documents to a given suspicious text. For this approach we have experimented varying different parameters. The second applies a document similarity function to pairs of documents in order to extract the most similar ones. For this second approach we have experimented using textual similarity, and also adding semantic knowledge.

3.1 Full-Text Search Engine

The approach that we have developed based on a full-text search engine contains an indexing phase and the filtering itself. To implement this we needed to allow the storage of our corpus' contents into an inverted index, and querying against it, and for this purpose we chose Lucene [Gospodnetic et al., 2009].

Indexing. The first step in order to produce a filtered set is to index the documents from our corpus of candidate documents to be the source of the plagiarism, given that we will have a large amount and we require fast retrieval. Lucene creates an inverted index that will allow efficient multi-word queries.

Filtering. After the indexing part is done, we calculate the similarity between a given suspicious document and every candidate that we have stored in the aforementioned full-text search engine's index. Then, we will extract those documents that have the highest similarity scores, and these will compose the corresponding filtered set for the given suspicious text.

The main difference between our approach and the one described in [Kasprzak et al., 2009] is the document scoring function, apart from the fact that we use a full-text search engine and they implemented their own indexing and querying approach. The aforementioned paper proposes a method to calculate this score based on the number of words that appear in both documents. In our case, however, we apply a more complex function that takes into consideration several factors. Concretely, we use the document scoring function implemented by Lucene, called *Lucene's Practical Scoring Function*, which is applied over a query and a document, and is defined as [Gospodnetic et al., 2009, Manning et al., 2008]:

$$score(q,d) = C(q,d) \cdot QN(q) \cdot \sum_{t \in q} \left(tf(t) \cdot idf(t)^2 \cdot B(t) \cdot N(t,d) \right) \qquad (1)$$

where q is a query, d is a document, $tf(t)$ is the term frequency of term t in document d, $idf(t)$ is the inverse document frequency of term t, $C(q,d)$ is a score factor based on how many of the query terms are found in the specified document, $QN(q)$ is a normalizing factor used to make scores between queries comparable, $B(t)$ is a search time boost of term t in the query q, and $N(t,d)$ encapsulates a few boost and length factors. The result of this function will be a normalized similarity value between 0 and 1.

For every suspicious document, we will extract all non-overlapping n-grams of a certain length, and use them as queries against Lucene, obtaining, for each of them, a list of documents that contain those n-grams, as well as a similarity score. For every source document we will keep the highest similarity score for any of its n-grams, and store those documents that returned the highest scores in the generated filtered set.

In the aforementioned algorithm, apart from the corpus of suspicious documents, the procedure takes the n-gram size as input, which will be used to generate the queries that will be executed against Lucene's index. Another input parameter is the criteria, which will define how to determine the documents that should be stored in the filtered set. In our case we have defined two different criteria, that are similarity score threshold and filtered set size. In our experiments we will explore different values for the similarity score threshold and the filtered set size. In addition, we analyze how different n-gram sizes affect the accuracy of our system.

3.2   Document Similarity Measure

We have experimented using only a textual similarity, which means comparing words in the document as they appear, and also using semantic information by expanding words into their equivalent terms to detect higher levels of obfuscation.

Textual Similarity. In order to apply the document similarity measure that we have developed, we first weight the words in every text in our corpora and then compare the weights of those terms that appear in both the suspicious and the source documents being compared. The similarity score between the two documents will be the sum of their common term weights. More concretely, we use a variant of the *term frequency inverse document frequency* function [Spärck Jones, 1972], or *tf-idf*, commonly used in information retrieval. It is applied over a term, t_i, in a given document, d_j, and is defined as:

$$tf\text{-}idf_{i,j} = tf_{i,j} \cdot idf_i \tag{2}$$

where the term frequency of t_i in document d_j, or $tf_{i,j}$, is:

$$tf_{i,j} = \frac{n_{i,j}}{\sum_k n_{k,j}} \tag{3}$$

being $n_{i,j}$ the number of occurrences of term t_i in document d_j, and the denominator is the sum of the number of occurrences of all terms in document d_j. Furthermore, the inverse document frequency of term t_i, or idf_i, is defined as:

$$idf_i = \log\left(\frac{|D|}{|\{d : t_i \in d\}|}\right) \tag{4}$$

where $|D|$ is the number of documents in our corpus.

For our purpose this definition of *tf-idf* is not optimal. The *tf* value is normalized by the length of a document to prevent longer documents from having a higher weight. This makes sense in information retrieval applications such as search engines. However, in our case it would be better to skip this normalization given that the more times a word appears in a document, the higher likelihood it will have to contain a plagiarized fragment, regardless of its size. Therefore, *tf* in our case will be defined as $tf_{i,j} = n_{i,j}$.

With regards to *idf*, given that in our case the number of documents in the corpus is constant and we will only need the relative *tf-idf* value, not the absolute, we can simplify this function by removing the $|D|$ and the logarithm, as:

$$idf_i = \frac{1}{|\{d : t_i \in d\}|} \tag{5}$$

Finally, the similarity score of two documents, being d_j the suspicious and d_k the source one, will be defined as follows:

$$sim_{d_j,d_k} = \sum_{t_w \in d_j \cap d_k} (tf_{w,j} \cdot idf_{w,j}) = \sum_{t_w \in d_j \cap d_k} \left(n_{w,j} \cdot \frac{1}{|\{d : t_w \in d\}|}\right) \tag{6}$$

Therefore, the similarity of two documents will be higher the more words they have in common, and also the fewer documents those terms appear in.

The approach that we propose to apply the variant of the *tf-idf* measure previously mentioned is applied as a two-pass algorithm. First of all, we calculate the *tf-idf* of the words of every source and suspicious documents, and afterwards store these values. Next, we compare every suspicious document against the corpus of source texts, using their stored *tf-idf* values and the similarity function previously defined. Finally, this algorithm will return, for every suspicious document, a set of the most similar source documents.

Semantic Similarity. The approach based on textual similarity is very strict and doesn't perform well for higher levels of obfuscation, where equivalent terms are used. To solve this issue we need to add semantic knowledge to our system, and for this purpose we use the WordNet::Similarity tool [Pedersen et al., 2004] to extract words with a similar meaning and consider them as well.

We have implemented two different approaches to take advantage of semantic information. In the first of them what we do is, in the indexing phase, extract all equivalent terms[1] for every word in a source document, calculate their *tf-idf* and store these values as well. Next, in the filtering phase, we expand every word in a suspicious document into its equivalent terms, considering them as if they appeared in the document as well. After this is done, we apply the textual similarity function over the original words and their expanded equivalent terms. This approach has the potential disadvantage that it will change the corpus' structure, and therefore our *tf-idf* measure might not be representative anymore. On the other hand, in the second approach we perform the indexing phase as we do in the textual similarity method, and use the semantic knowledge only during the evaluation step. More concretely, we extract the equivalent terms for every suspicious document word, and check if they exist in the source text being compared. If this is the case, we will add to the final similarity score the *tf-idf* of the aforemetioned equivalent term.

4   Experimentation and Results

To evaluate our system we used the external plagiarism corpora from the *1st International Competition on Plagiarism Detection* [Potthast et al., 2009]. This corpora contains a source documents corpus composed of 14,429 texts, and a suspicious documents corpus composed of 14,428 elements. Half of the suspicious texts contain a plagiarism, which ranges between 0% and 100% of the corresponding source document. In addition, the plagiarism length is evenly distributed between 50 and 5,000 words.

Given that there is a trade-off in our experiments between recall and precision, we used the F-score measure with $\beta = 1$ to determine the best result, as this measure is a combination of the previous two.

4.1   Full-Text Search Engine

We performed three sets of experiments based on the Lucene full-text search engine, varying different parameters.

---

[1] We used synonyms as equivalent terms.

Similarity Score Threshold. Generating filtered sets using a full-text search engine similarity score threshold, only including those candidate documents that had the highest scores for any given suspicious document n-gram query. We considered different thresholds, including in the filtered set only those documents that had a similarity score equal or greater than this value. In this experiment we used a fixed n-gram size of 25 words, given that the minimum plagiarism size of the corpora we experimented with was 50 words [Potthast et al., 2009], and therefore if we use a size of 25 we ensure that for every plagiarisms we will at least have one query with all tokens plagiarized. The results from this experiment are shown in Table 1, where column *Captured* represents the number of plagiarisms that are contained within the filtered set of source documents, and *Missed* those that are not included in this set. The other columns correspond to the metrics previously mentioned.

Table 1. Metrics using different similarity score thresholds and based on n-gram queries of size 25

Threshold	Size	Captured	Missed	Recall	Precision	F-score
0.0	14429	332	0	1.0000	0.0000	0.0000
0.1	9450.19	332	0	1.0000	0.3451	0.5131
0.2	4443.89	328	4	0.9880	0.6920	0.8139
0.3	1520.45	304	28	0.9157	0.8946	0.9050
0.4	348.48	271	61	0.8163	0.9758	0.8890
0.5	69.12	194	138	0.5843	0.9952	0.7363
0.6	22.11	118	214	0.3554	0.9985	0.5242
0.7	5.68	69	263	0.2078	0.9996	0.3441
0.8	1.81	55	277	0.1657	0.9999	0.2842
0.9	0.60	28	304	0.0843	1.0000	0.1556
1.0	0.42	17	315	0.0512	1.0000	0.0974

As we can see in Table 1, recall decreases as the threshold grows, and the opposite happens for precision. This is because higher thresholds are more restrictive and therefore will lead to smaller filtered sets. The F-score measure has a maximum value when the threshold is 0.3 for the corpora we used, so this would be the approximate optimal value for a similarity score threshold. However, we see that the number of filtered documents cannot be controlled by changing this parameter, as there is no strict correlation between the filtered set size and the similarity score threshold.

Filtered Set Size. Our second experiment consisted in fixing the filtered set size and observing what is the impact of this value in recall, precision and F-score. The filtered set will be filled with the documents that contain the highest similarity scores. The queries used by the information retrieval system are sentences extracted from every suspicious document. We identify sentences based on spacing and punctuation symbols. The results from this experiment are shown in Table 2, where we see that recall increases with the filtered set size, and precision decreases (by definition). We find that the maximum F-score value happens

Table 2. Metrics using different filtered set sizes and based on n-gram queries of size 25

Size	Captured	Missed	Recall	Precision	F-score
10	153	179	0.4608	0.9993	0.6308
20	199	133	0.5994	0.9986	0.7491
30	226	106	0.6807	0.9979	0.8094
40	232	100	0.6988	0.9972	0.8218
50	256	76	0.7711	0.9965	0.8694
60	260	72	0.7831	0.9958	0.8768
70	277	55	0.8343	0.9951	0.9077
80	277	55	0.8343	0.9945	0.9074
90	278	54	0.8373	0.9938	0.9089
100	283	49	0.8524	0.9931	0.9174
200	285	47	0.8584	0.9861	0.9179
500	292	40	0.8795	0.9653	0.9204
1000	302	30	0.9096	0.9307	0.9200

when we have a filtered set of around 500 document, and then starts to decrease given that the recall increase doesn't compensate the precision drop. Therefore, this will be an approximate optimal value.

One big advantage of this approach is that it transforms the complexity of the plagiarism detection task from linear to constant, based on the number of elements that will compose the filtered set. This will allow us to ensure that our plagiarism detection software finishes within a given time window, making it suitable for real-life use cases. Other authors, such as [Grozea et al., 2009]), have also chosen to convert this problem into linear complexity as a way of making it computationally tractable.

N-gram Size. Finally, we experimented using different n-gram sizes as information retrieval queries. In addition, we kept the 0.3 threshold restriction. The results from this experiment are shown in Table 3, where we see that recall decreases when the n-gram size increases, and the opposite applies to precision. This is because longer n-grams produce lower similarity scores due to the fact that, specially for smaller plagiarisms, more non-matching terms will be included in the query, and the normalization factors described in Lucene's scoring function will lower the overall score. However, and as it already happened when we experimented with the similarity score threshold, we cannot control the size of the filtered set by adjusting the n-gram size, and therefore this is a big constraint to take into account.

4.2 Document Similarity Measure

With regards to our document similarity measure based on a variant of *tf-idf*, we have performed two sets of experiments, described in what follows.

Table 3. Metrics using different n-gram sizes as queries and a score threshold of 0.3

N	Size	Captured	Missed	Recall	Precision	F-score
10	6435.88	325	7	0.9789	0.5540	0.7075
25	1520.45	304	28	0.9157	0.8946	0.9050
50	490.89	299	33	0.9006	0.9660	0.9321
100	226.63	299	33	0.9006	0.9843	0.9406
200	253.98	276	56	0.8313	0.9824	0.9006
500	329.31	176	156	0.5301	0.9772	0.6874
1000	363.23	110	222	0.3313	0.9748	0.4946

Textual Similarity. Our first set of experiments used the textual similarity measure, obtaining the results shown in Table 4, where we see that recall grows with the filtered set size, and precision drops by definition. We also find that F-score peaks when the filtered set size is 90.

Table 4. Metrics obtained applying the textual similarity measure

Size	Captured	Missed	Recall	Precision	F-score
10	238	94	0.7169	0.9993	0.8348
20	249	83	0.7500	0.9986	0.8566
30	255	77	0.7681	0.9979	0.8680
40	260	72	0.7831	0.9972	0.8773
50	260	72	0.7831	0.9965	0.8770
60	269	63	0.8102	0.9958	0.8935
70	273	59	0.8223	0.9951	0.9005
80	277	55	0.8343	0.9945	0.9074
90	279	53	0.8404	0.9938	0.9106
100	279	53	0.8404	0.9931	0.9104

Semantic Similarity. The second set of experiments used the semantic similarity measure, and more concretely the two variants detailed in Section 3.2. For the first of these variants, which expands a word into its equivalent terms both in the indexing and in the evaluation phases, we obtained the results shown in Table 5. The trend of the values shown in this table are similar to those presented in Table 4, although the values themselves are much lower in the former than in the latter. This means that the semantic measure adds a considerable amount of noise to the system, as we previously expected, given that it changes the nature of our corpus since it adds additional words in the indexing phase.

Next, Table 6 shows the results obtained using the second variant of the semantic similarity measure, which doesn't expand a word into its equivalent terms in the indexing phase, but only during evaluation time. As we see in this table, the results are much better than with the first variant, and are fairly similar to those obtained applying the textual similarity measure (Table 4), although never exceed them.

Table 5. Metrics obtained applying the first variant of the semantic similarity measure

Size	Captured	Missed	Recall	Precision	F-score
10	85	247	0.2560	0.9993	0.4076
20	95	237	0.2861	0.9986	0.4448
30	129	203	0.3886	0.9979	0.5593
40	153	179	0.4608	0.9972	0.6304
50	164	168	0.4940	0.9965	0.6605
60	181	151	0.5452	0.9958	0.7046
70	189	143	0.5693	0.9951	0.7242
80	198	134	0.5964	0.9945	0.7456
90	206	126	0.6205	0.9938	0.7640
100	209	123	0.6295	0.9931	0.7706

Table 6. Metrics obtained applying the second variant of the semantic similarity measure

Size	Captured	Missed	Recall	Precision	F-score
10	238	94	0.7169	0.9993	0.8348
20	249	83	0.7500	0.9986	0.8566
30	255	77	0.7681	0.9979	0.8680
40	260	72	0.7831	0.9972	0.8773
50	260	72	0.7831	0.9965	0.8770
60	266	66	0.8012	0.9958	0.8880
70	273	59	0.8223	0.9951	0.9005
80	277	55	0.8343	0.9945	0.9074
90	277	55	0.8343	0.9938	0.9071
100	277	55	0.8343	0.9931	0.9068

4.3 Technique Comparison

For recall, we observe that the first variant of the semantic similarity measure is clearly worse than any other approach. With regards to the full-text search engine method, it is also clearly worse than the document similarity approaches, but only for small filtered set sizes. For larger sizes it is comparable and even better some times. In addition, we observe that, even though most of the time both measures have the same performance, in some situations the textual similarity measure is better than the semantic one. Finally, when it comes to F-score, we see that the results are similar to those presented above for the recall metric, given that precision is the same for all approaches as we are fixing the filtered set size, and therefore F-score strongly correlates with variations in recall.

5 Conclusions and Future Work

In this paper we have presented a set of different approaches for document corpus filtering in the context of external plagiarism detection. All of these have in common that they are related to information retrieval techniques. Some of them use a full-text search engine, and the rest use a variant of the *tf-idf* function that is commonly used in information retrieval tasks.

Two of the approaches that we have proposed are less suitable given that they don't allow the user to control the number of resulting documents, which is the main goal of what is described in this paper. These are the usage of a full-text search engine using a similarity score threshold, and also varying the n-gram size. Therefore, although they produce good results, we would not consider them in future research given that a plagiarism detection system that uses these approaches for filtered set generation would require an undefined amount of time to complete.

When we compare the usage of a full-text search engine fixing the number of documents in the filtered set, and the textual similarity measure, we find that the results are comparable. For larger amounts of documents (e.g. 100) the former performs better, and for smaller amounts, the latter seems more suitable.

Finally, we have seen that the usage of semantic knowledge can add a considerable amount of noise, worsening the results. On the other hand, we haven't seen any gains that would justify their application. In addition, these techniques require many additional computations, with the corresponding performance impact. Therefore, from our experimentation we would not recommend to use semantic techniques like the ones described in this paper. Instead, the usage of the textual similarity measure, or the full-text search engine fixing the filtered set size, seem to be the best options, being the former a more likely choice given that it is more efficient than the latter.

As future work we would like to keep investigating the addition of semantic knowledge to solve certain kinds of obfuscations. This would include the recognition of the word's part-of-speech, and using different resources such as VerbNet and FrameNet, focusing more in altering different kinds of words rather than expanding every token into its equivalent terms.

Acknowledgements

This research has been partially funded by the Spanish Ministry of Science and Innovation (grant TIN2009-13391-C04-01) and the Conselleria d'Educació of the Spanish Generalitat Valenciana (grants ACOMP/2010/286 and PROME-TEO/2009/119).

References

[Barrón-Cedeño et al., 2009] Barrón-Cedeño, A., Rosso, P., Benedí, J.-M.: Reducing the Plagiarism Detection Search Space on the Basis of the Kullback-Leibler Distance. In: Gelbukh, A. (ed.) CICLing 2009. LNCS, vol. 5449, pp. 523–534. Springer, Heidelberg (2009)

[Basile et al., 2009] Basile, C., Benedetto, D., Caglioti, E., Cristadoro, G., Esposti, M.D.: A Plagiarism Detection Procedure in Three Steps: Selection, Matches and "Squares". In: Proceedings of the SEPLN 2009 Workshop on Uncovering Plagiarism, Authorship and Social Software Misuse, San Sebastian, Spain, pp. 19–23 (2009)

[Basile et al., 2008] Basile, C., Benedetto, D., Caglioti, E., Esposti, M.D.: An example of mathematical authorship attribution. Journal of Mathematical Physics 49, 125211–125230 (2008)

[Broder, 1997] Broder, A.: On the resemblance and containment of documents. In: Proceedings of the Compression and Complexity of Sequences 1997, SEQUENCES 1997, pp. 21–29. IEEE Computer Society Press, Washington, DC, USA (1997)

[Gospodnetic et al., 2009] Gospodnetic, O., Hatcher, E., McCandless, M.: Lucene in Action, 2nd edn. Manning Publications (2009)

[Grozea et al., 2009] Grozea, C., Gehl, C., Popescu, M.: ENCOPLOT: Pairwise Sequence Matching in Linear Time Applied to Plagiarism Detection. In: Proceedings of the SEPLN 2009 Workshop on Uncovering Plagiarism, Authorship and Social Software Misuse, San Sebastian, Spain, pp. 10–18 (2009)

[Kasprzak et al., 2009] Kasprzak, J., Brandejs, M., Křipač, M.: Finding Plagiarism by Evaluating Document Similarities. In: Proceedings of the SEPLN 2009 Workshop on Uncovering Plagiarism, Authorship and Social Software Misuse, San Sebastian, Spain, pp. 24–28 (2009)

[Manning et al., 2008] Manning, C.D., Raghavan, P., Schütze, H.: Introduction to Information Retrieval. Cambridge University Press, Cambridge (2008)

[Pedersen et al., 2004] Pedersen, T., Patwardhan, S., Michelizzi, J.: WordNet: Similarity - Measuring the Relatedness of Concepts. In: Proceedings of the North American Chapter of the Association for Computational Linguistics, pp. 1024–1025 (2004)

[Potthast et al., 2009] Potthast, M., Stein, B., Eiselt, A., Cedeño, A.B., Rosso, P.: Overview of the 1st International Competition on Plagiarism Detection. In: Proceedings of the SEPLN 2009 Workshop on Uncovering Plagiarism, Authorship and Social Software Misuse, San Sebastian, Spain, pp. 1–9 (2009)

[Si et al., 1997] Si, A., Leong, H.V., Lau, R.W.H.: CHECK: a document plagiarism detection system. In: Proceedings of the ACM Symposium on Applied Computing, San Jose, California, pp. 70–77 (1997)

[Spärck Jones, 1972] Spärck Jones, K.: A statistical interpretation of term specificity and its application in retrieval. Journal of Documentation 28(1), 11–21 (1972)

[Stein et al., 2007] Stein, B., Zu Eissen, S.M., Potthast, M.: Strategies for retrieving plagiarized documents. In: Proceedings of the 30th Annual International ACM SIGIR Conference on Research and Development in Information Retrieval, pp. 825–826 (2007)

Word Sense Disambiguation: A Graph-Based Approach Using N-Cliques Partitioning Technique

Yoan Gutiérrez[1], Sonia Vázquez[2], and Andrés Montoyo[2]

[1] Department of Informatics
University of Matanzas, Cuba
{yoan.gutierrez}@umcc.cu
[2] Research Group of Language Processing and Information Systems
Department of Software and Computing Systems
University of Alicante, Spain
{svazquez,montoyo}@dlsi.ua.es

Abstract. This paper presents a new approach to solve semantic ambiguity using an adaptation of the Cliques Partitioning Technique to N distance. This new approach is able to identify sets of strongly related senses using a multidimensional graph based on different resources: WordNet Domains, SUMO and WordNet Affects. As a result, each Clique will contain relevant information used to extract the correct sense of each word. In order to evaluate our approach there have been conducted several experiments using the data set of the "English All Words" task of Senseval-2 obtaining promising results.

Keywords: Word Sense Disambiguation, Cliques, N-Cliques, WordNet.

1 Introduction and Motivation

In Natural Language Processing (NLP) one of the main problems is the ambiguity. One word has different meanings according to the context where it appears and it is difficult to decide which sense is the most appropriate. To deal with this problem and determine the correct meanings of words a specific task has been created: Word Sense Disambiguation (WSD). One of the firsts methods to solve WSD was that based on Lesk algorithm [1]. It used the coincidence of descriptions to determinate the word senses. Others have used lexicons, list of concepts, taxonomies of concepts, and lexical dictionaries among different lexical resources. Among all of these different approaches the research community has also experimented with graph-based methods obtaining promising results. We can mention those approaches using structural interconnections such as Structural Semantic Interconnections (SSI) which creates structural specifications of the possible senses for each word in a context [2], another is "Exploring the integration of WordNet [3] and FrameNet" [4] proposal by Laparra and those using page-rank such as Sinha and Mihalcea [5], and Agirre and Soroa [6].

At the beginning, one of the main problems of WSD was how to evaluate different systems based on a variety of different repositories, corpus and language resources.

R. Muñoz et al. (Eds.): NLDB 2011, LNCS 6716, pp. 112–124, 2011.
© Springer-Verlag Berlin Heidelberg 2011

To solve this problem an international workshop on evaluating WSD Systems (Senseval[1]) was created. In Senseval different NLP tasks are defined and the participant systems are evaluated using the same repositories and corpus. In Senseval-2 [7] the best system obtained a 69% of accuracy in the WSD task, three years later in Senseval-3 [8] the better results were around 65.2% of accuracy and six years later in Semeval-2 [9] 55.5% of accuracy was obtained. As we can notice, despite of there have been developed lots of different systems, WSD is still an open research field due to the low accuracy of the results obtained.

In this paper we present a new graph-based approach for the resolution of semantic ambiguity of words in different contexts. The method proposed is an adaptation of the N-Cliques model [10] applying the Clique Partitioning Technique [11] to N distance for WSD. It is based on the multidimensional interrelations obtained from ISR-WN (Integration of Semantic Resources based on WordNet) [12] resource.

The organization of this paper is as follows: after this introduction, in section 2 we describe the techniques and resources used. In section 3 we present our new approach and how it works. Next, in section 4 we present the evaluation and a comparative study of results contrasted with the NLP systems that participated in the Senseval-2 competition of the Association for Computational Linguistics (ACL) and previous works. Finally, conclusions and a further works are detailed.

2   Models, Technique and Resources

In this work we present several models, algorithm and resources that have been combined to solve semantic ambiguities. Next sections describe each one and how we have integrated and adapted all of them in a new approach to WSD.

2.1   Cliques

Cliques are formally defined by Luce and Perry [13] in terms of friendship as "A Clique is a set of more than two people if they are all mutual friends of one another." In order to understand what is a Clique, we can follow the explanation by Cavique [14]: "Given an undirected graph $G = (V, E)$ where V denotes the set of vertices and E the set of edges, the graph $G_1 = (V_1, E_1)$ is called a sub-graph of G if $V_1 \subseteq V$, $E_1 \subseteq E$ and for every edge $(v_i, v_j) \in E_1$ the vertices $v_i, v_j \in V_1$. A sub-graph G_1 is said to be complete if there is an edge for each pair of vertices". A complete sub-graph is also called a Clique.

In order to build sets of elements from a graph structure many authors have worked with other alternatives of Cliques or cluster concepts of graphs. We can see [10, 15] with N-Cliques, [15] with K-Plex and [16] with Clubs and Clans. Also, we can find many others authors that have analyzed the Clique drawback as a NP-Complete (Not Polynomial Complete) problem, such as [17] and many others. In our approach we have chosen the N-Cliques model.

---

[1] http://www.senseval.org

2.2 N-Cliques

The N-Cliques model was introduced by Luce [10], where a N-Clique of a graph G is a subgraph of G induced by a set of points V associated with a maximal complete subgraph of the power graph G_n. In order to know the meaning of n we follow the Alba paper definition [18] that says: "…we will assume that the graph G is connected and has a diameter strictly greater than n. The nth power G^n of G is a graph with $V(G^n)=V(G)$ and such that u and v are adjacent in G^n iff $dG(u, v) \leq n$; that is, two points are adjacent in G^n whenever the distance between them in G is n or less."

The 1-Clique is identical to a Clique, because the distance between the vertices is one edge. The 2-Clique is the maximal complete sub-graph with a path length of one or two edges. Therefore, with the N-Cliques graph application we can obtain different subsets strongly integrated [19]. These models have been applied to Social Networks for different purposes. For example, they have been used to model broadcast networks(cellular networks) [20], to biomedical process [21], in NLP to lexical acquisition [22] and so on. To apply the N-Cliques model we have provided our own modification using the heuristic algorithm called Clique Partitioning.

2.3 Clique Partitioning Algorithm

This algorithm was introduced by Tseng and Siewiorek [11] only for Clique model or 1-Clique. We have included a little modification in order to apply this algorithm to N-Cliques. The original algorithm is shown in the following table and our contributions are underlined:

Table 1. Clique Partitioning Algorithm to N edges distances

```
/*Create super graph G'(S,E) S: SuperNodes, E: edges */
/*V: vertex, N:distance between pair nodes*/
S = ∅; E' = ∅;
for each vᵢ ∈ V do sᵢ = {vᵢ}; S = S ∈ {sᵢ}; endfor
for each eᵢ,ⱼ ∈ E do E' = E' ∈ {e'ᵢ,ⱼ}; endfor
while E'≠ ∅ do
   /* Find S_index1, S_index2 having most common neighbors*/
   MostCommons = -1;
   for each e'ᵢ,ⱼ ∈ E' do
      cᵢ,ⱼ = |COMMON-NEIGHBOR(G', sᵢ , sⱼ, N )|;
      if cᵢ,ⱼ > MostCommons then
      MostCommons = cᵢ,ⱼ ;  Index1 = i;  Index2 = j; endif
   endfor
   CommonSet = COMMON-NEIGHBOR (G',S_index1 , S_index2, 1);
   /*delete all edges linking S_Index1 or S_Index2 */
   E' = DELETE-EDGE(E', S_Index1);  E' = DELETE-EDGE(E', S_Index2);
   /*Merge S_index1 and  S_index2 into S_index1index2 */
            S_index1Index2 = S_index1 U S_index2;
   S = S - S_index1 - S_index2 ; S = S U{ S_index1Index2 };
   /* add edge from S_Index1Index2 to super-nodes in CommonSet */
   for each sᵢ ∈ CommonSet do
   E' = E' U {e'ᵢ.Index1Index2}; Enfor sᵢ
   /* Decrease in 1 the N value */
   If  N > 1 then  N = N-1; endif
Endwhile  Return S;
```

This modification creates a Clique with N distance and next $N-1$, $N-2$,...$N=1$ distances. The *COMMON-NEIGHBOR* $(G_1, s_i, s_j, \underline{N})$ returns the set of supernodes that are common neighbors to N distances of each s_i and s_j in G'. The procedure *DELETE-EDGE* (E_1, s_i) deletes all edges in E' which have s_i as their end super-node. To understand this algorithm we show an example at Fig. 1 using $N = 2$ edges distances. Note that *CommonSet* has the set of supernodes that are common neighbors to '*1*' distance of s_{index1} and s_{index2} in G'.

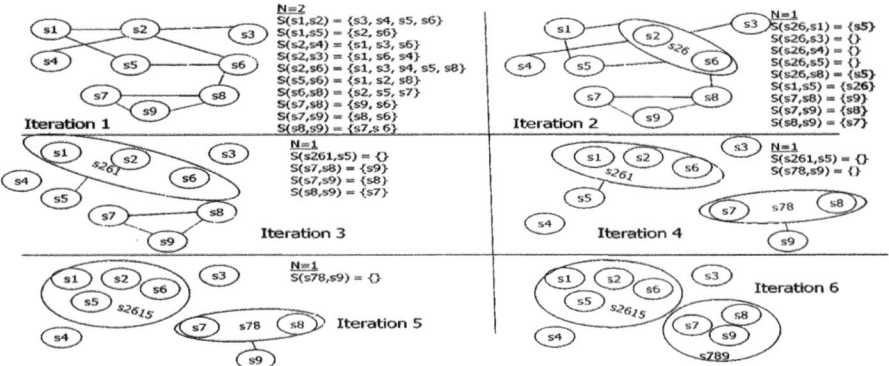

Fig. 1. Example of Clique Partitioning Algorithm to $N=2$ edges distances

As we can see this heuristic algorithm is able to obtain one set of nodes with maximal distance among all nodes <= 2 edges, because $N = 2$. And it continues obtaining Cliques with $N = N - 1$ where $N \geq 1$ for each iteration while $E' \neq \emptyset$. Next section describes the new resource used to apply this algorithm.

2.4 ISR-WN Resource

The main purpose of the ISR-WN resource is to map several semantic resources to obtain a more extensively network of related terms. Many efforts have been focused on the idea of building semantic networks like MultiWordNet (MWN) [23], EuroWordNet (EWN) [24], Multilingual Central Repository (MCR) [25], ISR-WN [12] among others. Each resource has different characteristics.

MWN is able to align the Italian and English lexical dictionaries conceptualized by Domain labels. EWN was developed to align Dutch, Italian, Spanish, German, French, Czech, English, and other lexical dictionaries. MCR integrates into the EWN framework an upgraded version of the EWN Top Concept ontology, the MWN Domains, the Suggested Upper Merged Ontology (SUMO) [26] and hundreds of thousands of new semantic relations and properties automatically acquired from corpora. ISR-WN takes into account different kind of labels linked to WN: Level Upper Concepts (SUMO), Domains and Emotion labels.

As we can see, each resource provides different semantic relations. However, ISR-WN has the highest quantity of semantic dimensions aligned, so it is a suitable resource to run our algorithm. Using ISR-WN we are able to extract important information from the interrelations of four ontological resources: WordNet[2], WordNet-Domains[3] (WND) [27], WordNet-Affects[4] (WNA) [28] and SUMO[5] through the links of these resources with WN and taking into account different semantic relations. This resource is based on WN1.6 and provides a tool which allows navigating from any sense of WN, Domain labels or SUMO categories, across internal relations such as Fig. 2 shows.

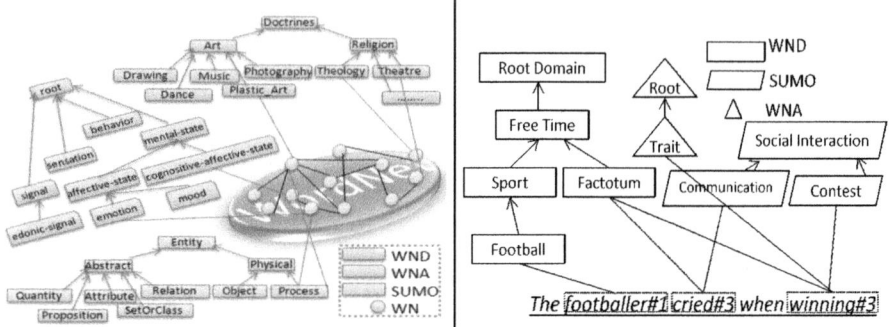

Fig. 2. ISR-WN Logical model   **Fig. 3.** Multidimensional features

At this point we can discover the multidimensionality of concepts that exists in each sentence. Fig.3 shows how the concepts characterize one sentence and how words are related through the semantic network. In order to establish the Concepts associated to each sentence we apply an adaptation of Reuters Vector [29] using the provided links of ISR-WN.

2.5 Reuters Vector

Using Reuters Vector method we are able to build an initial graph with the most relevant Concepts (labels of the semantic resources of ISR-WN) related to context. This method consists of the acquisition procedure to identify which Concept or semantic labels are relevant in the Reuters corpus [6] for a synset.

Such as Magnini said "As a First step a Relevant Lemma List was built as the union of the synonyms and of the gloss for that synset. This list represents the context of the synset in WordNet, and will be used to estimate the probability of a domain given a synset in the corpus. This information is collected in a vector, called **Reuters Vector**, with a dimension in each domain" [29]. In our proposal we have only used synonyms to build the Relevant Lemmas List. Each dimension has the Concept probability calculated with the Formula (2). For a better understanding the term $P(D/f)$ is explained in Formula (1).

---

[2] http://www.cogsci.princeton.edu/~wn/

[3] http://wndomains.fbk.eu

[4] http://wndomains.fbk.eu/wnaffect.html

[5] http://www.ontologyportal.org

[6] http://about.reuters.com/researchan/dstandards/corpus/

$$P(D|f) = \sum_{k=1}^{} \sum_{j=1}^{} P(D|s_j)_k \tag{1}$$

Where D is the given WordNet Domain or Concept and f (phrase) is a list of lemmas. $P(D|f)$ is joint probability distribution, k represents the $k\text{-}th$ lemma, s_j: represents the $j\text{-}th$ sense of the $k\text{-}th$ lemma and $P(D|s_j)$ is the probability of Domain D related to sense s, this probability is represented by formula (2) [29].

$$P(D|s) = \frac{P(s|D)P(D)}{\sum_{i=1}^{n} P(s|D_i)P(D_i)} \tag{2}$$

Where the probability of the synset for a Domain is assumed to be conditioned by the probability of its most related lemmas according to the following formula:

$$P(s|D) = P(l_1 \dots l_m|D) = \prod_{i=1}^{m} P(l_i|D) \tag{3}$$

Where l is a lemma in the list of Relevant Lemmas. The probability of a lemma given a Domain is its relative frequency in that Domain. To obtain the list of lemmas we get all lemmas from a given synset using its list of synonyms. (e.g.WN1.6 (*entity%1:03:00::* (List of lemmas (*entity, something*), gloss{*anything having existence (living or nonliving)*)))) (see formula (4) extracted from [29])

$$P(l_i|D) = \frac{c(l_i, D) + \epsilon}{c(D) + \epsilon|L|} \tag{4}$$

"Where $c(l, D)$ is the number of occurrences of lemma l in a Domain, $c(D)$ is the total number of occurrences of the domain and $|L|$ is the total number of the lemmas in the considered corpus. For each domain, $P(D)$ is assumed to be 1, because we have not special Domain requirements to fit" [29].

According to Reuters Vector it is recommended to use Corpus, however if we replace the Corpus for WordNet Domains, WordNet Affects or SUMO, we will be able to obtain Reuters Vector in the same way, with different categories or domains. Then D could be a SUMO category, Domain Label or Emotion Label. The links established among synsets and each resource are used to obtain $P(s|D)$. This method serves for the same purpose that the following [30, 31], so we can replace the use of Reuter Vectors if we want. The goal of this step is to obtain the most relevant Concepts associated to the sentence.

3 Method Structure

Once each constituent of our proposal have been described we present our method. It consists of three steps: 1st. Creating an initial semantic graph. 2th. Applying N-Cliques Partitioning Technique. 3th. Selecting the correct senses

As we can see in Fig.4, everything begins when a sentence is introduced. After that, the lemmas of the words are obtained and next the three steps are applied. Following, we describe the three steps.

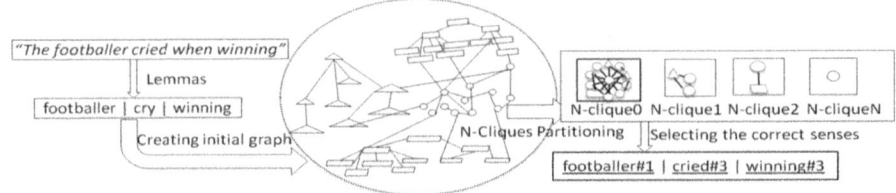

Fig. 4. General model of N-Cliques method

3.1   Creating an Initial Semantic Graph

The aim of this step consists of building a semantic graph from all senses of the words in each sentence. The connections among these senses are established using the shortest path with a fast variant of Floyd algorithm [32] between all senses vs all selected Concepts of Reuters Vector obtained (based on IRS-WN).

The process of building the initial graph on Fig. 5 guarantees that the graph built will be more centralized and the diameter between the corners will be the shortest. The Reuters Vector is able to obtain the common concepts of the sentences.

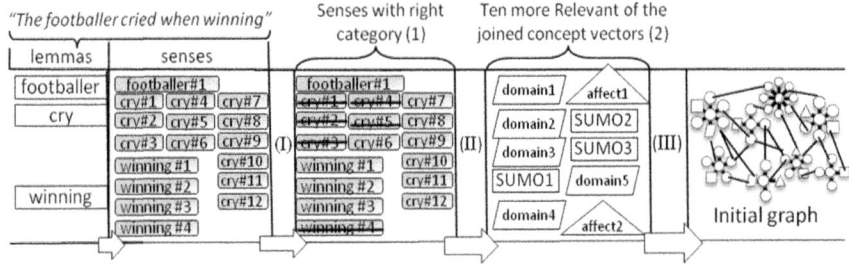

Fig. 5. Initial graph creation with all senses vs ten more relevant concepts

These are the steps to build the initial graph:

(I)   Eliminating the senses that not match with Freeling grammatical category suggested.
(II)   Obtaining Reuters Vectors.
(III) Obtaining the shortest path between all Senses (1) and all Concepts (2); and then creating the Initial Graph with the path without repeated nodes.

It is important to remark that the initial graph is composed by synsets (senses), Domains, Affects and SUMO categories, besides the Factotum domain for generic synsets will be removed of the Reuters Vector obtained.

3.2   Applying N-Cliques Partitioning Technique

Once we have obtained the initial graph it is used by N-Cliques Partitioning Technique as input. Then we apply this algorithm to obtain the N-Cliques which are composed by any element type that appears in the initial graph.

3.3 Selecting the Correct Senses

For this step we sort the N-Cliques by quantity of nodes. Later for each corresponding lemma of the sentence it is searched the synset that matches it. If it does not exist in the first N-Clique we pass to the next, and so on. The synset found in this process will be considered as correct. If in a N-Clique we find two or more synsets associated to the searched lemma, we select as correct the synset that the Freeling tool [33] notifies as the most frequent in the sentence where it appears. Notice that if the lemma to disambiguate is the verb 'be' we always select the Most Frequent Sense for this lemma (*be%2:42:03::*). The reason of doing this is because the verb 'be' is one of the most polisemous words and moreover, it is related in almost any type of contexts. If we took into account the verb 'be', we would obtain worst results because N-Cliques model applied on WSD tries to find semantic nodes on similar contexts.

In this approach we have considered as criterium for picking up a N-cliques the quantity of nodes however, it could be used another criterium such as applying PageRank technique described in previous sections. If we had used PageRank, the N-Cliques selected would have been those with more rank achieved. At present, we have chosen the inclusion of N-Cliques model with the simplest way but in further works we plan to use different techniques.

4 Evaluation

This method has been evaluated in the "English All Words" task of Senseval-2 [7]. The analysis has been divided in seven experiments that will be analyzed in detail to evaluate the influence of different combinations of resources. The experimental distance choice to apply the partition was *N=3*. The experiments are described next and the results are shows on Fig 6.

1. Applying the proposal with a graph composed by WN, WND, WNA and SUMO.
2. Applying the proposal with a graph composed by WN and WND.
3. Applying the proposal with a graph composed by WN and SUMO.
4. Applying the proposal with a graph composed by WN and WNA.
5. Applying the proposal with a graph composed by WN, WND and SUMO.
6. Applying the proposal with a graph composed by WN, WND and WNA.
7. Applying the proposal with a graph composed by WN, WNA and SUMO.

4.1 Analysis of the Behaviour of the Accuracies of Grammatical Categories

In each experiment conducted it has been analyzed the behaviour of each grammatical category. In Fig. 6 we can observe how our proposal has the highest accuracy for adverbs determination in WSD, reaching 64%. As we can notice nouns, adjectives and verbs have a low accuracy respectively. We can deduce that our method is very powerful to disambiguate adverbs. Related to nouns and adjectives the accuracy reaches 50% and 35% respectively. Finally, verbs obtain the worst results around 25% because they are the most ambiguous categories.

4.2 Analysis of Correct Senses According to Different N-Cliques

Once obtained the N-Cliques each one could contain a series of proposed senses. In order to evaluate the precision of each N-Clique we have compared our proposed senses to the provided senses of Senseval-2 "English All words" task. Table 2 shows two types of information, in one hand the N-Cliques with more quantity of right senses proposed by our method and in the other hand the N-Cliques with the right senses provided by Senseval-2. These N-Cliques are presented in decreasing order. As we can see the most relevant N-Cliques overlap in 100%, in both cases. This demonstrates that the most quantity of right senses and proposed senses are located at the same N-Cliques. Notice that our proposal provides that the right senses are focused on the first N-Cliques.

Table 2. Cliques that contain right senses and provided senses

Our proposal		Senseval-2 provided	
N-Cliques	Quantity of proposed senses	N-Cliques	Quantity of right senses
Clique0	361	Clique0	195
Clique2	268	Clique3	158
Clique3	251	Clique1	146
Clique1	251	Clique2	143
Clique4	229	Clique4	140

4.3 General Analysis

The integration of WordNet resources in WSD is a new way to obtain better results. The proposed method is one of several variants that could be used. Such as Fig. 6, the most relevant experiment was that use all resources where the recall obtained was 43.3%. This result could locate our proposal in the 11th place of Senseval-2 ranking as Table 3 shows. If we applied only WNA linked WN the precision obtained would be 69.3%. Our proposal is considered to work very well in detecting the right sense for adverbs, and not too accurate for verbs such as Fig.6 shows.

	Total P	Total R	Nouns R	Adjectives R	Verbs R	Adverbs R
Experiment 1	0.444	0.433	0.489	0.359	0.239	0.646
Experiment 2	0.432	0.422	0.493	0.332	0.220	0.639
Experiment 3	0.432	0.409	0.486	0.349	0.216	0.584
Experiment 4	0.693	0.397	0.471	0.349	0.230	0.546
Experiment 5	0.438	0.429	0.500	0.341	0.225	0.649
Experiment 6	0.434	0.425	0.496	0.339	0.222	0.638
Experiment 7	0.434	0.411	0.493	0.355	0.207	0.580

Fig. 6. Accuracy of the experiments: Precision (P) and Recall (R)

Table 3. Senseval-2 ranking

English All words - Fine-grained Scoring							
Rank	Precision	Recall	System	Rank	Precision	Recall	System
1	0.690	0.690	SMUaw	13	0.748	0.357	IRST
2	0.636	0.636	CNTS-Antwerp	14	0.345	0.338	USM 1
3	0.618	0.618	S-LIA-HMM	15	0.336	0.336	USM 3
4	0.575	0.569	UNED-AW-U2	16	0.572	0.291	BCU
5	0.556	0.550	UNED-AW-U	17	0.440	0.200	Sheffield
6	0.475	0.454	UCLA - gchao2	18	0.566	0.169	S - sel-ospd
7	0.474	0.453	UCLA - gchao3	19	0.545	0.169	S - sel-ospd-ana
8	0.416	0.451	CLR-DIMAP	20	0.598	0.140	S - sel
9	0.451	0.451	CLR-DIMAP (R)	21	0.328	0.038	IIT 2
10	0.500	0.449	UCLA - gchao	22	0.294	0.034	IIT 3
11	**0.444**	**0.433**	**Our Proposal**	23	0.287	0.033	IIT 1
12	0.360	0.360	USM 2				

4.4 Comparison with Previous WSD Approaches Graph-Based

In this section we present a comparison with some graph-based WSD methods mentioned on section 1. These proposals were tested using "English All Words" task corpus from Senseval-2. First, we describe Page-Rank method. It has been used to determinate the centrality of structural lexical network by Agirre and Soroa [6], and Sinha and Mihalcea [5]. Both proposals build sub-graphs related to the context, similar to ours, with the difference that they are able to assign weights to the labels ((nodes) senses) using the semantic relation of WordNet. Then, to disambiguate each word it is chosen the most weighted sense. A difference between Mihalcea and Agirre is that Mihalcea has an experimental use, first with six different similarity measures assigned to the edges and then applies four different measures of centrality. These approaches obtained respectively 58.6% and 56.37% of recall; however, ours does not improve these results obtaining only a 43.3% of recall. The lower recall is due to that our proposal non-apply weight techniques to obtain sort senses by scores, instead we are able to obtain many densely connected sub-graphs and select as right sense such as MFS suggests. Moreover, we do not provide a heuristic method to decide among senses that are integrating a same Clique which are the most engage with the context.

5 Conclusions and Further Works

In this paper we present a new proposal that combines different techniques, methods and resources. Our goal is to use the N-Cliques model in WSD based on ISR-WN and the Reuters Vector method. N-Cliques have been used in several research areas, but never in WSD, therefore it was necessary to find the correct context to apply this technique. Our proposal has been applied to Senseval-2 corpus where many experiments have been conducted. One of them was to study the behaviour of each grammatical category in different ways. We have demonstrated that the integration of different semantic resources influence in the accuracy of WSD results. The use of SUMO, WND and WNA independently obtained worse results than with the integration of all of them. Moreover, to apply the N-Clique model it is necessary to define a propitious scene. So, we have provided a wide semantic net with several resources to be executed. In this approach it was revealed, that we are able to define the right senses for the adverbs grammatical category with highest accuracy, and a little less for nouns. However, the accuracy of adjectives and verbs is a bit lower. The accuracy of precision and recall obtained could locate our proposal in 11th place on Senseval-2 ranking.

As further works, we want to replace the initial graph obtained from Reuters Vector using other options to determine if changing this step could improve the results. Besides, in order to decide the correct senses instead of using the Freeling frequencies we could use other possibilities. Moreover, the inclusion of more resources is not excluded for further works but this option depends on the enrichment of ISR-WN. Also, we could vary the N distance that we use in this work ($N=3$) with others. Finally, in order to improve our results we will integrate the Mihalcea and Agirre proposals to obtain more accurate results.

Acknowledgments

This paper has been supported partially by Ministerio de Ciencia e Innovación - Spanish Government (grant no. TIN2009-13391-C04-01), and Conselleria d'Educación - Generalitat Valenciana (grant no. PROMETEO/2009/119, ACOMP/2010/288 and ACOMP/2011/001).

References

1. Lesk, M.E.: Automatic sense disambiguation using machine readable dictionaries: How to tell a pine cone from a ice cream cone. In: Proceedings of the ACM SIGDOC Conference, Toronto, Ontario, pp. 24–26 (1986)
2. Navigli, R., Velardi, P.: Structural Semantic Interconnections: a Knowledge-Based Approach to Word Sense Disambiguation. IEEE Transactions on Pattern Analysis and Machine Intelligence (TPAMI) 27 (2005)
3. Fellbaum, C.: WordNet. An Electronic Lexical Database. University of Cambridge, Cambridge (1998)
4. Laparra, E., Rigau, G., Cuadros, M.: Exploring the integration of WordNet and FrameNet. In: Proceedings of the 5th Global WordNet Conference (GWC 2010), Mumbai, India (2010)
5. Sinha, R., Mihalcea, R.: Unsupervised Graph-based Word Sense Disambiguation Using Measures ofWord Semantic Similarity. In: Proceedings of the IEEE International Conference on Semantic Computing (ICSC 2007), Irvine, CA (2007)
6. Agirre, E., Soroa, A.: Personalizing PageRank for Word Sense Disambiguation. In: Proceedings of the 12th Conference of the European Chapter of the Association for Computational Linguistics (EACL-2009), Athens, Greece (2009)
7. Cotton, S., Edmonds, P., Kilgarriff, A., Palmer, M.: English All word. In: Linguistics, A.f.C. (ed.) SENSEVAL-2: Second International Workshop on Evaluating Word Sense Disambiguation Systems. Association for Computational Linguistics, Toulouse (2001)
8. Snyder, B., Palmer, M.: The English All Word Task. In: Linguistics, A.f.C. (ed.) SENSEVAL-3: Third International Workshop on the Evaluation of System of the Semantic Analysis of Text. Association for Computational Linguistics, Barcelona (2004)

9. Agirre, E., Lacalle, O.L.d., Fellbaum, C., Hsieh, S.-k., Tesconi, M., Monachini, M., Segers, P.V.a.R.: SemEval-2010 Task 17: All-words Word Sense Disambiguation on a Specific Domain. In: Proceedings of the 5th International Workshop on Semantic Evaluations (SemEval-2010), Association for Computational Linguistics, Uppsala (2010)
10. Luce, R.D.: Connectivity and generalized cliques in sociometric group structure. Psychometrika 15, 159–190 (1950)
11. Tseng, C.-J., Siewiorek, D.P.: Automated Synthesis of Data Paths in Digital Systems. IEEE Trans. on CAD of Integrated Circuits and Systems 5, 379–395 (1986)
12. Gutiérrez, Y., Fernández, A., Montoyo, A., Vázquez, S.: Integration of semantic resources based on WordNet. In: XXVI Congreso de la Sociedad Española Para el Procesamiento del Lenguaje Natural, SEPLN 2010, vol. 45, pp. 161–168. Universidad Politécnica de Valencia, Valencia (2010)
13. Luce, R.D., Perry, A.D.: A Method of Matrix Analysis of Group Structure. Psychometrie 14, 95–116 (1949)
14. Cavique, L., Mendes, A.B., Santos, J.M.A.: An Algorithm to Discover the k-Clique Cover in Networks. In: Lopes, L.S., Lau, N., Mariano, P., Rocha, L.M. (eds.) EPIA 2009. LNCS, vol. 5816, pp. 363–373. Springer, Heidelberg (2009)
15. Balasundaram, B., Butenko, S., Hicks, I.V., Sachdeva, S.: Clique relaxations in Social Network Analisis: The Maximun k-plex Problem (2006)
16. Mokken, R.J.: Cliques, Clubs and Clans, vol. 13. Elsevier Scientific Publishing Company, Amsterdam (1979)
17. Wood, D.R.: An algorithm for finding a maximum clique in a graph. Operations Research Letters, 211–217 (1997)
18. Alba, R.D.: A Graph-Theoretic Definition of a Sociometric Clique. Journal of Mathematical Sociology 3, 113–126 (1973)
19. Friedkin, N.E.: Structural Cohesion and Equivalence Explanations of Social Homogeneity. Sociological Methods & Research 12, 235–261 (1984)
20. Clark, B.N., Colbourn, C.J., Johnson, D.S.: Unit Disk Graphs, vol. 86. Elsevier Science Publishers B.V., North-Holland (1990)
21. Kose, F., Weckwerth, W., Linke, T., Feihn, O.: Visualizing plant metabolomic correlation networks using clique-metabolite matrices. Bioinformatics 17, 1198–1208 (2001)
22. Widdows, D., Dorow, B.: A graph model for unsupervised lexical acquisition. In: Proceedings of the 19th International Conference on Computational Linguistics. ACL, Morristown (2002)
23. Pianta, E., Bentivogli, L., Girardi, C.: MultiWordNet. Developing an aligned multilingual database. In: Proceedings of the 1st International WordNet Conference, Mysore, India, pp. 293–302 (2002)
24. Dorr, B.J., Castellón, M.A.M.a.I.: Spanish EuroWordNet and LCS-Based Interlingual MT. In: AMTA/SIG-IL First Workshop on Interlinguas, San Diego, CA (1997)
25. Atserias, J., Villarejo, L., Rigau, G., Agirre, E., Carroll, J., Magnini, B., Vossen, P.: The MEANING Multilingual Central Repository. In: Proceedings of the Second International Global WordNet Conference (GWC 2004), Brno, Czech Republic (2004)
26. Suggested Upper Merged Ontology (SUMO). Ontology Portal (2009)
27. Magnini, B., Cavaglia, G.: Integrating Subject Field Codes into WordNet. In: Proceedings of Third International Conference on Language Resources and Evaluation (LREC-2000), pp. 1413–1418 (2000)
28. Strapparava, C., Valitutti, A.: WordNet-Affect: an affective extension of WordNet. In: Proceedings of the 4th International Conference on Language Resources and Evaluation (LREC 2004), pp. 1083–1086. Lisbon (2004)
29. Magnini, B., Strapparava, C., Pezzulo, G., Gliozzo, A.: Comparing Ontology-Based and Corpus-Based Domain Annotations in WordNet. In: Proceedings of the First International WordNet Conference, Mysore, India, pp. 21–25 (2002)

30. Vázquez, S., Montoyo, A., Rigau, G.: Using Relevant Domains Resource for Word Sense Disambiguation. In: Proceedings of the International Conference on Artificial Intelligence, IC-AI 2004. CSREA Press, Las Vegas (2004)
31. Gutiérrez, Y., Fernández, A., Montoyo, A., Vázquez, S.: UMCC-DLSI: Integrative resource for disambiguation task. In: Proceedings of the 5th International Workshop on Semantic Evaluation, pp. 427–432. Association for Computational Linguistics, Uppsala (2010)
32. Floyd, R.W.: Algothm 97: Shortest path. C. ACM 5, 345 (1963)
33. Atserias, J., Casas, B., Comelles, E., González, M., Padró, L., Padró, M.: FreeLing 1.3: Syntactic and semantic services in an opensource NLP library. In: Proceedings of LREC 2006, Genoa, Italy (2006)

OntoFIS as a NLP Resource in the Drug-Therapy Domain: Design Issues and Solutions Applied

María Teresa Romá-Ferri[1], Jesús M. Hermida[2], and Manuel Palomar[2]

[1] Department of Nursing
[2] Department of Software and Computing Systems,
University of Alicante, Aptdo. de Correos 99, E-08030 Alicante, Spain
`mtr.ferri@ua.es`, `{jhermida,mpalomar}@dlsi.ua.es`

Abstract. In the Health domain, and specifically in the drug-therapy domain, in order to improve the access to the information of different types of users, several informational resources, semantically annotated, are under development. One of the existing development lines is oriented to reusing the effort spent on the design of the existing resources on the Web and obtaining knowledge-based resources for natural language processing (NLP) tasks. In this line, OntoFIS was designed as a NLP resource aimed at filling the gap of multilingual knowledge-based resources within the domain. The design process used for building OntoFIS merges the best approaches of several ontology design methodologies. However, given the characteristics of the drug-therapy domain, whose needs of knowledge are very precise, the process of formalisation of the domain knowledge led to a set of issues. Thus, this paper discusses the main issues found and the solutions analysed and applied in each case.

Keywords: OntoFIS, ontologies, ontology design methodology.

1 Introduction

At present, the amount of information and the types of sources available on the Web constantly grow year by year. Despite this, it is a fact that this information does not fulfil the actual needs of all the users, either with a professional or lay profile. This overflow of information is an issue that directly affects the normal behaviour of many citizens, who tend to accept as veridical any of the first pieces of information found on the Web that seems appropriate to their needs. The application of this mechanism of action in specialised domains, such as health, leads to several risks.

In order to minimise or even overcome these risks, one of the strategies consisted in the development of several terminological and knowledge-based resources, such as ontologies, which can provide non-ambiguous semantics to the informational content. However, there is still a shortfall in the creation of resources in the health domain, except for those created in English. In this domain, there still exist some scenarios where this shortfall is even more notable, such as the drug therapy. On one hand, within this specific scenario, the information is based on text and normally expressed in the users' mother tongue (in our scenario, in Spanish), whereas the information based on scientific evidences is usually in English.

R. Muñoz et al. (Eds.): NLDB 2011, LNCS 6716, pp. 125–136, 2011.
© Springer-Verlag Berlin Heidelberg 2011

On the other hand, the existing terminological knowledge sources are deficient in terms for identifying pharmaceutical specialties (GELOCATIL® Comp. 650 mg is a registered trademark) and the main active ingredients. This fact might be motivated by the large number of specialties and their diversity (only in Spain, in 2008, 17,204 pharmaceutical specialties were available [1]). Furthermore, these specialties are not generally sold in all the countries with the same name or pharmaceutical form (in Portugal: GELOCATIL tablet 650 mg; in Belgium: WITTE KRUIS MONO tablet 500 mg; or in the United Kingdom: ZANZA tablet 500 mg). Even active ingredients can be called using different standard names (in Spanish *paracetamol*, while in American English *acetaminophen*).

In order to reduce these limitations, several informational resources, semantically annotated, are under development. One of the approaches followed aims at reusing the effort spent on the design of the existing resources on the Web and obtaining knowledge-based resources for natural language processing (NLP) tasks [2, 3]. This specific approach was applied in the design of OntoFIS (in Spanish, *Ontología Farmacoterapéutica e Información para el Seguimiento*, Drug-Therapy and Information Follow-up Ontology, [4, 5, 6]), whose goal is to fill the gap of multilingual knowledge-based resources. Once built, it was populated with 55,000 instances approximately, with terminology used within the Spanish National Health System, specified in Spanish.

The design methodology employed in the development of OntoFIS is based on common aspects of the existing approaches [7, 8, 9, 10]. However, due to the characteristics of the drug-therapy domain, whose needs of knowledge are very precise, the process of formalisation of OntoFIS led to a set of issues. In this context, the aim of this paper is to analyse the main issues found and discuss the solutions applied in each case.

This paper is organised as follows: Section 2 presents the main features of OntoFIS and describes the design process followed during its development. Section 3 analyses the different issues of the design process and describes the solutions taken into consideration in each case. Subsequently, Section 4 discusses the main benefits of the design methodology and the last section draws the principal conclusions and describes the lines of future work.

2   OntoFIS: Characteristics and Design Process

Considering the initial requirements of information, we designed a new ontology called OntoFIS, which is aimed at representing knowledge from the clinical practice, focused on the rational use of medicines. Based on the ontology principles expressed in [11], OntoFIS is a conceptualization of common and shared knowledge from the drug-therapy domain. It was created as a potential multilingual resource to help in the task of semantic annotation of Web resources and NLP tasks such as automatic retrieval and classification of documents, or information extraction from narrative texts related with this domain.

There exist 23 main concepts related to four knowledge cores [4]: *(i)* drugs for prescription (name, composition, classification), *(ii)* administration method (period, frequency, quantity, forms, route, preparation, method), *(iii)* control of results related to drug safety and effectiveness (side effects, time, interactions, adverse reactions, precautions), and *(iv)* medical diagnosis (diseases, syndromes, signs or symptoms). Each of the concepts includes synonyms and expressions in five of the main languages of the European Union (Spanish, English, French, Italian and German). These concepts are interconnected by means of a set of 647 semantic relations, obtained from a taxonomy of 36 types of relation.

The ontology is populated with 55,000 instances approximately. Among these, because of their terminological value, we would like to highlight the following: 17,204 pharmaceutical specialties/registered trademarks (commercialised in Spain), active ingredients (4,456) and chemical groups (3,200).

2.1 Design Process

The methodology chosen for the design of OntoFIS was based on common aspects of current proposals [7, 8, 9, 10]. The process is divided in three stages: *(i)* systematic capture of domain knowledge using an informal formalisation; *(ii)* description of captured knowledge by means of a conceptual modelling language, in this case, UML; and *(iii)* formalisation of knowledge by using a knowledge representation language, OWL.

Among the existing representations we have focused on reusing those UML class diagrams. Obtaining ontologies from these diagrams allows us to carry out other tasks, such as, semantically annotating the content of Web applications [5, 12].

The following subsections describe the stages of the development of OntoFIS.

2.1.1 Knowledge Capture

In this first stage, we carried out a set of tasks with the purpose of finding information sources tailored to the domain of interest: *(i)* search in bibliographic databases; and *(ii)* analysis of domain-specific information sources: manuals about Internal Medicine, Pharmacology and Nursing cares and inserts of different types of drugs.

In a first step, from these information sources we extracted the most significant terms and definitions, and a collection of questions usually asked by users with different profiles, expressing their needs of information, which would be the basic knowledge core of our ontology. In the end, 144 representative terms were collected and, simultaneously, each term was described with the available dictionaries and manuals, thus creating a gloss of terms.

In a second step, the selected terms were classified in different groups: essential terms, secondary terms (high specificity or attributes) and synonyms (also translations).

2.1.2 Knowledge Representation Using UML

In this stage, we used the mechanisms of knowledge representation provided by UML class diagrams to describe the conceptual model of OntoFIS. This second stage began with the analysis of the gloss of essential terms in order to establish the type of concept they represent, according to our criteria of functionality and level of specification of our ontology (medium granularity). From this collection of terms we extracted the main ontology concepts and detected the relation triples *(subject, relation, object)*.

Once analysed the captured information from the domain, we built the UML class diagram of OntoFIS. Fig. 1 depicts a fragment of this class diagram, which contains three of the main classes and a subset of the relations defined, and offers a vision of the complexity of the present scenario.

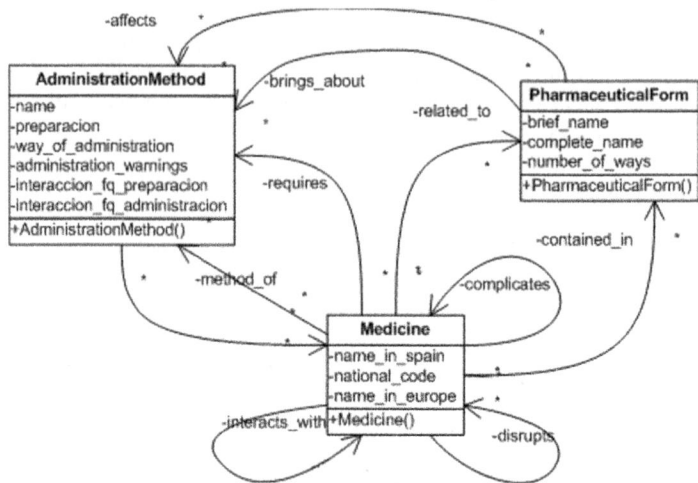

Fig. 1. Fragment of OntoFIS UML class diagram

2.1.3 Knowledge Formalisation: UML-OWL Transformations

This section describes the last process of the design method, i.e. the transformation process that turns the different elements of the UML class diagram into elements of an OWL ontology (for a more detailed description, see [5]). This process is divided in two phases: content transformation and name transformation.

The first task to be carried out in the design of the process was the definition of a collection of transformation rules capable to transform a UML class diagram into an OWL representation of our ontology from a semantic point of view. The purpose was to obtain elements or mechanisms in both diagrams capable of describing equivalent knowledge. Table 1 briefly summarises our collection of transformations, turning class diagram elements into OWL elements. The final transformation rules were applied, sorted by their complexity level, beginning with the simplest mechanisms (classes and hierarchy) and finishing with the most complex ones (aggregations and compositions).

Another important aspect in the transformation process was the design of rules for name conversion between both representations to avoid conflicts. There is also a clear naming gap between UML and OWL identifiers. While an OWL ontology use URIs to identify uniquely each of its elements (global identification), some elements from different class diagram may have the same name (local identification). Table 2 shows a summary of these name conversions. This is the second phase of the transformation process. Once obtained an element with equivalent semantics in OWL, our method assigns an URI to it.

Table 1. Equivalences between UML class diagram elements and OWL ontology elements

Elements of UML Class Diagram	Elements of OWL
Class	`owl:Class`
Class Description	`rdfs:comment`
Representative Terms	`rdfs:label`
Attribute (Simple datatype)	`owl:DatatypeProperty`
Attribute (Complex datatype)	`owl:ObjectProperty`
Inheritance	`rdfs:subClassOf`
Rol de Asociación	`owl:ObjectProperty`
Role Multiplicity	`owl:Restriction` + `owl:cardinality`,...
Aggregation	`owl:UnionOf`
Composition	Generic properties `partOf` and `hasPart` + subproperties

Table 2. Name building rules for OWL ontologies

Target Element	URI Generation Rule
`owl:Class`	URI_ONTODEF[1] + '#' + CLASS_NAME[2]
`owl:DatatypeProperty`	URI_ONTODEF + '#' + CLASS_NAME + { '-','''..''',':',';' } + ATTR_NAME[3]
`owl:ObjectProperty`	URI_ONTODEF + '#' + ASSOC_NAME[4] + { '-','''..''',':',';' } + ROLE_NAME[5]
Instance	URI_INSTANCE + '#' + CLASS_NAME + { '-','''..''',':',';' } + INSTANCE_ID

[1]URI_ONTODEF   URL of the file which contains the definition of the ontology.
[2]CLASS_NAME   Name of the UML class from which comes the element.
[3]ATTR_NAME   Name of the UML attribute from which comes the element.
[4]ASSOC_NAME   Name of the UML association from which comes the element.
[5]ROLE_NAME   Name of the UML rol of association from which comes the element.

3 Design Issues and Solutions Applied

Despite the fact that our design method reuses aspects of the best-known ontology design methodologies, the specific characteristics of the drug-therapy domain (i.e. the large quantity of semantic relations needed to describe knowledge) introduced new issues not taken into consideration before. This section addresses the description of the issues found and the solutions proposed in order to solve them.

The main issues were found in the last two stages of our design processes: Knowledge Representation and Formalisation. In this sense, it is worth mentioning that no issues were found during the first stage, Knowledge Capture, because as introduced in the first section, in this domain there exist multiple and diverse information sources that considerably facilitate the initial process of selection of the domain knowledge. Having said that, the following subsections aim at describing the main issues found and the different approaches followed to solve them.

3.1 Issues during the Representation Process

As described in Section 2.1, our approach aims at reusing of UML class diagrams. According to Cranefield and Purvis [12], UML can also be used for the representation of ontologies because of its simplicity and the intensive use in several software development processes.

(Issue 1, I1) However, **UML** has some known shortcomings when representing knowledge in a general way: *(i)* small number of representation mechanisms; *(ii)* this representation expresses knowledge local either to the model or the designer, which is difficult to share and reuse among the community members; and *(iii)* it is complicated to establish relations between elements of different models stored in different physical locations, even using XMI [13]. Therefore, it is not a suitable candidate as knowledge formalisation language to be used in semantic annotation tasks.

(I1-Solution 1) In order to overcome these shortcomings the design method proposes a new stage, Knowledge Formalisation using OWL-DL. The main advantage of OWL-DL regarding UML is that it was designed to be used by computing systems and its expressivity level can be adapted to different types of applications. This solution has been followed by other approaches within the Health domain, such as SNOMED-CT®.

(I2) UML Visualisation. The number of elements contained in our resulting diagram (with 647 associations between 23 classes, illustrated in Fig. 2) raised an issue of usability and maintainability of the final model.

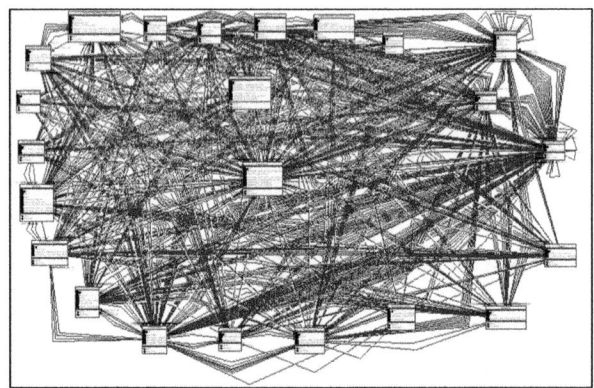

Fig. 2. Visualisation of the class diagram resulting from the second stage

(I2-Solution 1) Use a UML editor able to create a non-visual representation. This implied losing one of the main benefits, a global visual representation, which is highly recommendable for experts from others domains, different from Computer Science, e.g. the Health domain. This solution was therefore discarded.

(I2-Solution 2) Extend the UML metamodel with the metaclass "role container". In order to minimise the number of associations between classes, we decided to group those associations between the same classes within the same element. Role containers (see Fig. 3) are divided in three parts: name, direct roles (from class A to class B), and

inverse roles (from class B to class A). This approach allows the designers to simplify the diagram and thus improve its usability (legibility). Nonetheless, it is a fact that users can lose some visual information such as role cardinality or inverse roles (within the same association). Although it would be possible to add these features to our visual representation, in this case this information is not completely relevant to the user this diagram might be addressed at, mainly with a professional profile in the Health domain.

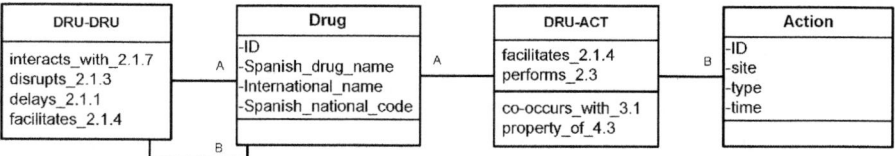

Fig. 3. Visual representation of the *DRU-ACT* and *DRU-DRU* role containers

3.2 Issues during the Formalisation Process

As can be noticed, there is a representation gap between UML class diagrams and OWL documents. This gap can be divided into two aspects: the equivalence between the semantics of the elements and the generation of identifiers.

(I3) There are not direct transformation mechanisms for all the elements of the UML class diagrams. On one hand, OWL has a higher degree of expressivity and there are therefore some characteristics that UML cannot represent (without using OCL, Object Constraint Language). For instance, UML has no direct mechanism for the representation of some characteristics of OWL Object Properties, such as transitivity or symmetry. On the other hand, UML provide some representation mechanisms which have no trivial or no possible equivalence. Although the concept of association is one of the elements most utilised in class diagrams, it does not have an equivalent representation in OWL. Instead, association roles are used to obtain the different OWL properties of an ontology. Moreover, aggregations and compositions are the elements of a class diagram whose capture and processing is more complex, since their semantics vary according to each use case.

(I3-Solution 1) This is an issue intrinsic to the languages themselves. Therefore, there is no effective method of solving this problem. We opted to extend the knowledge included in the representation during the formalization stage when necessary.

(I4) Elements whose semantics depends on the interpretation of the designer. Aggregations and compositions are the elements of a class diagram whose capture and processing is more complex, since their semantics vary according to each use case. It is worth mentioning that [14, 15] provide generic solutions for these problems. However, they still depend on the interpretation of the designer.

(I4-Solution-1) The use of NLP techniques would be a need in order to disambiguate the semantics of aggregations in the context of a certain model. However, the solution of this issue was out of the scope of this work. In this case, due

to domain restrictions, our final model did not contain any aggregation nor composition.

(I5) Semantic incompatibility of the relations caused during the process of transformation. An important aspect in the transformation process was the specification of rules for name conversion between both representations that avoid conflicts. As mentioned before, there is a naming gap between UML and OWL identifiers as well. Whereas OWL ontologies use URIs to uniquely identify each of its elements (global identification), some elements from different class diagrams may have the same name (local identification) such as attributes or association roles. For instance, the "*Spanish_national_code*" class attribute and the "*facilitates*" association role will be turned into the "*Drug..Spanish_national_code*" Datatype Property and the "*AS3..facilitates*" Object Property, respectively.

The algorithm proposed transforms each attribute or role following the same rules (see Tables 1 and 2). At the end of the process, the transformation generates 647 types of relations (Object Properties) when the number of types of actual relations, formally specified during the Knowledge Capture stage, is just 36 types. This can lead to problems when carrying out inferences over the knowledge contained, since properties with different identification and no relation between them are obviously treated like different properties. In this domain, the certainty of the knowledge contained in the ontology is a key aspect for further works.

It is usual that in UML class diagrams the designer specifies association roles with the same name. However, the semantics of these roles is usually different, so that it is possible to transform each of them into a new Object property. In this domain, the number of possible role names is limited and the names are associated to a well-defined semantics, which cannot be changed. Therefore, the *AssociationRole* transformation should be modified to obtain just 36 OWL Object properties, which should be used 647 times between classes. Fig. 4 illustrates a part of the UML class diagram obtained from the Knowledge representation stage. Each of these association roles ("facilitates") share the same semantics even though they are in different associations.

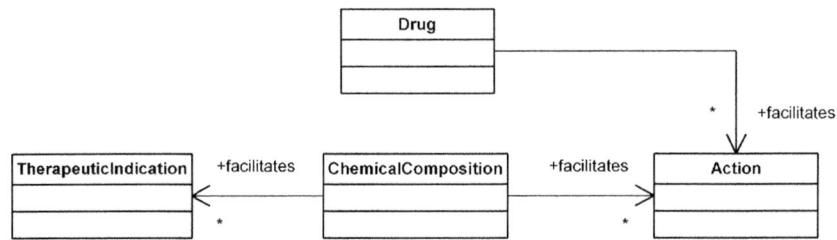

Fig. 4. Part of the UML class diagram resulting from the Knowledge Representation stage

(I5-Solution 1) The first proposed solution consisted in changing the naming rule for those conflicting elements (association roles). The new rule combined the names of the origin and destination class with the name of the role, using the following pattern: `URIbase/OriginClass-AssocRole-DestinationClass`, e.g.

http://www.dlsi.ua.es/ontofis/Drug-consists_of-ActiveSubstace. In case this combination was already employed, the transformation algorithm would reuse the same ObjectProperty definition.

The solution acted as a quality evaluation process, which helps either humans or software processes to manage them and find duplicates or incorrect associations, mainly caused by human errors during the Knowledge Representation phase. Nonetheless, the underlying representation problem is not resolved yet.

(I5-Solution 2) The second solution directly addresses the representation issue changing the conflicting element transformations. In this sense, the solution applied consists in the removal of the domain-range constraints from the transformation, thus generating generic Object properties (with *owl:Thing* as domain and range) that group the semantics of a set of role names from different associations. Although this solution simplifies and solves the representation, it can also lead to imprecision in the knowledge contained in the ontology, since any instantiation of the Object properties could be valid. This reduces the reusability of the ontology, since the knowledge description is incomplete. Therefore, it is necessary to document the information of the domain captured during the first stage in order to populate it correctly. Furthermore, after applying the solution, the resulting ontology (and its instances) should be evaluated by the designers (such evaluation can be found in [6]).

(I5-Solution 3) To solve the issue detected in the last solution, we propose a change on the transformation that constrains domain and range values. The designed approach is the following;

1) For each type of relationship, an Object property is created with generic domain and range (class *owl:Thing*), as in Solution 2. For instance, the relationship called "facilitates" is represented in OWL as follows:
   ```
   <owl:ObjectProperty rdf:ID="facilitates"/>
   ```
2) Subsequently, in each class defined as origin of a role, the transformation process includes an *AllValuesFrom* restriction in its definition. In this way, the possible range values can be constrained to the instances of the destination class.

The OWL ontology (using the XML syntax) resulting from the application of the new transformation over the model elements of Fig. 4 will be the following:

```
<owl:Class rdf:ID="Action"/>

<owl:Class rdf:ID="Drug">
  <rdfs:subClassOf>
    <owl:Restriction>
      <owl:onProperty rdf:resource="#facilitates"/>
      <owl:allValuesFrom rdf:resource="#Action"/>
    </owl:Restriction>
  </rdfs:subClassOf>
</owl:Class>
```

When a class is origin of two instances of the relation, the *AllValuesFrom* restriction is defined from the union of the range classes.

```
<owl:Class rdf:ID="TherapeuticIndication"/>

<owl:Class rdf:ID="ChemicalComposition">
  <rdfs:subClassOf>
```

```
<owl:Restriction>
  <owl:allValuesFrom>
    <owl:Class>
      <owl:unionOf rdf:parseType="Collection">
        <owl:Class rdf:about="#TherapeuticIndication"/>
        <owl:Class rdf:about="#Action"/>
      </owl:unionOf>
    </owl:Class>
  </owl:allValuesFrom>

  <owl:onProperty rdf:resource="#facilitates"/>
</owl:Restriction>

  </rdfs:subClassOf>
</owl:Class>
```

This solution has been successfully tested with part of the knowledge of OntoFIS. In order to ensure its appropriateness, we are carrying out a set of experiments for evaluating the knowledge contained in the OWL formalisation of the whole ontology.

4 Discussion

The paper presents the issues detected during the design of OntoFIS and the solutions applied to each of them in order to obtain a resource with terminological knowledge in the Drug-Therapy domain. The approach followed in the design methodology was based on the use of UML as knowledge representation language. This approach aims at reducing the existing bottleneck knowledge acquisition problem, either in the design of ontologies or the creation of new semantically-annotated resources.

In addition, the design of a UML diagram with the domain knowledge is an interesting option when working in a multidisciplinary environment. UML diagrams facilitate the visualisation of the knowledge in a way that it can boost the interchange of information between experts in diverse domains (e.g. computer science-health). In this domain, the use of UML-like diagrams is usual. For instance, the HL7 (Health Level Seven International, http://www.hl7.org) defines a standard with an UML-like notation used for identifying and describing data formats and informational exchanges between health information systems. This representation allows the verification of the contained knowledge (represented concepts, their features and the semantic network created from the relations between them) in a visual manner.

We could have used other similar approaches such as the OMG's Ontology Design Metamodel (ODM) [16], which offers a set of modelling primitives to represent OWL ontologies (a complete description of this approach can be found in [17]). However, we finally chose classic UML class diagrams because of their simplicity and visual expressivity. In addition, while the ODM approach is bound to OWL (at present just in version 1.1), with UML, we can generate different types of formalisations in the last stage of our design process.

When applying our generic ontology design methodology in this domain, we found the main problems in the transformation process between the Knowledge Representation and Formalisation stages. We are aware that the presented solutions are not perfect. They have different shortcomings that we have been commenting before. Nevertheless, the solutions are still valid if the correct control mechanisms are introduced when indicated. The design of OntoFIS did take these aspects into consideration and, therefore, the ontology is valid for the purposes it was aimed at: being the key resource in the task of semantic annotation of Web resources and in NLP tasks – automatic retrieval and classification of documents, and information extraction from narrative texts related with this domain.

5   Conclusions and Future Work

OntoFIS is an ontology that captures knowledge of the Drug-therapy domain that can be used as a resource in NLP tasks. The design methodology followed in the process of development combines the best processes of several existing methodologies. In this paper, we described the main problems found when we applied a general design methodology to the design of an ontology, OntoFIS, in a domain with a very characteristic needs of knowledge. For each of these issues we proposed a solution and analysed its benefits and shortcomings.

Our methodology has three stages: Knowledge Capture, Representation (in UML class diagrams) and Formalisation (in OWL ontologies). The main issues were found in the last two stages. Due to the characteristics of the domain (low number of concepts and high number of relations between them), the main detected issues are related to the visualisation of the diagrams and the gap between the UML representation and the OWL formalisation. The solutions proposed are not perfect, but they are always valid when they are applied together with control mechanisms.

Future work aims at *(i)* applying and evaluating the solution 3 of the Issue 5 on the whole OntoFIS; *(ii)* applying the methodology and the designed solutions to another specific subdomain within the health domain; and *(iii)* proposing an adaptation of the design methodology taking into consideration the experiences obtained during the development of ontologies in the health domain.

Acknowledgments. This paper has been supported by the Spanish Ministry of Education under the Doctoral Fellowship Program (FPU) (AP2007-03076), the Spanish Ministry of Science and Innovation under the TEXTMESS 2.0 project (TIN2009-13391-C04-01), and the Valencian Government (Generalitat Valenciana) under the Prometeo program (PROMETEO/2009/199) and the ACOMP/2011/001 research action.

References

1. Consejo General de Colegios Oficiales de Farmacéuticos: Nuevos medicamentos 2008. Farmacéuticos 345, 54-61 (2009)
2. Guarino, N.: Formal Ontology in information Systems. In: Proceedings of Formal Ontology in Information Systems, FOIS 1998, pp. 3–15. IOS Press, Amsterdam (1998), http://www.loa-cnr.it/Papers/FOIS98.pdf
3. Bodenreider, O.: Biomedical ontologies in action: role in knowledge management, data integration and decision support. Yearb. Med. Inform., 67–79 (2008)
4. Romá-Ferri, M.T., Hermida, J., Montoyo, A., Palomar, M.: Representación del conocimiento farmacoterapéutico: diseño de una ontología. In: Actas del XI Congreso Nacional de Informática de la Salud (INFORSALUD 2008), Sociedad Española de Informática de la Salud (SEIS), Madrid, pp. 240–247 (2008)

5. Hermida, J.M., Romá-Ferri, M.T., Montoyo, A., Palomar, M.: Reusing UML Class Models to Generate OWL Ontologies. A Use Case in the Pharmacotherapeutic Domain. In: Proceedings of the KEOD 2009, Madeira, Portugal, pp. 281–286 (2009)
6. Romá-Ferri, M.T., Cruanes, J., Palomar, M.: Quality indicators of the OntoFIS pharmacotherapeutic ontology for semantic interoperability. In: Proceedings of the IADIS International Conference e-Health 2009, Algarve, Portugal, pp. 107–114 (2009)
7. Uschold, M., Grüninger, M.: Ontologies: principles, methods and applications. Knowledge Engineering Review 11(2), 93–155 (1996),
   https://eprints.kfupm.edu.sa/55793/
8. Fernández-López, M., Gómez-Pérez, A., Pazos-Sierra, A., Pazos-Sierra, J.: Building a chemical ontology using METHONTOLOGY and the ontology design environment. IEEE Intelligent Systems & Their Applications 4(1), 37–46 (1999)
9. Noy, N.F., McGuinness, D.L.: Ontology development 101: a guide to creating your first ontology. Technical report KSL-01-05, Stanford Knowledge Systems Laboratory and technical report SMI-2001-0880, Stanford Medical Informatics (2001),
   http://protege.stanford.edu/publications/
   ontology_development/ontology101.pdf
10. Gómez-Pérez, A., Fernández-López, M., Corcho, O.: Ontological engineering: with examples from the areas of knowledge management, e-commerce and the semantic web. Springer, London (2004)
11. Gruber, T.R.: Toward principles for the design of ontologies used for knowledge sharing. Technical report KSL-93-04, Stanford Knowledge Systems Laboratory (1993),
    http://ksl-web.stanford.edu/knowledge-sharing/papers/
    onto-design.rtf
12. Cranefield, S., Purvis, M.K.: UML as an ontology modelling language. In: Proceedings of the Workshop on Intelligent Information Integration, 16th International Joint Conference on Artificial Intelligence. CEUR Workshop Proceedings, Germany, pp. 46–53 (1999)
13. Lundell, B., Lings, B., Persson, A., Mattsson, A.: UML Model Interchange in Heterogeneous Tool Environments: An Analysis of Adoptions of XMI 2. In: Wang, J., Whittle, J., Harel, D., Reggio, G. (eds.) MoDELS 2006. LNCS, vol. 4199, pp. 619–630. Springer, Heidelberg (2006)
14. Rector, A., Welty, C.: Simple part-whole relations in OWL ontologies. W3C Editor's Draft (August 11, 2005),
    http://www.w3.org/2001/sw/BestPractices/OEP/SimplePart/
    Whole/index.html
15. Veres, C.: Aggregation in ontologies: Practical implementations in OWL. In: Lowe, D.G., Gaedke, M. (eds.) ICWE 2005. LNCS, vol. 3579, pp. 285–295. Springer, Heidelberg (2005)
16. Object Management Group: Ontology Definition Metamodel Version 1.0. Object Modeling Group (May 2009), http://www.omg.org/spec/ODM/1.0/PDF
17. Gasěvić, D., Djurić, D., Devedžic, V.: Model Driven Engineering and Ontology Development, 2nd edn. Springer, Berlin (2009)

Exploiting Unlabeled Data for Question Classification*

David Tomás[1] and Claudio Giuliano[2]

[1] Department of Software and Computing Systems, University of Alicante, Spain
[2] Human Language Technology Group, FBK-Irst, Italy

Abstract. In this paper, we introduce a kernel-based approach to question classification. We employed a kernel function based on latent semantic information acquired from Wikipedia. This kernel allows including external semantic knowledge into the supervised learning process. We obtained a highly effective question classifier combining this knowledge with a bag-of-words approach by means of composite kernels. As the semantic information is acquired from unlabeled text, our system can be easily adapted to different languages and domains. We tested it on a parallel corpus of English and Spanish questions.

Keywords: question classification, semi-supervised learning, kernel methods.

1 Introduction

Question classification is one of the main tasks carried out in a question answering (QA) system [1]. The goal is to assign labels to questions based on the expected answer type. For example, a question like "Who was the first man to fly across the Pacific Ocean?" could be classified as *person*, while "Where is the Orinoco river?" could be classified as *location*. In a QA system, this information allows narrowing down the set of expected answers to those matching the class identified (a name of a person or a name of a location in the previous examples).

Most of the state-of-the-art approaches to question classification rely on linguistic information obtained from chunkers, parsers, named entity recognizers or lexical databases [2]. The use of this information binds the systems to particular tools and manually constructed linguistic resources, creating dependencies that make them difficult to move to new languages and domains. In addition, these tools and resources are not always available in all languages, and their use could require an excessive computational cost for real-time systems.

In this paper, we present a semi-supervised machine learning approach to question classification based on kernel methods. We extend the traditional bag-of-words representation, offering an effective way to integrate external semantic information in the question classification process by means of semantic kernels [3]. The advantage of our approach when compared with traditional semi-supervised learning [4], is that we do not need a collection of unlabeled questions (which are difficult to obtain) but a generic

* This research has been partially funded by the Spanish Government under project TEXT-MESS 2.0 (TIN2009-13391-C04-01) and Prometeo (PROMETEO/2009/199), and by the Italian Ministry of University and Research and by the Autonomous Province of Trento under project ITCH (RBIN045PXH).

R. Muñoz et al. (Eds.): NLDB 2011, LNCS 6716, pp. 137–144, 2011.
© Springer-Verlag Berlin Heidelberg 2011

corpus of texts. The result is a flexible system easily adaptable to different languages and domains. We tested the robustness of our approach on a parallel corpus of questions in English and Spanish.

The rest of the paper is organized as follows: Section 2 describes the kernels defined for this task and how the semantic information is included in the system; Section 3 shows the experiments carried out and the results obtained; in Section 4 related work is presented; finally, conclusions and future work are discussed in Section 5.

2   Kernels for Question Classification

Kernel methods are a popular machine learning approach within the natural language processing community [3]. The strategy adopted by kernel methods consists of splitting the learning problem in two parts. They first embed the input data in a suitable feature space, and then use a linear algorithm to discover nonlinear patterns in the input space. Typically, the mapping is performed implicitly by a so-called *kernel function*. Potentially any kernel function can work with any kernel-based algorithm. We used *support vector machines* (SVM) [5] in our approach.

Formally, the kernel is a function $k : X \times X \rightarrow \mathbb{R}$ that takes as input two data objects and outputs a real number characterizing their similarity, with the property that the function is symmetric and positive semi-definite. That is, for all x_i, $x_j \in X$, it satisfies

$$k(x_i, x_j) = \langle \phi(x_i), \phi(x_j) \rangle \,, \tag{1}$$

where ϕ is an explicit mapping from X to an (inner product) feature space $\mathcal{F}$.

The simplest method to estimate the similarity between two questions is to compute the inner product of their vector representations in the VSM. Formally, we define a space of dimensionality N in which each dimension is associated with one word from the dictionary, and the question q is represented by a row vector

$$\phi(q) = (f(t_1, q), f(t_2, q), \ldots, f(t_N, q)) \,, \tag{2}$$

where the function $f(t_i, q)$ records whether a particular token t_i is used in q. Using this representation we define the *bag-of-words kernel* between questions as

$$k_{BOW}(q_i, q_j) = \langle \phi(q_i), \phi(q_j) \rangle = \sum_{l=1}^{N} f(t_l, q_i) f(t_l, q_j) \,. \tag{3}$$

However, such an approach does not deal well with lexical variability and ambiguity. To address the shortcomings of the bag-of-words representation, we introduce the class of semantic kernels and show how to define an effective semantic VSM using unlabeled data.

2.1   Semantic Kernels

It has been shown that semantic information is fundamental for improving question classification accuracy [2]. In the context of kernel methods, semantic information can be integrated considering linear transformations of the type $\tilde{\phi}(q_j) = \phi(q_j)\mathbf{S}$, where $\mathbf{S}$ is a $N \times k$ matrix [3]. The matrix $\mathbf{S}$ can be rewritten as $\mathbf{S} = \mathbf{WP}$, where $\mathbf{W}$ is a diagonal matrix determining the word weights, while $\mathbf{P}$ is the *word proximity matrix* capturing the semantic relations between words. The proximity matrix $\mathbf{P}$ can be defined by setting non-zero entries between those words whose semantic relation is inferred from an external source of domain knowledge. The *semantic kernel* takes the general form

$$\tilde{k}(q_i, q_j) = \phi(q_i)\mathbf{SS}'\phi(q_j)' = \tilde{\phi}(q_i)\tilde{\phi}(q_j)' . \tag{4}$$

In the next paragraphs, we describe two alternatives approaches to define the proximity matrix. The first makes use of manually built lists of semantically related words and it is defined for comparative purposes only, while the second exploits unlabeled data and represents the main contribution of this work.

Explicit Semantic Kernel. WordNet and manually constructed lists of semantically related words typically provide a simple and effective way to introduce semantic information into the kernel. To define a semantic kernel from such resources, we can explicitly construct the proximity matrix $\mathbf{P}$ by setting its entries to reflect the semantic proximity between the words i and j in the specific lexical resource. In our experiments, we used the class-specific word lists manually constructed by [2] to define $\mathbf{P}$.[1] In the resulting semantic matrix, rows are indexed by words and columns are indexed by answer type. The (i, j)th entry is 1 if the word w_i is contained in the list l_j; 0 otherwise. The corresponding kernel, called *explicit semantic kernel*, is defined as follows:

$$k_{ES}(q_i, q_j) = \phi(q_i)\mathbf{PP}'\phi(q_j)' = \tilde{\phi}(q_i)\tilde{\phi}(q_j)' . \tag{5}$$

Latent Semantic Kernel. An alternative approach to define a proximity matrix is by looking at co-occurrence information in a large corpus. Two words are considered semantically related if they frequently co-occur in the same texts. *Latent semantic indexing* (LSI) is an effective vector space representation of corpora being able to acquire semantic information using co-occurrence information. This second approach is more attractive because it allows us to automatically define semantic models for different languages and domains.

We use singular value decomposition (SVD) to automatically define the proximity matrix $\mathbf{P}$ from Wikipedia, represented by its term-by-document matrix $\mathbf{D}$, where the $\mathbf{D}_{i,j}$ entry gives the frequency of term t_i in document d_j [6]. SVD decomposes the term-by-document matrix $\mathbf{D}$ into three matrixes $\mathbf{D} = \mathbf{U}\mathbf{\Sigma}\mathbf{V}'$, where $\mathbf{U}$ and $\mathbf{V}$ are orthogonal matrices whose columns are the eigenvectors of $\mathbf{DD}'$ and $\mathbf{D}'\mathbf{D}$ respectively, and $\mathbf{\Sigma}$ is the diagonal matrix containing the singular values of $\mathbf{D}$.

The selection of a representative corpus is an important part of the process of defining a semantic space. The use of Wikipedia allows us to define a open-domain statistical model. Under this setting, we define the proximity matrix $\mathbf{P}$ as follows:

$$\mathbf{P} = \mathbf{U}_k\mathbf{\Sigma}_k^{-1} , \tag{6}$$

---

[1] Available at `http://l2r.cs.uiuc.edu/~cogcomp/Data/QA/QC/`

where $\mathbf{U}_k$ is the matrix containing the first k columns of $\mathbf{U}$ and k is the dimensionality of the latent semantic space and can be fixed in advance.

The matrix $\mathbf{P}$ is used to define a linear transformation $\rho : \mathbb{R}^N \rightarrow \mathbb{R}^k$, that maps the vector $\phi(q_j)$, represented in the standard VSM, into the vector $\tilde{\phi}(q_j)$ in the latent semantic space. Formally, ρ is defined as follows:

$$\rho(\phi(q_j)) = \phi(q_j)(\mathbf{WP}) = \tilde{\phi}(q_j) , \tag{7}$$

where q_j is represented as a row vector, and $\mathbf{W}$ is a $N \times N$ diagonal matrix determining the word weights such that $\mathbf{W}_{i,i} = idf(w_i)$, where $idf(w_i)$ is the *inverse document frequency* of word w_i. The *latent semantic kernel* is explicitly defined as follows

$$k_{LS}(q_i, q_j) = \langle \rho(\phi(q_i)), \rho(\phi(q_j)) \rangle . \tag{8}$$

2.2 Composite Kernels

Once we have defined all the individual kernels representing syntactic and semantic information, we can define the composite kernel to combine and extend the individual ones based on the closure properties of the kernel functions. We have defined three composite kernels for question classification: $k_{BOW} + k_{LS}$ combines the bag-of-words (k_{BOW}) with semantic information automatically acquired from Wikipedia (k_{LS}); $k_{BOW} + k_{ES}$ combines the bag-of-words with semantic information acquired from manually constructed lists of words semantically related to specific answer types (k_{ES}); $k_{BOW} + k_{LS} + k_{ES}$ combines the bag-of-words with semantic information automatically acquired from Wikipedia and semantic information acquired from manually constructed lists of words semantically related to specific answer types.

3 Experiments

In this section, we describe the evaluation framework and the results obtained with our approach. The purpose of these experiments is to test the effect of the different semantic kernels in the classification, and the robustness of this approach in dealing with different languages. We compare our results with other state-of-the-art systems.

3.1 Description of the Data Set

The UIUC corpus[2] has become a *de facto* standard for the evaluation of question classification systems. It was first described in [2] and contains a training set of 5,452 questions and a test set of 500 questions in English.

These questions are labeled with a two level hierarchical taxonomy of classes. The first level consists of 6 coarse-grained classes (like *human*, *location* or *numeric*) that are subclassified on a second level of 50 fine-grained classes (with refinements like *city*, *country* or *mountain* for the coarse class *location*).

We evaluated our system on both coarse- and fine-grained classification. The latter is a touchstone for machine learning approaches, as the number of samples per question

---

[2] Available at http://l2r.cs.uiuc.edu/~cogcomp/Data/QA/QC/.

class is drastically reduced with respect to the coarse classification. It is in this framework that we can really test the generalization capabilities of our proposal and the effect of semantic information induced by the semantic kernels.

In order to test the flexibility and robustness of our approach with different languages, we manually translated the whole UIUC corpus into Spanish. We performed the same experiments on both English and Spanish corpora.

3.2   Experimental Setup

All the experiments were performed using the SVM package LIBSVM[3] customized to embed our own kernels, and the SVD library LIBSVDC.[4] To define the English and Spanish proximity matrices, we performed the SVD using 400 dimensions ($k = 400$) on the term-by-document matrices obtained from 50,000 pages randomly selected from the English and Spanish versions of Wikipedia. For English, we took as input the preprocessed version of Wikipedia made available by [7], without considering the syntactic and semantic annotation. For Spanish, we wrote a script to remove the HTML and wiki markups from the pages. The statistical significance of all the results was checked by means of the *approximate randomization* procedure [8], with significance levels of 0.05 and 0.01.

3.3   Experimental Results

Table 1 shows the accuracy on the English and Spanish benchmark. The results obtained in English employing only the latent semantic kernel k_{LS}, show that the semantic information induced by this kernel is not enough for the task of question classification. Using the composite kernel $k_{BOW} + k_{LS}$, we improve the results when compared with the baseline k_{BOW}, achieving 90.0% on coarse classes and 83.2% on fine classes. This difference is statistically significant ($p < 0.01$) in both coarse and fine classification. In this case, the composite kernel allows to successfully complementing the information from the bag-of-words with the semantic knowledge obtained by means of the k_{LS}. On the other hand, the difference between $k_{BOW} + k_{LS}$ and $k_{BOW} + k_{ES}$ is not statistically significant in both coarse and fine classification. This means that the improvement achieved with both resources is equivalent. The advantage of the approach with the composite kernel $k_{BOW} + k_{LS}$ is that we do not need any handcrafted resources.

Finally, the composite kernel $k_{BOW} + k_{LS} + k_{ES}$ further improves the results obtained with the previous kernels at a significance level of $p < 0.05$ for the fine-grained classification, obtaining 85.6% precision. This result reveal that k_{LS} and k_{ES} capture different semantic relations and can complement each other. Figure 1 shows the learning curves for the experiments on the English corpus. The use of the composite kernel $k_{LS} + k_{BOW}$ increases the generalization capabilities of the system, achieving the same accuracy that the baseline k_{BOW} with just half of the training samples.

In order to test the performance of the system in different languages, we repeated the previous experiments on the Spanish corpus. In this case we did not build the kernel k_{ES}, as it implied the translation of the whole list of related words. We leave this as

---

[3] http://www.csie.ntu.edu.tw/~cjlin/libsvm/
[4] http://tedlab.mit.edu/~dr/svdlibc/

Table 1. Results for English and Spanish

Kernel	English		Spanish	
	Coarse	Fine	Coarse	Fine
k_{BOW}	86.4	80.8	80.2	73.4
k_{LS}	70.4	71.2	73.2	58.4
$k_{BOW} + k_{LS}$	90.0	83.2	**85.0**	**75.6**
$k_{BOW} + k_{ES}$	89.4	84.0	-	-
$k_{BOW} + k_{LS} + k_{ES}$	**90.8**	**85.6**	-	-

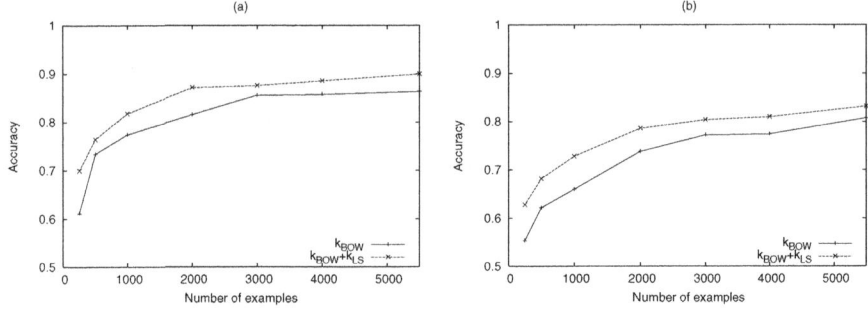

Fig. 1. Learning curves for coarse (a) and fine (b) classes for the English task

future work. The results in Table 1 confirm again that the composite kernel $k_{BOW} + k_{LS}$ obtains statistically significant improvement with respect to the baseline k_{BOW} in both coarse ($p < 0.01$) and fine ($p < 0.05$) classification. We obtained overall lower results when compared with the English versions of the experiments. The main cause may be that the Spanish Wikipedia corpus employed to build the proximity matrix was obtained employing simple heuristics to clean the documents. Figure 2 shows the learning curves for the Spanish system. Compared with English, they show less improvement in terms of training data reduction for the experiments on fine classes.

4   Related Work

The work by [2] was one of the first serious attempts to evaluate the performance of question classification systems in isolation. They developed a hierarchical classifier based on SNoW. They employed several resources to obtain a linguistically rich feature space, including head chunks, named entities and a handcrafted list of semantically related words. Their system obtained 91% precision for coarse classes and 84.2% for fine classes.

In [9], they experimented and compared five different machine learning algorithms, demonstrating that SVM outperforms all the other algorithms using a bag-of-ngrams representation. They employed a tree kernel to take advantage of the syntactic structures of the questions, obtaining 90% precision for coarse classes. The tree kernel did not

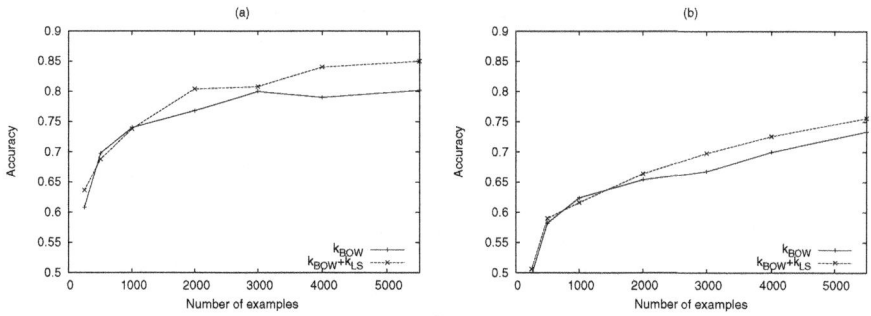

Fig. 2. Learning curves for coarse (a) and fine (b) classes for the Spanish task

improve the results of the n-gram model for fine classes, concluding that the syntactic tree does not contain the information required to distinguish between the various fine categories within a coarse category.

The work developed in [10] was a first attempt to apply SVD for dimensionality reduction in question classification, but they did not obtain any improvement with this technique. They achieve 79.8% precision for fine-grained classification with SVD and 2000 dimensions, obtaining worse results than the original n-gram representation. The main difference with our approach is that they built the statistical model using a very small corpus of questions (i.e., the training and test sets), instead of exploiting a large unlabeled corpus of documents. Even though they defined a representative corpus, the very small size of the corpus compromised the results.

The work in [4] presented an approach to semi-supervised question classification using the Tri-training algorithm [11] and part-of-speech features. They performed experiments on different subsets of the UIUC training corpus, while the rest of the training instances were employed as unlabeled data. These subsets of the corpus are the same that we presented in the learning curve in Figure 1 (b). The results obtained with our kernel $k_{BOW} + k_{LS}$ outperform their experiments for all the subsets defined.

5   Conclusions and Future Work

In this paper we presented an approach to question classification based on kernel methods. We employed composite kernels to incorporate semantic relations between words and extend the bag-of-words representation. We defined a latent semantic kernel to obtain a generalized similarity function between questions. The model was acquired from unlabeled documents from Wikipedia, resulting in a flexible system easily adaptable to different languages and domains. We further improved the system including an explicit semantic kernel based on lists of semantically related words.

The system was tested on the UIUC data set, a corpus of English questions widely employed in question classification research. In fine-grained classification, our semantic kernel outperformed previous approaches based on tree kernels and parsing information. In coarse classification, our results are comparable with the state-of-the-art, surpassing many other systems that make an intensive use of linguistic resources and tools. In both cases our approach significantly improves the bag-of-word representation. Our kernel $k_{BOW} + k_{LS}$ employs unlabeled documents from Wikipedia to improve the bag-of-words representation. This semi-supervised approach demonstrated to outperform previous approaches to semi-supervised question classification based on unlabeled corpus of questions. Moreover, we obtained a parallel corpus in Spanish to test the system with different languages and demonstrate the robustness and flexibility of our approach. As in the experiments on the English corpus, we obtained a significant improvement by including the latent semantic kernel.

For future work, we want to investigate the effect of varying the corpus, the number of documents, and dimensions used to define the semantic space. Besides that, we plan to review the Spanish documents gathered from Wikipedia to solve the problems mentioned in Section 3.3. We also want to translate the list of semantically related words in order to perform a comparative experiment with the kernel k_{ES} on the Spanish corpus.

References

1. Voorhees, E.M.: The trec-8 question answering track report. In: Eighth Text REtrieval Conference, vol. 500-246, pp. 77–82. National Institute of Standards and Technology, Gaithersburg (1999)
2. Li, X., Roth, D.: Learning question classifiers. In: 19th International Conference on Computational Linguistics. pp. 1–7. Association for Computational Linguistics, Morristown (2002)
3. Shawe-Taylor, J., Cristianini, N.: Kernel Methods for Pattern Analysis. Cambridge University Press, Cambridge (2004)
4. Nguyen, T.T., Nguyen, L.M., Shimazu, A.: Using semi-supervised learning for question classification. Information and Media Technologies 3(1), 112–130 (2008)
5. Cristianini, N., Shawe-Taylor, J.: An introduction to Support Vector Machines. Cambridge University Press, Cambridge (2000)
6. Deerwester, S.C., Dumais, S.T., Landauer, T.K., Furnas, G.W., Harshman, R.A.: Indexing by latent semantic analysis. Journal of the American Society of Information Science 41(6), 391–407 (1990)
7. Zaragoza, H., Atserias, J., Ciaramita, M., Attardi, G.: Semantically annotated snapshot of the english wikipedia, vol.1 (2007), http://www.yr-bcn.es/semanticWikipedia
8. Noreen, E.W.: Computer-Intensive Methods for Testing Hypotheses. John Wiley & Sons, New York (1989)
9. Zhang, D., Lee, W.S.: Question classification using support vector machines. In: 26th Annual International ACM SIGIR Conference, pp. 26–32. ACM, New York (2003)
10. Hacioglu, K., Ward, W.: Question classification with support vector machines and error correcting codes. In: North American Chapter of the Association for Computational Linguistics, pp. 28–30. Association for Computational Linguistics, Morristown (2003)
11. Zhou, Z.H., Li, M.: Tri-training: Exploiting unlabeled data using three classifiers. IEEE Transactions on Knowledge and Data Engineering 17(11), 1529–1541 (2005)

A System for Adaptive Information Extraction from Highly Informal Text

Laura Alonso i Alemany and Rafael Carrascosa

NLP group
FaMAF-UNC
Córdoba, Argentina
alemany@famaf.unc.edu.ar

Abstract. We present a first version of ADO, a system for **A**daptive **D**ata **O**rganization, that is, information extraction from highly informal text: short text messages, classified ads, tweets, etc. It is built on a modular architecture that integrates in a transparent way off-the-shelf NLP tools, general procedures on strings and machine learning and processes tailored to a domain.

The system is called adaptive because it implements a semi-supervised approach. Knowledge resources are initially built by hand, and they are updated automatically by feeds from the corpus. This allows ADO to adapt to the rapidly changing user-generated language.

In order to estimate the impact of future developments, we have carried out an orientative evaluation of the system with a small corpus of classified advertisements of the real estate domain in Spanish. This evaluation shows that tokenization and chunking can be well resolved by simple techniques, but normalization, morphosyntactic and semantic tagging require either more complex techniques or a bigger training corpus.

Keywords: user-generated text, information extraction, natural language processing suites.

1 Introduction and Motivation

In recent times there has been a significant increase in on-line, publicly available text like community-centered blogs, short text messages by mobile phones, user states in social networks, short advertisements or auction listings. This user-generated content provides a growing amount of privileged information for a variety of goals, ranging from product or service reviews to epidemic surveillance.

We have developed ADO, a system to deal with this kind of text. The ultimate goal of ADO (**A**daptive **D**ata **O**rganization) is to perform information extraction on these pieces of textual information, providing an end-to-end facility from raw text feeds to a knowledge base. However, in many application domains this ultimate goal is unattainable for the very simple reason that knowledge of the domain is not known and organized so that it can be expressed formally. That is why the system is highly flexible and customizable, easily adapting to different processing needs.

R. Muñoz et al. (Eds.): NLDB 2011, LNCS 6716, pp. 145–152, 2011.
© Springer-Verlag Berlin Heidelberg 2011

One of the crucial features of ADO is its adaptability. On-line, typically user-generated text is rapidly changing, at rates that often exceed human processing capability. To tackle this, ADO implements a semi-supervised approach to updating its knowledge-based resources. Based on bootstrapping techniques, initial, human-annotated resources are automatically updated and enriched with upcoming new variants of words and structures.

In this paper we will present a proof of concept of ADO in processing classified ads of the real estate domain, an application similar to [10]. This kind of corpus presents the advantage of providing plentiful, time-stamped corpora and belonging to a domain with a clear organization of knowledge that can be easily translated into a knowledge base.

The rest of the paper is organized as follows. In the following section, we describe the general architecture of the system, providing detail on its modules. Then, we present the evaluation corpus of real estate classified advertisements and its manual annotation. In Section 4 we describe an evaluation of ADO with this corpus, with a detail for each of the levels, together with an analysis of error that will direct future developments. We finish with a summary of contributions and an outline of future work.

2   Architecture of the System

ADO is based on a highly modular architecture that makes it easier to incorporate improvements and to port it to different domains or genres. Processing is organized in a layered approach typical of classical NLP architectures:

1. tokenization
2. normalization
3. morpho-syntactic tagging
4. chunking
5. semantic tagging
6. integration in knowledge base

Standard NLP and ML modules and procedures are used whenever available, mainly from the NLTK toolkit [8]. The information flow deals with module-specific procedures in a transparent way, using parameters to access their high-level features. Since this is a prototype system, it has been implemented in python, but each module may interface with tools that have been implemented in any other language.

The input to ADO is an XML collection of documents. Two main kinds of output are envisaged. The system can provide a knowledge base containing the relevant information extracted from documents. Alternatively, it can also provide the same collection, with each document associated to stand-off annotation with information from the processing layers by which the collection has been analyzed.

For all processing layers, at least two basic approaches are supported: ADO can process hand-written rules in intuitive regular expressions. Alternatively, an annotated corpus can be provided, from which a statistical classifier is learnt.

In the rest of the section we describe the functionality of some of the processing layers.

2.1 Normalization

In order to provide some generalization to the data and reduce data sparseness, normalization assigns a normal form to the many different variants by which a word can be realized. In standard language, this normal form is the lemma. In user-generated language or in other kinds of highly informal or specific language, normalization is more complex than lemmatization. We have to deal with flective morphemes but also with a heavy use abbreviations, misspellings, writing by the sound or even usage of numbers as letters [14].

Very often, vocabulary normalization for user-generated text is done by hand-made dictionaries of abbreviations and variants. However, the variation in this kind of text is very high, and new variants of words and entirely new ways of expression are produced rapidly.

We propose a bootstrapping approach to alleviate human intervention in this process. We substitute dictionary-based normalization by clusters of words with the same canonical form. Then, given a new word, an automatic classifier finds the most similar word in the manually created clusters. Similarity is calculated using a string edit distance, so that a pair of strings are closer if less edit operations are needed to transform one into the other. If similarity between these two strings is over a given threshold, the new word is clustered together with the most similar; if not, it constitutes a new cluster. Thus, the task of the human annotator is reduced to creating an initial set of clusters and eventually validating automatic associations, because clusters are updated automatically or semi-automatically from corpus feeds.

The distance to be used is a parameter, so that the user can use standard edit distances available with the system (like Levenshtein [7], Jaro-Winkler [15]) or else a tailored edit distance can be used. Several approaches have been proposed to learn the costs of edit operations for string edit distances, from stochastic transducers [13,5,2,11] to conditional random fields [9], maximum entropy approaches [12], noisy channel models [1,3] or pair-Hidden Markov Models [4].

The clustering criterion of a new word with a pre-existing cluster is also a parameter, it can be chosen among single link, complete link or average link. The distance threshold for a word to create its own cluster or to be appended to a pre-existing cluster is also a parameter.

2.2 Semantic Tagging

In this step, chunks determined in the previous layer are assigned one of a class of semantic categories, possibly organized in an ontology. If the ontology is organized taxonomically (with a clear *is-a* backbone), ADO can provide semantic classification at different levels of granularity.

2.3 Integration in a Knowledge Base

As a final output, ADO can provide a knowledge base where all the relevant information that has been extracted from documents is organized. To do that, the system needs to be provided with the structure of the knowledge base to be populated, and also with templates with which to extract information from documents. Exploiting these two data structures, ADO will extract the relevant information from each document with a template filling procedure based on Hungarian matching [6], and then insert it into the corresponding slots in the knowledge base.

3   An Annotated Corpus of Classified Advertisements

We wanted to carry out a preliminary evaluation of the system that indicates the main sources of error and allows us to establish priorities with respect to the following steps of development.

We manually annotated a small corpus of 100 classified advertisements of the real state domain in Spanish. These advertisements were randomly extracted from a 1 million word corpus of advertisements compiled from the on-line version of the Argentinean newspaper *"La Voz del Interior"*. The annotated corpus has 2500 words, with an average of 16 words per advertisement.

The manual annotation replicated all the processing steps in the automatic system, except template matching, with the following results:

tokenization: there were 2650 character strings between white spaces in the original corpus, in contrast with 2514 words identified manually.

normalization: only a reduction of dimensionality of 10% was achieved by normalization: 559 different forms were reduced to 455 different lemmas (excluding numbers and punctuation). The 10 most frequent forms cover 9% of the words, while the 10 most frequent lemmas cover 12% of the words.

morpho-syntactic: only two categories were defined: *head* or *modifier*, and they were equally distributed (50% of the words were tagged as head, the rest as modifier). We expect that these tags are sufficient to recover the structure of chunks and assign them a semantic class, the following processing steps.

chunking: 1121 chunks were identified, with an average of 2.5 words per chunk, with 83% of the chunks having 1, 2 or 3 words, and only 6% having 5 or more words.

semantics: 43 fine-grained semantic categories were defined. The 10 most frequent categories (*telephone, price, zone, address, bedroom, seller*, etc.) cover 57% of the chunks, while the 10 least frequent account only for 2% of the chunks. Coarser semantic categories were also defined, creating a taxonomy of categories. In the lower level there are 43 categories, 27 in the following, 7 in the next and 3 in the topmost level: *administrative, physical, miscellaneous*.

4   Evaluation

For this first evaluation, very simple, unexpensive procedures have been implemented in the system. We expect that the evaluation will shed light onto which aspects of the system should be improved first with a bigger impact in performance. The last step, integration in a knowledge base, has not been carried out because the knowledge base and corresponding templates have not been designed for this domain.

4.1   Evaluation Metrics

Evaluation of the system is mainly based on accuracy. In each level, units can be assigned to various classes, and accuracy gives the proportion of times units have been assigned to the correct class. For tokenization and chunking, we have calculated accuracy based on the BIO model, so units can be assigned to one of three classes: beginning of span, inside of span or outside of span. Since the corpus was very small, evaluation was carried out via ten-fold cross validation.

Additionally, we provide the evaluation of basic baselines for each level of processing, to assess the benefit provided by complex procedures:

tokenization tokens are strings between white spaces
normalization each token is its own normal form
morphosyntactic tags are assigned randomly, weighted by their frequency in the gold standard
chunking each *head* constitutes a chunk, together with all *modifiers* to its left
semantic each chunk is assigned the most frequent semantic class in the gold standard

4.2   Configuration of the System

In this preliminary implementation, we have skipped the step of integration in a Knowledge Base, focusing in previous processing steps.

The system was configured as follows:

1. **tokenization** regular expression rules have been used to tokenize the input text, separating punctuation signs and numbers from words, and grouping telephone numbers.
2. **normalization** normalization has been carried out by applying simple levenshtein string edit distance with a single-link clustering criterion and a threshold of 5 for appending new words to a pre-existing cluster. Numbers and punctuation signs were not normalized. The clusters of variants of the same word were manually created from a single day of advertisements, totalling 3359 ads.
3. **morpho-syntax** a back-off combination of Markov Models of different lengths (from 2 to 0) was used, as provided by NLTK, with the categories *head* and *modifier*. The features used by the tagger are the normal form of the word and the previous word, and the tag assigned to the previous word.
4. **chunking** the NLTK Naive Bayes chunker was used, using the morphosyntactic tag assigned to a word and to the two words to its left.
5. **semantics** the NLTK Naive Bayes classifier was used, using as features:

- number of words in the chunk.
- morpho-syntactic tag, normal form and token form of words in chunk.
- morpho-syntactic tag, normal form and token form of the two words at the left of the chunk.
- morpho-syntactic tag, normal form and token form of the two words at the right of the chunk.

4.3 Analysis of Results

Results can be seen in Table 1. In order to better analyze the performance of each module, we have evaluated the performance when perfect input from the gold standard is provided (table at the left). We have also evaluated the system when each layer gets input from the automatic output of the inmediately preceding layer and when it gets input from the baseline (table at the right). This allows us to roughly assess the benefit of each procedure.

Very good results are achieved for tokenization (96% accuracy), which is remarkable given the rule-based, unexpensive approach. The baseline for this task, considering whitespace separated strings as tokens, is high (90%), but rules outperform it. Probably a couple of rules more can help in achieving near-perfect performance, as noted by [14].

For chunking, implementing a very simple machine learning approach, results are also very good (90%), although the baseline is also high (87%). Automatic analysis does not seem to damage performance in this level either. This seems to indicate that a rule-based approach, a bigger corpus or more complex features can achieve better results.

The performance for morphosyntactic tagging, with 20% error, needs to be improved. In this layer, the baseline is much lower than the machine-learning approach, which is not surprising because categories are equally distributed.

Table 1. Results of ten-fold cross-validation for the different processing layers of the basic ADO as applied to newspaper classified ads of the real estate domain (mean accuracy and standard deviation between folds, between parentheses). In the left, each layer is provided by perfect input, from the gold standard, in the right, each layer is provided by automatic or baseline input.

	perfect input for each layer				automatic input for each layer		
	accuracy	baseline			accuracy	baseline
Tokenization	96	–	90 (2)	Tokenization	96	–	90 (2)
Normalization (lemma)	11 (3)	38 (6)	Normalization (lemma)	11 (3)	38 (6)
Normalization (cluster)	55 (10)	– –	Normalization (cluster)	55 (10)	– –
Morphosyntactic	90 (4)	49 (2)	Morphosyntactic	81 (9)	49 (2)
Chunking	90 (1)	87 (2)	Chunking	90 (1)	87 (2)
Semantic (finest)	55 (4)	10 (1)	Semantic (finest)	33 (8)	14 (9)
Semantic (fine)	62 (5)	13 (5)	Semantic (fine)	53 (10)	17 (14)
Semantic (coarse)	77 (2)	33 (4)	Semantic (coarse)	65 (13)	36 (14)
Semantic (top)	85 (3)	66 (2)	Semantic (top)	72 (13)	66 (15)

We can also see that errors in normalization produce an increase of 10% more error in tagging. The machine learning approach taken seems adequate, but the learning corpus is small, highly unstructured and the domain presents vocabulary sparseness. All this makes it highly probable that, given a word, one has never seen it and has to treat it as a new word, guessing its tag from the previous sequence of words, which does not provide much certainty because language in this domain is highly unstructured. To tackle these problems, we plan to increase the size of the training corpus, but given the sparseness of the data, we will apply active learning techniques, selecting examples of those sequences in the markov model with higher entropy, contributing to increase the certainty of the model.

Semantic tagging has a good level of performance, given the size of the training corpus. It is notable that performance is well above the baseline for all granularities. For coarser distinctions, performance consistently improves, which seems to validate the structure of the domain ontology. It is notable that errors from automatic processing previous layers significantly affect the performance at this layer, more acutely for finer-grained distinctions and for intelligent classifiers (as opposed to the baseline). The high standard deviation between folds from automatic tagging also indicates that no regularities can be found, in contrast with low deviations for gold standard input. Therefore, to improve performance of semantic tagging it seems necessary to address performance issues in previous layers, particularly in normalization. In any case, a bigger corpus would be beneficial for this task.

The level with worst performance is normalization. For pairing a word with the gold standard lemma, performance is 11%, which is even below the performance of the baseline, considering each token as its own normal form. Performance improves when words are assigned to clusters, reaching 55%. We found no significant differences when changing parameters for agglomeration threshold, clustering criterion and distance. We expect that a gain in performance can be achieved by training a string edit distance, where each edit operation has a corpus-specific cost, instead of the uniform cost of the levenshtein distance. As mentioned in Section 2.1 in previous work, this approach is promising.

5   Conclusions and Future Work

We have presented a preliminary version ADO, a system for information extraction from informal text. A detailed evaluation indicates that the normalization process is the most critical, and that improving it would have an impact on the overall performance of the system. The features of text in this domain, with high sparseness of normal forms, make this difficult, but training a string edit distance seems like a promising approach to tackle this problem.

As future work, besides exploring training edit distances, we will enlarge the training corpus by active learning techniques. We will evaluate the impact of automatically updating normalization dictionaries with feeds from the corpus, a crucial aspect of the system. We will also implement the integration in a knowledge base and evaluate the end-to-end performance of the system.

Finally, we will perform a portability test by adapting ADO to the processing of short text messages sent by cell phone. This adaptation will concern mainly the normalization module, and information extraction facilities will be event-focused, used only to extract named entities and milestones like dates or places.

Acknowledgements. This work has been funded by the Argentinean Fon-CyT Programme, via the PAE-PICT-2007-02290 project *Data Mining in Semi-structured Text*, and by the Spanish Science and Innovation Ministry, via the *KNOW2* project (TIN2009-14715-C04).

References

1. AiTi, A., Min, Z., PohKhim, Y., ZhenZhen, F., Jian, S.: Input normalization for an english-to-chinese sms translation system. In: The Tenth Machine Translation Summit (2005)
2. Bilenko, M., Mooney, R.J.: Adaptive duplicate detection using learnable string similarity measures. In: Proceedings of the Ninth ACM SIGKDD (2003)
3. Cook, P., Stevenson, S.: An unsupervised model for text message normalization. In: Workshop on Computational Approaches to Linguistic Creativity. NAACL (2009)
4. Durbin, R., Eddy, S., Drogh, A., Mitchison, G.: Biological sequence analysis: Probabilistic models of proteins and nucleic acids. Cambridge University Press, Cambridge (1998)
5. Gómez-Ballester, E., Forcada-Zubizarreta, M.L., Micó-Andrés, M.L.: A gradient-descent method to adapt the edit-distance to a classification task. IOS Press, Amsterdam (2000)
6. Kuhn, H.W.: The hungarian method for the assignment problem. Naval Research Logistic Quarterly 2, 83–97 (1955)
7. Levenshtein, V.I.: Binary codes capable of correcting deletions, insertions, and reversals. Cybernetics and Control Theory 10(8), 707–710 (1966)
8. Loper, E., Bird, S.: Nltk: The natural language toolkit. In: Proceedings of the ACL Demonstration Session, pp. 214–217 (2004)
9. McCallum, A., Bellare, K., Pereira, F.: A conditional random field for discriminatively-trained finite-state string edit distance. In: Proceedings of the Twenty-First Annual Conference on Uncertainty in Artificial Intelligence (UAI 2005), pp. 388–395. AUAI Press, Arlington (2005)
10. Michelson, M., Knoblock, C.A.: Phoebus: a system for extracting and integrating data from unstructured and ungrammatical sources. In: AAAI 2006 (2006)
11. Oncina, J., Sebban, M.: Learning stochastic edit distance: Application in handwritten character recognition. Pattern Recognition 39(9), 1575–1587 (2006)
12. Pakhomov, S.: Semi-supervised maximum entropy based approach to acronym and abbreviation normalization in medical texts. In: ACL, pp. 160–167 (2002)
13. Ristad, E.S., Yanilos, P.N.: Learning string edit distance. IEEE Transactions on Pattern Analysis and Machine Intelligence 20, 522–532 (1998)
14. Sproat, R., Black, A., Chen, S., Kumar, S., Ostendorf, M., Richards, C.: Normalization of non-standard words. Computer Speech and Language 15(3) (2001)
15. Winkler, W.E.: The state of record linkage and current research problems. Tech. Rep. R99/04, Statistics of Income Division (1999)

Pythia: Compositional Meaning Construction for Ontology-Based Question Answering on the Semantic Web

Christina Unger and Philipp Cimiano

CITEC, Bielefeld University, Germany

Abstract. In this paper we present the ontology-based question answering system Pythia. It compositionally constructs meaning representations using a vocabulary aligned to the vocabulary of a given ontology. In doing so it relies on a deep linguistic analysis, which allows to construct formal queries even for complex natural language questions (e.g. involving quantification and superlatives).

Keywords: ontology-based question answering, compositionality.

1 Introduction

The growing Semantic Web provides a large amount of ontology-based semantic markup that question answering systems can exploit in order to interpret and answer natural language questions. This means that user questions can be interpreted with respect to a particular ontology which provides natural language expressions with a well-defined meaning, thereby allowing to retrieve precise answers.

In this paper we present the ontology-based question answering system Pythia. It is based on the following two main ideas. First, it uses principled linguistic representations in order to compositionally construct general meaning representations that can subsequently be translated into formal queries. Such a deep linguistic analysis allows Pythia to construct formal queries even for complex natural language questions, e.g. involving quantification and superlatives. And second, it relies on a specification of the lexicon-ontology interface that explicates possible linguistic realizations of ontology concepts. This allows to build meaning representations that use a vocabulary aligned to the vocabulary of a given ontology, thereby ensuring a precise and correct mapping of natural language terms to corresponding ontology concepts.

In the following sections we present the system, its architecture, and report on evaluation results with respect to a subset of DBPedia and compare our system with related work.

2 Approach

The architecture of our question answering system Pythia can be depicted very roughly as follows:

R. Muñoz et al. (Eds.): NLDB 2011, LNCS 6716, pp. 153–160, 2011.
© Springer-Verlag Berlin Heidelberg 2011

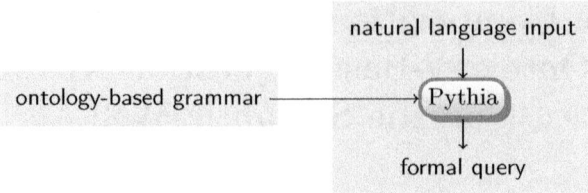

Natural language input is tranformed into a formal query by means of a linguistic analysis that is driven by an ontology-based grammar. Before explicating this transformation, we will briefly describe the grammar and its generation. A more detailed account of grammar generation and of the motivation to use ontology-specific grammars is given in [4].

In Pythia, natural language expressions are parsed and interpreted with respect to a grammar which we assume to be composed of two parts: an ontology-specific part and an ontology-independent part. The ontology-specific part contains lexical entries that refer to individuals, concepts, and properties of the underlying ontology. It is generated automatically from an ontology-lexicon model, as will be described below. The ontology-independent part comprises functional expressions like auxiliary verbs, determiners, wh-words and so on. The overall picture can be sketched as follows:

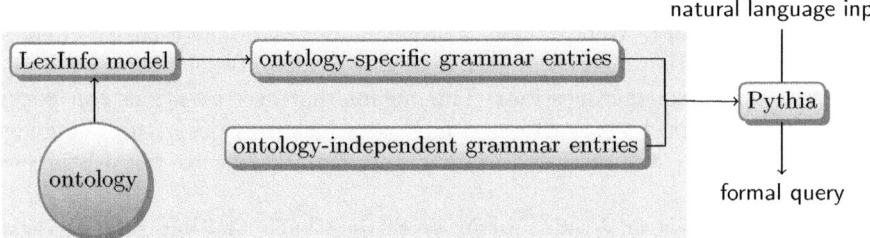

Both parts of the grammar use principled linguistic representations. More specifically, we assume grammar entries to be pairs of a syntactic and a semantic representation. As syntactic representation we take trees from Lexicalized Tree Adjoining Grammar (LTAG [5]). LTAG is very well-suited for ontology-based grammar generation because it allows for flexible basic units; we can, for example, assume complex grammar entries for examples like population of or has...inhabitants. As semantic representations we take DUDEs [2], a kind of Underspecified Discourse Representation Structures (UDRS [6]) augmented with information that allows for a flexible semantic composition.

The first step in generating a grammar from a given ontology is to enrich the ontology with information about its verbalization. The framework we use for this is LexInfo[1] [3], which offers a general frame for creating a declarative specification of the lexicon-ontology interface by connecting concepts of the ontology to information about their linguistic realization, i.e. word forms, morphology, subcategoriziation frames and how syntactic and semantic arguments correspond to

---

[1] http://lexinfo.net

each other. The lexical entries specified by LexInfo are then input to a general mechanism for generating grammar entries, i.e. pairs of syntactic and semantic representations.

For example, the object property **borders** in the ontology is first specified to be verbalized as the transitive verb **to border** together with the relevant linguistic information (inflection, subcategorization, and so on). The resulting lexical entry is then input to a grammar generation mechanism, which specifies general templates for mapping LexInfo entries to grammar entries. Applied to the entry for **to border** it gives rise to a family of elementary LTAG trees, two of them – one for active and one for passive use – are given in 1a. They are both paired with the UDRT-like semantic representation in 1b.

1. (a)

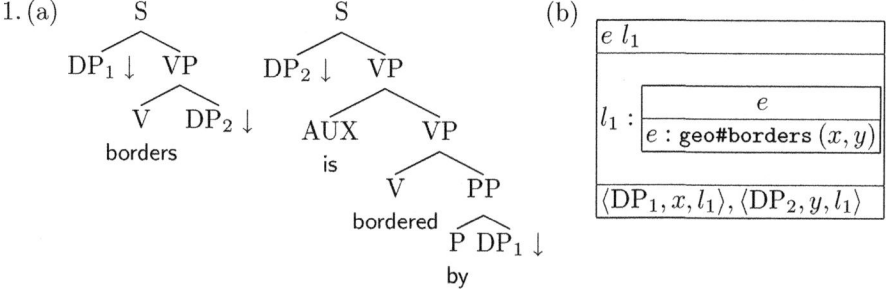

The syntactic structure encoded in the elementary trees captures the lexical material that is needed for verbalizing the property **borders**. The semantic representation contains a DRS labelled l_1, which provides the predicate **geo#borders** corresponding to the intended concept in the ontology (the prefix **geo#** abbreviates the namespace of the ontology), as well as information about the semantic arguments (x and y) and about which substitution node in the syntactic structure will provide them.

These linguistic representations are then used for parsing and interpreting natural language questions. The process of mapping natural language input to formal queries can be depicted as follows:

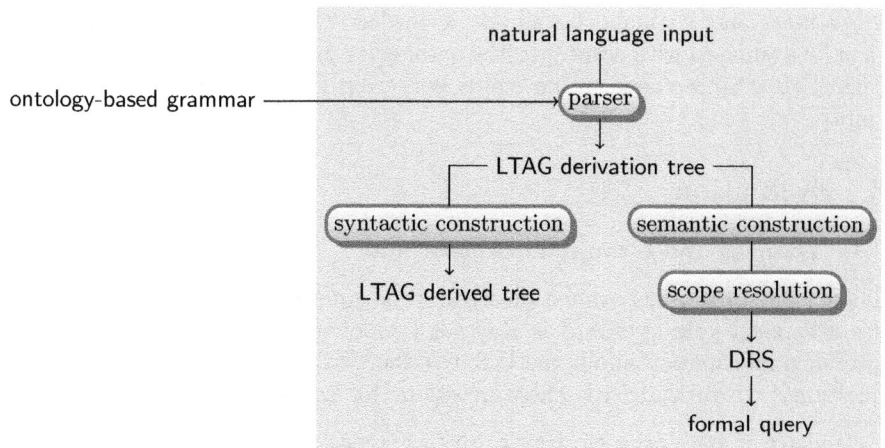

It involves three main steps. First, the input is handed to a parser, which works along the lines of the Earley-type parser devised by Schabes & Joshi [7]. It constructs an LTAG derivation tree, considering only the syntactic part of the grammar entries involved. Next, syntactic and semantic composition rules apply in tandem in order to construct a derived tree together with an according DUDE. The syntactic composition rules are LTAG's standard substitution and adjoin operations, and the semantic composition rules are parallel operations on DUDEs: an argument saturating operation (much like function application) that interprets substitution, and a union operation that interprets adjoin. Once all argument slots are filled, the constructed DUDE corresponds to an equivalent UDRS, which is then subject to scope resolution, resulting in a set of disambiguated Discourse Representation Structures (DRS [8]). Those are subsequently translated into a formal query. In the presented version of the system we use queries formulated in FLogic [9], but any other query language, e.g. SPARQL, could be used as well.

As an example, consider the input question Which states border Hawaii?. The parser produces a derivation tree that yields the derived tree in 2a. Parallel to this, a UDRT-like semantic representation is built which resolves to the DRS in 2b (the question mark serves to point out those variables whose values should be provided as an answer).

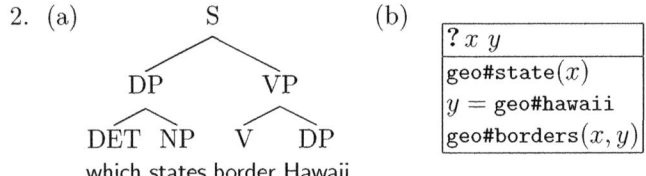

2. (a)   S   (b)
         DP   VP
     DET NP   V   DP
     which states border Hawaii

$? x \; y$
$geo\#state(x)$
$y = geo\#hawaii$
$geo\#borders(x, y)$

Subsequently, the DRS is translated into the following FLogic query:

```
FORALL X,Y <- X:geo#state AND equal(Y,geo#hawaii) AND
              X[geo#border -> Y]. orderedby X
```

It reads similarly to a first-order formula: for all bindings of X and Y such that X is a state and Y equals Hawaii and X borders Y, return all X. The query can then be evaluated with respect to the ontology, e.g. by means of the OntoBroker Engine [10]. Since there are no states bordering Hawaii, the returned result is empty.

3   Evaluation

3.1   Dataset and Grammar Generation

There is no established evaluation standard for question answering systems, but an often-used gold standard is Raymond Mooney's ontology comprising geographical information about the U.S. together with a set of 880 annotated user questions[2]. In order to use these questions for evaluation, we extracted from

---

[2] Available at ftp://ftp.cs.utexas.edu/pub/mooney/nl-ilp-data/geosystem/

DBpedia a subset containing all U.S. states, cities, mountains, lakes, rivers and roads. Furthermore, we annotated 865 of the 880 questions with corresponding FLogic query results. (The remaining 15 questions are questions which are out of scope of the ontology, such as Which rivers do not run through USA?).

In order to customize our question answering system to the domain of U.S. geography, we first constructed a LexInfo model for the extracted subset of DBpedia, which specifies how the concepts and relations of this ontology are verbalized. The constructed LexInfo model contains 678 lexical entries, of which 600 correspond to common nouns representing individuals and could be constructed automatically. The remaining 78 entries were built manually, using LexInfo's API. The effort to do so amounted to less than two minutes per entry, leading to a total amount of approximately two and a half hours. Next, those LexInfo entries were input to automatic grammar generation, yielding 2785 grammar entries (pairs of syntactic and semantic representations). Additionally, we manually specified 149 grammar entries for domain-independent elements such as determiners, wh-words, auxiliary words, and so on. The complete set of grammar entries are then used by Pythia for processing user questions. All mentioned resources – the dataset, the questions annotated with FLogic queries, the lexicon model, and the grammar files – are available at http://www.sc.cit-ec.uni-bielefeld.de/pythia.

3.2   Evaluation Results and Discussion

Running Pythia on the above mentioned 865 user questions and comparing the results of the constructed queries with the results given by the gold standard queries, we reach a recall of 67 % and precision of 82 %, leading to an F-measure of 73,7%.

Cases in which the system fails to construct an appropriate query can be pinned down to reasons that can roughly be categorized as Pythia-internal and Pythia-external failures. Pythia-external failures mean failures for which Pythia is not to blame. On the one hand side, these comprise questions that are ill-formed:

- *syntactically ill-formed questions:*
  questions that are incomplete or ungrammatical and therefore are not parsed (e.g. What is capital of Iowa?, What are the capital city in Texas?)
- *semantically ill-formed questions:*
  questions that violate sortal restrictions (e.g. Which states border the Missouri river?)

On the other hand side, they comprise questions that fail due to data incompleteness. For example, the ontology concept highest_point should be extensionally equivalent to locations with maximal height, which however is not always the case. Thus, if the gold standard uses the concept highest_point, while Pythia constructs a query asking for the location with maximal height, the results of both sometimes do not match, although the meaning of both queries are intuitively equivalent.

By Pythia-internal failures we mean questions that could in principle be parsed and answered, but for one reason or the other the system fails to do so. There are mainly two reasons for such failure. One reason is incomplete coverage, i.e. there is lexical material (e.g. the verb washed by) or syntactic constructions (e.g. topicalizations and NP disjunctions) missing in our grammar. Examples of questions that cannot be parsed for these reasons are Of the states washed by the Mississippi river, which has the lowest point? and How many states have cities or towns named Springfield?. The other reason for failure is non-compositionality, i.e. cases where the logical form of the whole question is not exactly the composition of the meaning of the parts of the question. This involves components that do not contribute anything to the overall meaning (e.g. american, in the US, give me) and some specific cases of counting with respect to a restriction (e.g. in Which river flows through the most states?, where the number of states has to be counted for each river). However, we skip a deeper discussion of these cases.

The table in Fig. 1 gives an overview of the qualitative coverage of our approach. The categories are taken from Cimiano & Minock [16]). Full treatment is expressed by +, missing treatment by −, (+) denotes partial or ad hoc coverage, and (−) means something is not captured yet but could in principle be incorporated.

Question types	wh-questions	+
	how ADJ/many	+
	requests	(+)
	topicalized questions	(−)
	nominals	+
Ambiguities	lexical, syntactic, scope	+
Other phenomena	spatial propositions	+
	adjectival modifiers and superlatives	+
	aggregation and comparisions	+
	negation	+
	coordination	(−)
	non-compositionality	(+)
	variability	(+)
	handling out-of-scope questions	−
	temporal aspects	−

Fig. 1. Covered question types and phenomena

Finally, let us briefly compare Pythia's results (67% precision and 82% recall) with other systems that were evaluated on the Geobase dataset. C-Phrase[3] by Minock [11] reaches 80-90% precision and recall after 120 minutes of authoring (cf. [12]). Mooney's learned semantic parsers (cf. Mooney [13]), on the other hand, reach a precision between 70 and nearly 100%, and a recall between 60

---

[3] http://code.google.com/p/c-phrase/

and 80%. They require no manual effort but rely on a large enough training corpus (queries annotated with semantic representations). PRECISE [14], on the other hand, requires no customization as the needed lexicon is extracted automatically from the input database. Only semantically tractable questions are answered, thereby reaching 100% precision, while for semantically intractable questions the system requests a paraphrase. Of the 880 questions, about 80% of the questions turn out to be tractable.

So, in general Pythia's results are competetive, albeit slightly excelled by other systems. This is because Pythia comes with a trade-off between coverage and the manual effort that is required with respect to the LexInfo model. A majority of the yet uncovered cases could in principle be covered by manually extending the LexInfo model and by including new domain-independent constructions like coordination. The real benefit of Pythia, however, is that it is able to handle linguistically complex queries involving quantification, superlatives and comparisons, aggregation functions, and the like – that is, phenomena that most other (and especially shallow) systems cannot cope with and would be difficult to extend with.

4   Conclusion and Future Work

We presented an ontology-based question answering system that parses user questions with respect to a domain-specific lexicon built automatically from a specification of linguistic realizations of ontology concepts. It compositionally constructs meaning representations that are aligned to the vocabulary of the underlying ontology and can easily be translated into a formal query.

This approach has several advantages. Due to the use of principled linguistic representations, Pythia is able to handle a wide range of linguistically complex queries, involving quantifiers, numerals, comparisons and superlatives, negation, and so on. Furthermore, due to the explicit specification of the lexicon-ontology interface, it is able to correctly map natural language terms to corresponding ontology concepts, even if they are superficially different (e.g. mapping has... inhabitants to the property *population*).

The major challenge for such an approach concerns portability. Adapting Pythia to a new domain requires the creation of a new LexInfo model for that domain, from which domain-specific grammar entries can be generated. This means that in a linguistically rich approach like ours, ontological support comes with a price: scalability. Pythia works very well for a relatively small domain, but requires non-negligible effort for larger domains. Part of our research therefore focuses on replacing the manual construction of LexInfo models by a largely automatic mapping of natural language expression to entities and relations. This will become especially important when applying the approach to larger domains on the Semantic Web, e.g. the whole of DBpedia.

References

1. Bunt, H.: Semantic Underspecification: Which Technique For What Purpose? In: Computing Meaning, vol. 83, pp. 55–85. Springer, Netherlands (2007)
2. Cimiano, P.: Flexible semantic composition with DUDES. In: Proceedings of the 8th International Conference on Computational Semantics, IWCS, Tilburg (2009)
3. Cimiano, P., Buitelaar, P., McCrae, J., Sintek, M.: Lexinfo: A declarative model for the lexicon-ontology interface. Journal of Web Semantics: Science, Services and Agents on the World Wide Web 9(1), 29–51
4. Unger, C., Hieber, F., Cimiano, P.: Generating LTAG grammars from a lexicon-ontology interface. In: Bangalore, S., Frank, R., Romero, M. (eds.) 10th International Workshop on Tree Adjoining Grammars and Related Formalisms (TAG+10). Yale University, New Haven and London (2010)
5. Schabes, Y.: Mathematical and Computational Aspects of Lexicalized Grammars. Ph. D. thesis, University of Pennsylvania (1990)
6. Reyle, U.: Dealing with ambiguities by underspecification: Construction, representation and deduction. Journal of Semantics 10, 123–179 (1993)
7. Schabes, Y., Joshi, A.K.: An Earley-type parsing algorithm for Tree Adjoining Grammars. In: Proceedings of the 26th Annual Meeting of ACL, Buffalo, New York, pp. 258–269 (1988)
8. Kamp, H., Reyle, U.: From Discourse to Logic. Kluwer, Dordrecht (1993)
9. Kifer, M., Lausen, G.: F-logic: A higher-order language for reasoning about objects, inheritance, and scheme. Technical report, SIGMOD Record 18(2) (1989)
10. Decker, S., Erdmann, M., Fensel, D., Studer, R.: Ontobroker: Ontology based access to distributed and semi-structured information. Database Semantics: Semantic Issues in Multimedia Systems, 351–369 (1999)
11. Minock, M.: C-Phrase: A system for building robust natural language interfaces to databases. Data Knowl. Eng. 69(3), 290–302 (2010)
12. Minock, M., Olofsson, P., Näslund, A.: Towards building robust natural language interfaces to databases. In: Kapetanios, E., Sugumaran, V., Spiliopoulou, M. (eds.) NLDB 2008. LNCS, vol. 5039, pp. 187–198. Springer, Heidelberg (2008)
13. Mooney, R.: Learning for semantic parsing. In: Gelbukh, A. (ed.) CICLing 2007. LNCS, vol. 4394, pp. 311–324. Springer, Heidelberg (2007)
14. Popescu, A.-M., Etzioni, O., Kautz, H.: Towards a theory of natural language interfaces to databases. In: IUI 2003: Proceedings of the 8th International Conference on Intelligent User Interfaces, pp. 149–157. ACM, New York (2003)
15. Schiehlen, M.: Semantic Construction from Parse Forests. In: Proceedings of the 16th International Conference on Computational Linguistics, Copenhagen (1996)
16. Cimiano, P., Minock, M.: Natural Language Interfaces: What Is the Problem? – A Data-Driven Quantitative Analysis. In: Horacek, H., Métais, E., Muñoz, R., Wolska, M. (eds.) NLDB 2009. LNCS, vol. 5723, pp. 192–206. Springer, Heidelberg (2010)

'twazn me!!! ;('
Automatic Authorship Analysis of Micro-Blogging Messages

Rui Sousa Silva[1,3], Gustavo Laboreiro[2,4],
Luís Sarmento[2,4], Tim Grant[1],
Eugénio Oliveira[2], and Belinda Maia[3]

[1] Centre for Forensic Linguistics at Aston University
[2] Faculdade de Engenharia da Universidade do Porto - DEI - LIACC
[3] CLUP - Centro de Linguística da Universidade do Porto
[4] SAPO Labs Porto

Abstract. In this paper we propose a set of stylistic markers for automatically attributing authorship to micro-blogging messages. The proposed markers include highly personal and idiosyncratic editing options, such as 'emoticons', interjections, punctuation, abbreviations and other low-level features. We evaluate the ability of these features to help discriminate the authorship of Twitter messages among three authors. For that purpose, we train SVM classifiers to learn stylometric models for each author based on different combinations of the groups of stylistic features that we propose. Results show a relatively good-performance in attributing authorship of micro-blogging messages ($F = 0.63$) using this set of features, even when training the classifiers with as few as 60 examples from each author ($F = 0.54$). Additionally, we conclude that emoticons are the most discriminating features in these groups.

1 Introduction

In January 2010 the *New York Daily News* reported that a series of Twitter messages exchanged between two childhood friends led to one murdering the other. The set of Twitter messages exchanged between the victim and the accused was considered a potential key evidence in trial, but such evidence can be challenged if and when the alleged author *refutes* its authorship. *Authorship analysis* can, in this context, contribute to confirming or excluding the hypothesis that a given person is the true author of a queried message, *among several candidates*. However, the micro-blogging environment raises new, significant challenges as the messages are *extremely short* and fragmentary. For example, Twitter messages are limited to 140 characters, but very frequently have only 10 or even fewer words. Standard stylistic markers such as *lexical richness, frequency of function words*, or *syntactic measures* — which are known to perform well with longer, 'standard' language texts — perform worse with such short texts, whose

R. Muñoz et al. (Eds.): NLDB 2011, LNCS 6716, pp. 161–168, 2011.
© Springer-Verlag Berlin Heidelberg 2011

language is 'fragmentary' [1]. Traditional authorship analysis methods are considered unreliable for text excerpts smaller than 250-500 words, as the accuracy tends to drop significantly with text length decrease [9].

In this paper we use a text classification approach to investigate whether some 'non-traditional' stylistic markers, such as the type of emoticons, provide enough stylistic information to be used in authorship attribution. We focus specifically on *Twitter* for its popularity, and address Portuguese in particular, which is one of the most widely used languages in this medium[1].

2   Related Work

In recent years, there has been considerable research on authorship attribution of some *user-generated contents* — such as *e-mail* (e.g. [2]) and, more recently, *web logs* (e.g. [3,4,5]) and 'opinion spam' (e.g. [6]). However, research on authorship attribution of Twitter messages has been scarce, and raised robustness problems.

To tackle the problem of robustness in computational stylometric analysis, research (e.g. the 'Writeprints technique' [10]) was applied to four different text genres to discriminate authorship and detect similarity of online texts among 100 authors. The performance obtained was good, but (a) the procedure did not prove to be content-agnostic, and (b) did not analyse Twitter messages. Also, using structural features that are possibly due to editing and considering 'idiosyncratic features' usage anomalies to include misspellings and grammar mistakes it is bound to compromise the results.

More recently, it has been demonstrated that the authorship of twitter messages can be attributed with a certain degree of certainty [11]. Surprisingly, the authors concluded that authorship could be identified at 120 tweets per user, and that more messages would not improve accuracy significantly. However, their method compromises the authorship identification task of most unknown messages, as they reported a loss of 27% accuracy when information about the interlocutor's user data was removed.

It has also been demonstrated that authorship could be attributed using 'probabilistic context-free grammars' [12] by building complete models of each author's (3 to 6) syntax. Nevertheless, the authors used both syntactic and lexical information to determine each author's writing style.

Conversely, we propose a content-agnostic method, based on low-level features to identify authorship of unknown messages. This method is independent of user information, so not knowing the communication participants is irrelevant to the identification task. Moreover, although some of the features used have been studied independently, this method is innovative in that the specific combination of the different stylistic features has never been used before and has not been applied to such short texts.

---

[1] http://semiocast.com/downloads/Semiocast_Half_of_messages_on_Twitter_are_not_in_English_20100224.pdf

3   Method Description and Stylistic Features

Authorship attribution can be seen as a typical *text classification* task: given examples of messages written by a set of authors (classes), we aim to attribute authorship of messages of unknown authorship. In a forensic scenario, the task consists of discriminating the authorship of messages of a small number of potential authors (e.g. 2 to 5), or determining whether a message can be attributed to a certain ('suspect') author.

The key to framing authorship attribution as a text classification problem is the selection of the feature sets that best describe the *style* of the authors. We propose four groups of stylistic features for automatic authorship analysis, each dealing with a particular aspect of tweets. All features are *content-agnostic*; to ensure a robust authorship attribution and prevent the analysis from relying on topic-related clues, they do not contain lexical information.

Group 1: Quantitative Markers. These features attempt to grasp simple quantitative style markers from the message as a whole. The set includes message statistics, e.g. length (in characters) and number of tokens, as well as token-related statistics (e.g. average length, number of 1-character tokens, 2-consonant tokens, numeral tokens, choice of case, etc). We also consider other markers, e.g. use of dates, and words not found in the dictionary[2] to indicate possible spelling mistakes or potential use of specialised language.

As *Twitter*-specific features, we compute the number of user references (e.g. @user_123), number and position of *hashtags* (e.g. #music), in-message URLs and the URL shortening service used. We also take note of messages starting with a username (a reply), as the author may alter their writing style when addressing another person.

Group 2: Marks of Emotion. Another highly personal — and hence idiosyncratic stylistic marker — is the device used to convey emotion. There are mainly three non-verbal ways of expressing emotion in user-generated contents: (i) *smileys*; (ii) *'LOLs'*; and (iii) *interjections*.

Smileys (':-)') are used creatively to reflect human emotions by changing the combination of eyes, nose and mouth. This work explores three axes of idiosyncratic variation: *range* (e.g. number of happy smileys per message), *structure* (e.g. whether the smiley has a nose) and *direction* of the smiley.

Another form of expression is the prevalent 'LOL', which usually stands for *Laughing Out Loud*. Frequently users manipulate the basic 'LOL' and 'maximise' it in various other forms, e.g. by repeating its letters (e.g. 'LLOOOLLL') or creating a loop (e.g. 'LOLOL'). This subgroup describes several instances of *length*, *case* and *ratio* between 'L' / 'O', so as to distinguish between 'LOL' and the exaggeration in multiplying the 'O', as in 'LOOOOL'.

We identify interjections as tokens consisting of only two alternating letters that are not a 'LOL', such as 'haaahahahah'. Other popular and characteristic

---

[2] We use the GNU Aspell dictionary for European Portuguese.

examples are the typical Brazilian laughing 'rsrsrs' and the Spanish laughing 'je-jeje' — both of which are now commonly found in European Portuguese *Twitter*. We count the number of interjections used in a message, their average length and number of characters.

Group 3: Punctuation. The choice of punctuation is a case of writing style [13], mostly in languages whose syntax and morphology is highly flexible (such as Portuguese and Spanish). Some authors occasionally make use of expressive and non-standard punctuation, either by repeating ('!!!') or combining it ('?!?'). Others simply skip punctuation, assuming the meaning of the message will not be affected. Ellipsis in particular can be constructed in less usual ways (e.g. '..' or '......'). We count the frequency of these and other peculiar cases, such as the use of punctuation after a 'LOL' and at the end of a message (while ignoring URLs and *hashtags*).

Group 4: Abbreviations. Some abbreviations are highly idiolectal, thus depending on personal choice. We monitor the use of three types of abbreviations: 2-consonant tokens (e.g. 'bk' for 'back'), 1- or 2-letter tokens followed by '.' or '/' (e.g. 'p/') and 3-letter tokens ending in two consonants, with (possibly) a dot at the end (e.g. 'etc.').

4   Experimental Setup

This study is focused on the authorship identification of a message among three candidate authors. We consider only three possible authors as forensic linguistic scenarios usually imply a limited number of suspect authors, and is hence more realistic. We chose to use Support Vector Machines (SVM) [14] as the classification algorithm for its proven effectiveness in text classification tasks and robustness in handling a large number of features. The *SVM-Light* implementation [14] has been used, parametrised to a linear kernel. We employ a *1-vs-all* classification strategy; for each author, we use a SVM to learn the corresponding stylistic model, capable of discriminating each author's messages. Given a suspect message from each author, we use each SVM to predict the degree of likelihood that each author is the true author. The message authorship is attributed to the author of the highest scoring SVM. We also consider a threshold on the minimum value of the SVM score, so as to introduce a *confidence* parameter (the minimum score of the SVM classifier considered valid) in the authorship attribution process. When none of the SVM scores achieves the minimum value, authorship is left undefined.

Our data set consists of *Twitter* messages from authors in Portugal, collected in 2010 (January 12 to October 1). We counted over 200,000 users and over 4 million messages during this period (excluding messages posted automatically, such as news feeds). From these, we selected the 120 most prolific *Twitter* authors in the set, responsible for at least 2,000 *distinct* and *original* messages (i.e. excluding *retweets*), to extract the sets of messages for our experiments. We divide the 120 authors into 40 groups of 3 users at random, and maintain these groups throughout our experiments. The group of 3 authors forms the basic testing unit of our experiment.

We perform two sets of experiments. In *Experimental Set 1*, the classification procedure uses all possible groups of features to describe the messages. We use data sets of sizes 75, 250, 1,250 and 2,000 messages/author. In *Experimental Set 2*, we run the training and classification procedure using *only one* group of features at a time. We use the largest data set from the previous experiment (2,000 messages/author) for this analysis. We measure *Precision* (P), *Recall* (R) and F $(2PR/(P+R))$ considering:

$$P = \frac{\# \text{ messages correctly attributed}}{\# \text{ messages attributed}} \qquad R = \frac{\# \text{ messages correctly attributed}}{\# \text{ messages in the set}}$$

We run the training and classification procedures in each set of experiments and use the *confidence* parameter to draw *Precision vs. Recall graphs*. As these experiments consider *three* different authors, the baseline is $F = 0.33$ ($P = 0.33$ at $R = 0.33$). All experiments were conducted using a 5-fold cross validation, and run for all 40 groups of 3 authors. For varying levels of Recall (increments of 0.01) we calculate the maximum, minimum and average Precision that was obtained for all 40 groups. All F values are calculated using the average Precision.

5   Results and Analysis

Figure 1 shows the Precision vs Recall graphs for Experimental Set 1. Data set increases (from 75 to 2,000 messages/author) returns improvements in the

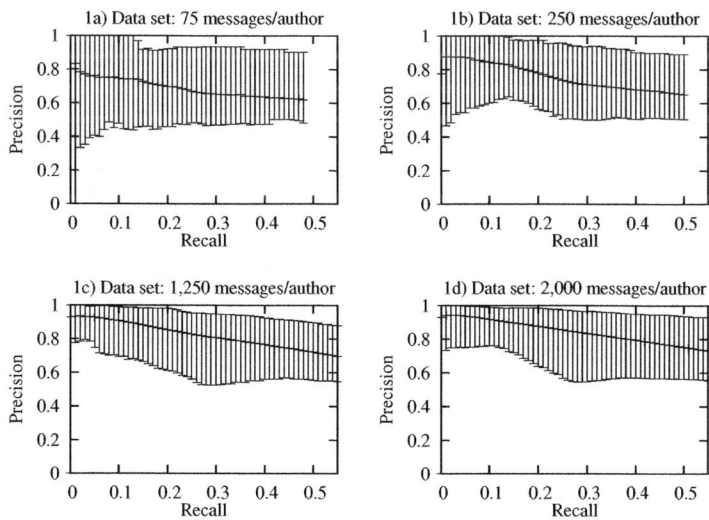

Fig. 1. Performance of each data set size. Each graph plots maximum, average and minimum Precision at varying levels of Recall (40 groups of 3 authors).

minimum, maximum and average Precision values. In addition, the robustness of the classifier also benefits from the added examples, as the most problem situations (corresponding to the minimum precision values) are handled correctly more frequently. The best F values are always obtained at the highest value of Recall, meaning that they too follow this improvement trend.

For the smaller data set (75 messages, Figure 1a), the minimum Precision curve is nearly constant, not showing a benefit from the decision threshold (at the cost of Recall). We speculate this is due to two reasons. First, given the large feature space (we use at least 5680 dimentions), and the relatively small number of non-negative feature component in each training example (most messages have between 64 and 70 features), a robust classification model can only be inferred using a larger training. Second, with such small sets it is highly probable that both the training and the test sets are atypical and distinct in terms of feature distribution. Still, the performance values obtained are far above the baseline, and an F value of 0.54 is reached. In the larger data sets (Figures 1c and 1d) we always obtain a Precision greater than 0.5. This means that even in the more difficult cases, the attribution process is correct more often than not. However, the contribution of the extra examples for the F values is lower when we go beyond 250 messages/author (where we get 0.59), even if they increase almost linearly up to 0.63 (for 2,000 messages/author).

Figure 2 presents the Performance vs Recall curves for authorship attribution with a classification procedure using *only one* group of features at a time. Quantitative Markers (Group 1) show an average performance, with minimum Precision

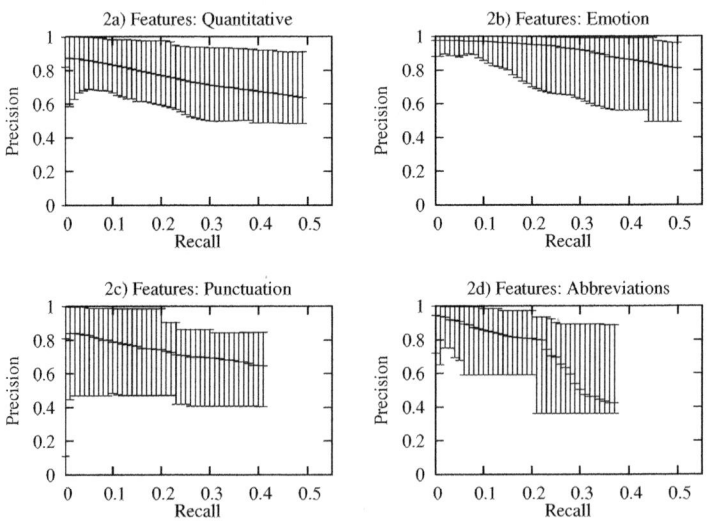

Fig. 2. Performance of each individual set of features. Each graph plots maximum, average and minimum Precision at varying levels of Recall (40 groups of 3 authors, 2,000 messages from each author).

and maximum Recall of 0.49, and maximum F value of 0.55 (Figure 2a). This shows that, albeit Twitter length constraints, there is room for stylistic choices like length of tokens, length of message posted, etc. Markers of *expression of emotion*, including *smileys*, *LOLs* and *interjections* (Group 2) achieve a relatively high performance, and clearly outperform all other feature groups (Figure 2b). It achieves an F value of 0.62 (where using *all* features togheter achieves 0.63). This is particularly interesting since these features are specific to user-generated contents, and to our knowledge their relevance and effectiveness in authorship attribution is now quantified for the first time. The difference between the average and minimum Precision values is an indicator that the low performance of this feature group is an infrequent event. The group of features including punctuation (Group 3) performs slightly worse than the previous groups, and scores only 0.50 on the F measure (Figure 2c). The difference between the best and worse case is significant, but the average Precision degrades as the Recall increases. Our evaluation demonstrates that our approach, although quite simplistic, is capable of detecting stylistic variation in the use of punctuation, and of successfully using this information for authorship attribution. This result is in line with those reported previously by [8] for punctuation-based features applied to automatic authorship attribution of sentences from newspapers. Group 4, containing features on the use of abbreviation, led to the worst results (maximum F value of 0.40). The shape of the curve rapidly approaches the baseline values, proving that this group is not robust (Figure 2d). Manual evaluation shows that these abbreviations are used rarely. However, as the low Recall/high Precision part of the curve suggests, they carry stylistic value, in spite of being used only in a relatively small number of cases. Finally, the performance when using *all* groups of features simultaneously (Figure 1f) is better than using any group of features *individually*, showing that all individual groups of features carry relevant stylistic information that can be combined, and suggesting that the investment in devising new groups of stylistic features may lead to additional global performance improvements — especially the recall.

6   Conclusions

Our experiment demonstrates that standard text classification techniques can be used in conjunction with a group of content-agnostic features to successfully attribute authorship of Twitter messages to three different authors. Automatic authorship attribution of such short text strings, using only *content-agnostic* stylistic features, had not been addressed before. Our classification approach requires a relatively small amount of training data (as little as 100 example messages) to achieve good performance in discriminating authorship.

Surprisingly, the group of emoticons outperforms all other feature groups tested, with a relatively high performance. The relevance and effectiveness of these features for automatic authorship attribution are now demonstrated for the first time. Quantitative and punctuation markers show average results, carrying some idiolectal information, despite the text length constraints. On balance, it can be argued that all features carry relevant information, since using all groups of features simultaneously allows inferring more robust authorship classifiers than using any group of features individually.

Acknowledgments. This work was partially supported by grant SFRH/BD/47890/2008 FCT-Portugal, co-financed by POPH/FSE.

References

1. Grant, T.: Txt 4n6: Idiolect free authorship analysis. In: Coulthard, M., Johnson, A. (eds.) Routledge Handbook of Forensic Linguistics. Routledge, New York (2010)
2. de Vel, O., Anderson, A., Corney, M., Mohay, G.: Mining e-mail content for author identification forensics, vol. 30, pp. 55–64. ACM, New York (2001)
3. Park, T., Li, J., Zhao, H., Chau, M.: Analyzing writing styles of bloggers with different opinions. In: Proceedings of the 19th Annual Workshop on Information Technologies and Systems (WITS 2009), Phoenix, Arizona, USA, December 14-15 (2009)
4. Goswami, S., Sarkar, S., Rustagi, M.: Stylometric analysis of bloggers' age and gender. In: International AAAI Conference on Weblogs and Social Media (2009)
5. Koppel, M., Schler, J., Argamon, S.: Computational methods in authorship attribution. Journal of the American Society for Information Science and Technology 60(1), 9–26 (2009)
6. Jindal, N., Liu, B.: Opinion spam and analysis. In: WSDM 2008: Proceedings of the International Conference on Web Search and Web Data Mining, pp. 219–230. ACM, New York (2008)
7. Pavelac, D., Justino, E., Olivera, L.S.: Author identification using stylometric features. Intelligencia Artificial,Revista Iberoamericana de IA 11(36), 59–66 (2007)
8. Sousa-Silva, R., Sarmento, L., Grant, T., Oliveira, E.C., Maia, B.: Comparing sentence-level features for authorship analysis in portuguese. In: PROPOR, pp. 51–54 (2010)
9. Hirst, G., Feiguina, O.: Bigrams of syntactic labels for authorship discrimination of short texts. Lit. Linguist. Computing 22(4), 405–417 (2007)
10. Abbasi, A., Chen, H.: Writeprints: A stylometric approach to identity-level identification and similarity detection in cyberspace. ACM Trans. Inf. Syst. 26(2), 1–29 (2008)
11. Layton, R., Watters, P., Dazeley, R.: Authorship attribution for twitter in 140 characters or less. In: Workshop Cybercrime and Trustworthy Computing, pp. 1–8 (2010)
12. Raghavan, S., Kovashka, A., Mooney, R.: Authorship attribution using probabilistic context-free grammars, pp. 38–42 (2010)
13. Eagleson, R.: Forensic analysis of personal written texts: a case study. In: Gibbons, J. (ed.) Forensic Linguistics: An Introduction to Language in the Justice System, pp. 362–373. Longman, Harlow (1994)
14. Joachims, T.: Text categorization with support vector machines: learning with many relevant features. In: Nédellec, C., Rouveirol, C. (eds.) ECML 1998. LNCS, vol. 1398, pp. 137–142. Springer, Heidelberg (1998)

Opinion Classification Techniques Applied to a Spanish Corpus

Eugenio Martínez-Cámara, M. Teresa Martín-Valdivia,
and L. Alfonso Ureña-López

University of Jaén
Department of Computer Science
SINAI - Sistemas Inteligentes de Acceso a la Información
{emcamara,maite,laurena}@ujaen.es

Abstract. Sentiment analysis is a new challenging task related to Text Mining and Natural Language Processing. Although there are some current works, most of them only focus on English texts. Web pages, information and opinions on the Internet are increasing every day, and English is not the only language used to write them. Other languages like Spanish are increasingly present so we have carried out some experiments over a Spanish film reviews corpus. In this paper we present several experiments using five classification algorithms (SVM, Nave Bayes, BBR, KNN, C4.5). The results obtained are very promising and encourage us to continue investigating in this line.

Keywords: Opinion mining, sentiment polarity classification, subjective corpora, machine learning algorithms.

1 Introduction

Opinion Mining (OM) is defined as the computational treatment of opinion, sentiment, and subjectivity in text. This new discipline combines Natural Language Processing (NLP) with data mining techniques and includes a large number of tasks [10]. This type of analysis is becoming increasingly important due mainly to the large amount of comments written on the Internet by millions of users all over the world through blogs, forums or social networks. The fundamental reason for the increasing presence of such sites is that the way people consume and relate has changed. The users, that will make a purchase, prefer to be as informed as possible before spending their money. So, the main search engines provide options that allow users to search opinions about the products they are querying. A clear example is the U.S. version of Bing[1].

On the other hand, although comments in the web are expressed at any language, especially after the explosion of the Web 2.0 and the social web, most of research in this field of NLP has been focus on English texts [10]. However, the number of subjective texts that are written in other languages such us Russian,

---

[1] http://www.bing.com

R. Muñoz et al. (Eds.): NLDB 2011, LNCS 6716, pp. 169–176, 2011.
© Springer-Verlag Berlin Heidelberg 2011

German, Spanish or Arabic is increasing every day [2]. Thus, OM should not focus on a unique language, but would also have to study other languages, and even research on sentiment analysis (SA) from a multilingual perspective.

Actually, the use of other languages on OM is one of the reasons of this work. In this paper we present the experiments performed over a corpus of Spanish film reviews. We study different classifiers to determine the polarity of comments written in Spanish. In addition, we also have conducted three different experiments for each of the classifiers studied using different types of information from film reviews to analyze how it affects the results.

This paper is organized as follows: Section 2 provides a brief description of related works with a special attention to those which use different language to English. Subsequently, we describe the corpus that we have used and the process of preparing the data to be used in the experiments. Then, we explain the algorithms that have been utilized in the experimentation. Section 5 describes the run of the selected algorithms, the results obtained and their interpretation. Finally, we propose the working lines to be followed in the future.

2   Opinion Mining in other Languages

Although, OM is a new discipline, there is a number of works related to this area, and more specifically to the classification of polarity. We can distinguish two main methodologies in dealing with the problem. On one hand, machine learning approach based on using a collection of data to train the classifiers. On the other hand, the approach based on semantic orientation does not need prior training, but it takes into account the orientation of words (positive or negative). Although, most of works carried out in this area use a set of data, mainly managing text written in English, there are also some researches studying the use of other languages.

An example of an experimentation in a different language than English that is presented in Spanish is [5]. It describes the creation of the Spanish corpus used in this paper. This work follows the semantic approach. In this paper we follows the machine learning approach to compare the results with Cruz et al. work [5].

In [6] a German corpus with Amazon products reviews is used to train a classifier to determine the polarity of the opinions. Denecke uses a machine translator to translate the comments from German into English and then applies SentiWordNet [14]. Zhang et al. [18] applies Chinese SA on two datasets. The experiments were run using rule-based and machine learning approaches (SVM, Nave Bayes, and Decision Tree).

On the other hand, some multilingual researches have been also accomplished. For example, Ahmad, Cheng and Almas [1] performed a local grammar approach for three languages: Arabic, Chinese and English using financial news, while Boldrini et al. [2] developed EmotiBlog which is a corpus that includes comments on several subjects in three languages: Spanish, English and Italian.

---

[2] http://www.internetworldstats.com/stats7.htm

3   Muchocine Corpus

For our experiments we have used a Spanish movie reviews corpus [5]. The collection contains 3,878 reviews of the website *muchocine*[3].

The corpus consists of documents not written by professionals writers, but rather by the web users. This may appear anecdotal, though it increases the difficulty of the task, because the texts may not be grammatically correct, or they can include spelling mistakes or informal expressions.

The opinions are rated on a scale from 1 to 5. One point means that the movie is very bad, and 5 means very good. Rated 3 films can be categorized as "neutral" which means the user consider the film is neither bad nor good. For this study, the "neutral" opinions have not been used, so the total number of documents on which the experiments have been performed is 2,625, with 1,274 negative reviews, and 1,351 positive reviews.

Each document has the name of the author who has written the review, the title of the movie, the score assigned, the higher mark that can be assigned to a film, the source of the data, a brief summary, and the reviews.

The corpus has been preprocessed as usually by removing stopwords and applying a root extraction algorithm (stemmer). Finally, it has been used the TF-IDF weighing scheme to form the vectors to be used in the training phase.

4   Classification Methods

Most of the researches on subjective classification of documents have been made over English texts, and the SVM algorithm is the most widely used. On Spanish there is not any machine learning research conducted on SA. In order to solve this deficiency, we have trained five classifiers (SVM, Naïve Bayes, BBR, KNN, C4.5) with the corpus *muchocine*. In the next subsections we briefly describe the learning algorithms that have been used.

4.1   SVM

Support Vector Machines algorithm (SVM) [17] is based on the structural risk minimization principle from the computational learning theory, and seeks a decision surface to separate the training data points into two classes and makes decisions based on the support vectors that are selected as the only effective elements in the training set. Due to its versatility, SVM has several configuration options. For this paper we have used the *libSVM*[4] implementation [4], type C-SVC, and with a lineal kernel. The SVM algorithm has been successful used in several OM researches [10].

---

[3] http://www.muchocine.net
[4] http://www.csie.ntu.edu.tw/ cjlin/libsvm/

4.2 Naïve Bayes

Naïve Bayes (NB) [8] is based on the Bayes theorem. Due to its complex calculation, the algorithm has to make two main assumptions: first, it considers the Bayes denominator invariant, and second, it assumes that the input variables are conditional independence. The algorithm estimates the value of the conditional likehoods. There are several estimation methods, which we have chosen the Kernel Density Estimation. Its behavior differs depending on the value of its parameter bandwidth, which has been set at 0.05.

The use of Naïve Bayes is justified by its good performance in SA problems [16].

4.3   BBR

Regression is a widely used technique that obtains good results in classification problems. BBR [7] is a Bayesian implementation of the logistic regression that avoids overfitting the training data. The algorithm is based on the calculation of the following conditional likelihood:

$$P(y|\beta, x_i) = \psi(\beta^T, x_i) = \psi(\sum_i \beta_j x_{i,j}) . \qquad (1)$$

One Bayesian approach is used to avoid the overfitting involves a prior distribution on β specifying that each β_j is likely to be near 0. We have used a Gaussian distribution because it is which returns the best results.

BBR has reached good results in other texts classification problems [7] [9].

4.4   KNN

K Nearest Neighbors (KNN) is a case-based learning method, which keeps all the training data for classification. KNN is very simple, for each new item to be classified KNN seeks in the training set the k closest items, and then it returns the major class in the "neighbors" set. There are several similarity measures which we have used the Euclidean distance because that measure fits good to our problem. KNN has been used in other OM works [16].

4.5   C4.5

C4.5 [12] is a tree decision algorithm. It is an improvement on IDE3 [11].

5   Experiment Results

As it has been already mentioned, each document with a film review has several fields, but for this experimentation, we have only used the film score, the summary, and the full review. As we have already indicated in Section 3, it has not been included in the experiments the films with a score of 3. Thus, the four remaining classes will be reduced to two ones, positive and negative. The reviews with a rating less than 3 was considered negative (-1), while the ones were rated with 4 or 5 are labeled as positive reviews (1).

We have decided to apply the classifiers to only the summaries (S), only to the body of the reviews (B) and to the summary and the body together (S_B). To evaluate the classifiers we apply a 10-cross-validation process. The evaluation has been carried out on the following measures [13]: Precision, Recall, F1, Accuracy and the Kappa value.

As we pointed in Subsection 4.1 the type SVM chosen is C-SVM with a linear kernel, and the value of its parameter C is 0. The results are shown in Table 1. The results obtained in the three corpora (S, B and S_B) show that SVM has a very good performance. On the other hand, the experiment "S_B" has reached the best result, due to the fact that it has more information than the others. However, the difference between "S_B" and "B" is quite limited. In contrast, the results obtained by the "S" experiment can not be considered low, although the information processed is less than the data managed in the other two experiments.

The results achieved with NB are shown in Table 2. The NB performance is similar to the SVM one. The best result has been obtained when the model is build with "S_B" corpus. The results are over 80%, but they are slightly worse than the ones obtained with SVM. The difference in precision between "B" and "S_B" increases with NB, becoming more important the information of the summary when it is analyzed the summary and the body together.

Table 3 presents the results reached by the algorithm BBR. The literature said that BBR is good for text categorization problems [7], and the results shown in Table 3 prove that it is worthy too for applying in OM problems. Meanwhile

Table 1. SVM results

Corpus	Prec.	Recall	Acc.	F1	Kappa
S	76.23%	76.07%	76.15%	76.14%	0.552
B	85.67%	85.49%	85.56%	85.57%	0.711
S_B	**86.84%**	**86.67%**	**86.74%**	**86.75%**	**0.736**

Table 2. Naïve Bayes results

Corpus	Prec.	Recall	Acc.	F1	Kappa
S	75.37%	74.63%	74.86%	74.99%	0.495
B	81.35%	81.32%	81.33%	81.33%	0.626
S_B	**83.16%**	**83.14%**	**83.16%**	**83.14%**	**0.663**

Table 3. BBR results

Corpus	Prec.	Recall	Acc.	F1	Kappa
S	73.51%	73.31%	73.41%	73.40%	0.467
B	81.42%	77.14%	76.61%	79.22%	0.537
S_B	**87.21%**	**87.01%**	**87.08%**	**87.10%**	**0.741**

the precision and the recall of "S_B" are very similar, but for "B" there is a significant difference that has not been given in the two previous algorithms. This could indicate that the precision for a class has been far better than for the other. However, the homogeneity between precision and recall returned when the summary is only considered. For this algorithm is also outstanding the contribution when the summary is studied with the body of the comment, because the difference between "B" and "S_B" is the 5.79 points. But, in this case, the summary information not only improves the average precision but also helps to homogenize the values obtained per class, and it can be seen this reflected in the values obtained for Accuracy and F1.

The results of KNN depends on the number of "neighbors" that are used to classify the new item (the parameter K). After applying different methods to calculate the optimal K value, the best results were reached with $K = 72$. The experiments are shown in Table 4. In this case, the results are not as good as the previous algorithms. The Recall, Accuracy and F1 score, even in the best case, is bellow 80%. The difference between Precision and Recall is closest the two points per each experimentation, so one class is classified better than the other one. In this case the difference between any evaluation measure of each of the three experiments is lower than in the previous algorithms. So, if it is considered that KNN classifies according to the similarity between items, it can be concluded that the summaries represent very well to the documents, and the importance of the body of the review is not proportional to its size.

Table 4. KNN results

Corpus	Prec.	Recall	Acc.	F1	Kappa
S	76%	75.53%	75.7%	75.76%	0.512
B	78.62%	76.92%	77.25%	77.76%	0.542
S_B	**80.03%**	**78.42%**	**78.74%**	**79.21%**	**0.572**

Table 5. C4.5 results

Corpus	Prec.	Recall	Acc.	F1	Kappa
S	65%	64.73%	64.87%	64.86%	0.295
B	66.90%	66.88%	66.86%	66.88%	0.337
S_B	**68.15%**	**68.20%**	**68.13%**	**68.17%**	**0.362**

Table 5 shows the results returned by C4.5, which are the lowest among all the algorithms but maintain the same trend. So the best performed is reached with "S_B" corpus. As happened with the other algorithms, the difference between "S" and "S_B" or between "S_B" and "B" is quite low, showing that the polarity of an opinion is not proportional to the amount of text.

In all the algorithms the best result is obtained with "S_B" corpus. Using "S_B" we can see that the regression algorithms perform better than the others. But, when the information to be analyzed is considerably reduced, "S", the

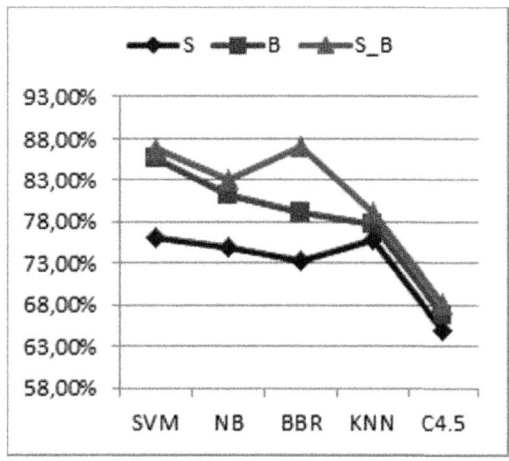

Fig. 1. F1 values of the three experiments

regression algorithms get worse results, so it can be due to the reduction of predictor variables in the regression model. Figure 1 shows the evolution of each algorithm.

It should be highlighted the fact that the short length of the summaries contains enough information of the review, because their results are above 73% in four of the five algorithms and it is also remarkable the improvement in the results when the summaries and the body are analyzed together.

6   Conclusion and Further Work

In this work we have carried out a first approach to SA in Spanish. We have applied supervised classifiers with results in the best case of 87.21% of accuracy.

Although the experimentation has focused on texts in Spanish, the objective is to deal the problem from a multilingual perspective to classify the polarity without taking into account the language in which it is written. So our next step is work in multilingual SA.

On the other hand, the semantic content in opinion texts and the knowledge of the domain is very important. So, our further work will be addressed to use external semantic resources like SentiWordNet [14],or General Inquirer[5] [15].

Acknowledgements. This work has been partially supported by a grant from the Fondo Europeo de Desarrollo Regional (FEDER), project TEXT-COOL 2.0 (TIN2009-13391-C04-02) from the Spanish Government, a grant from the Andalusian Government, project GeOasis (P08-TIC-41999) and Geocaching Urbano research project (RFC/IEG2010).

---

[5] http://www.wjh.harvard.edu/~inquierer

References

1. Ahmad, K., Cheng, D., Almas, Y.: Multi-lingual Sentiment Analysis of Financial News Streams. In: Proceedings of Science, GRID 2006 (2006)
2. Boldrini, E., Balahur, A., Martínez-Barco, P., Montoyo, A.: EmotiBlog: an annotation scheme for emotion detection and analysis in non-traditional textual genres. In: DMIN, pp. 491–497. CSREA Press
3. Carletta, J.: Assessing agreement on classification tasks: the kappa statistic. In: Computational Linguistics, vol. 22(2). MIT Press, Cambridge (1996)
4. Chang, C.C., Lin, C.J.: LIBSVM: a Library for Support Vector Machines (2001)
5. Cruz, F.L., Troyano, J.A., Enriquez, F., Ortega, J.: Clasificación de documentos basada en la opinión: experimentos con un corpus de críticas de cine en español. Procesamiento de Lenguaje Natural 41 (2008)
6. Denecke, K.: Using SentiWordNet for multilingual sentiment analysis. In: ICDE Workshops, pp. 507–512. IEEE Computer Society, Los Alamitos (2008)
7. Genkin, A., Lewis, D., Madigan, D.: Large-Scale Bayesian Logistic Regression for Text Categorization (2004)
8. Mitchell, T.: Machine Learning. McGraw-Hill, New York (1997)
9. Ortiz-Martos, A., Martín-Valdivia, M.T., Ureña-Lopez, L.A., Cumbreras-García, M.A.: Detección automática de Spam utilizando Regresión Logística Bayesiana. Procesamiento del Lenguaje Natural 35, 127–133 (2005)
10. Pang, B., Lee, L.: Opinion mining and sentiment analysis. Foundation and Trends in Information Retrieval 2(1-2), 1–135 (2008)
11. Quinlan, J.R.: Induction of Decision Trees. Machine Learning 1, 81–106 (1986)
12. Quinlan, J.R.: Programs for Machine Learning. Morgan Kaurfman, San Francisco (1993)
13. Sebastiani, F.: Machine Learning in automated text categorization. ACM Computing Surveys (CSUR) 34(1), 1–47 (2002)
14. Esuli, A., Sebastiani, F.: SentiWordNet: A publicly Available Lexical Resource for Opinion Mining. In: Proceedings of Language Resources and Evaluation, LREC (2006)
15. Stone, P.J.: The General Inquierer: A Computer Approach to Content Analysis. The MIT Press, Cambridge (1996)
16. Tan, S., Zhang, J.: An empirical study of sentiment analysis for Chinese documents. Expert System with Applications 34, 2622–2629 (2008)
17. Vapnik, V.: The Nature of Statistical Learning Theory. Springer-Verlag, New York (1995)
18. Zhang, C., Zeng, D., Li, J., Wang, F.-Y., Zuo, W.: Sentiment analysis of Chinese documents: From sentence to document level. JASIST 60, 2474–2487 (2008)

Prosody Analysis of Thai Emotion Utterances

Sukanya Yimngam[1], Wichian Premchaisawadi[2], and Worapoj Kreesuradej[1]

[1] Technology King Mongkut's Institute of Technology Ladkrabang,
Bangkok 10520, Thailand
[2] Graduate School of Information Technology, Siam University
235 Petchkasem Road, Phasi-charoen, Bangkok 10163, Thailand
{Sukanya.Yimngam,yimngam}@yahoo.com, {Wichian.Premchaisawadi,
wichian}@siam.edu,
{Worapoj.Kreesuradej,worapoj}@it.kmitl.ac.th

Abstract. Emotion speech synthesis is the most important process to generate the naturalness of utterances in text-to-speech system. The interjection utterances in Thai language are employed in express a number of emotions. This paper presents a study of the prosody parameters of the interjection utterances clipped from Thai utterances in the movies. The Thai emotional utterances from various movies have been analyzed and classified into 8 emotional types consisting of neutral, anger, happiness, sadness, fear, pleasant, unpleasant and surprise. The classification of prosodic features is based on fundamental frequency (F0), intensity and duration. This paper compares the prosodic features in the Thai language and other languages including English, Italian, French, Spanish and Arabic. The comparison results show that there are significant differences of prosodic features for each emotion in each language. Therefore, the quality of a text-to-speech system is based on the prosodic analysis of each language.

Keywords: emotion, prosody parameters, text-to-speech, speech synthesis.

1 Introduction

Previous research proposed by Yimngam et al. [8] has studied the classification of Thai emotional words by using interjection words, which were defined in the Thai Royal Institute Dictionary (1999). The results obtained in their research study show that the annotators' opinion of the emotional expression of the interjection words is significantly different from that defined in the dictionary. These results indicate that several factors such as sex and age have an effect on the classification of emotional expression. However, interjection words in text, is not really used in many situations. In this paper, we used interjection words in utterances defined in the Royal Institute Dictionary rather than using interjection words in text. We extracted several prosodic parameters of Thai Interjection utterances into 8 types of emotions, which consisted of neutral, anger, happiness, sadness, fear, pleasant, unpleasant and surprise. We emphasize the importance of prosodic features such as fundamental frequency (F0), intensity and duration. The comparison of these factors among the Thai language and other languages such as English, Italian, French, Spanish and Arabic was also conducted.

R. Muñoz et al. (Eds.): NLDB 2011, LNCS 6716, pp. 177–184, 2011.
© Springer-Verlag Berlin Heidelberg 2011

2 Emotional Speech Classification and Prosodic Parameters

In text to speech, many processes such as text analysis, letter to utterances, prosody generation and speech synthesis are involved in the construction of a text to speech system [14], [15]. Emotion speech synthesis is the most important process to generate the naturalness of utterances. There are two different approaches in the emotional speech classification. First, a categorical approach, the emotions are treated as distinct categories such as anger, joy, fear, etc. [12] Second, a dimensional approach [13], [16], it is a simplified representation of the essential properties of emotions such as evaluation (positive/negative), activation (active/passive) and power (dominant/ submissive). This paper uses the categorical approach to identify Thai emotion utterances from different movies. Each utterance was classified into 8 basic emotion types namely: neutral, anger, surprise, happiness, sadness, fear, pleasant and unpleasant. In the previous studies on the expression of the emotions, prosodic parameters are the specific features of speech that convey emotional information. In the initial step, some prosodic parameters such as fundamental frequency (F0), intensity and duration of the utterances were measured. These global prosodic parameters are often treated as universal cues for emotions. The model of each emotion in speech relies on a number of these parameters.

3 Related Works

3.1 In Thai Language

In Thai text-to-speech, various types of research works [7] such as speech analysis and tools, text analysis and tools, pronunciation modeling tools, Thai speech synthesis, linguistic/prosodic processing, waveform synthesis and language resources have been developed and implemented. Some of them are described in more detail below.

Tumtavitikul [1] presented the acoustics of Thai intonation in four types of emotional speech: anger, surprise, happiness and sadness. The results of her study indicate that the speaking rate, average pitch, and pitch range of intonation as well as the amplitude of the utterances varies according to the type of emotion involved. Anger and surprise are in agreement with Luksaneeyanawin [2]. Anger and sadness are in agreement with Cahn [3].

Chuenwattanapranithi [4] reported the findings of a somewhat unconventional investigation of emotional speech. Instead of looking for direct acoustic correlates of multiple emotions, they tested a specific theory, the size code hypothesis of emotional speech, about two emotions — anger and happiness. According to the hypothesis, anger and happiness are conveyed in speech by exaggerating or understating the body size of the speaker. In the six experiments from two studies, they synthesized vowels with a 3D articulatory synthesizer with parameter manipulations derived from the size code hypothesis, and asked Thai listeners to judge the body size and emotion of the speaker. Vowels synthesized with a longer vocal tract and lower F0 were mostly heard from a larger person if the length and F0 differences were stationary, but from an angry person if the vocal tract was dynamically lengthened and F0 was dynamically lowered. The opposite was true for the perception of small body size and happiness. These results provide preliminary support for the size code hypothesis. They also address the potential benefits of theory-driven investigations in emotion research.

Yimngam et al. [8] studied a comparison between the words defined in the dictionary and a survey of annotators' opinion. The results show that the annotators' opinion of the emotional expression of the interjection words is significantly different from that defined in dictionary. They were found that many factors such as sex and age had an effect on emotional expression. Therefore, these factors should be utilized to identify emotional expressions in text in order to improve the efficiency of Thai text-to-speech development in the future.

3.2 Other Languages

Burkhardt et al. [6] studied the effects of prosody changes on emotional speech in French, German, Greek and Turkish. Semantically identical sentences expressing emotional relevant content were translated into the target languages and were manipulated systematically with respect to pitch range, duration model, and jitter simulation. Nonetheless, there were differences between the different countries. These findings also indicate that results based on data-analysis from different cultures cannot be applied without reservations.

Dakkak et al. [9] developed an automated tool for emotional Arabic synthesis. They introduced a work done to incorporate emotions: anger, joy, sadness, fear and surprise, in an educational Arabic text-to-speech system. They presented a methodology to extract rules from prosodic parameters: pitch, duration and intensity for emotion generation.

Schröder [10] proposed an overview of existing approaches and techniques in the synthesis of emotional speech. First, formant synthesis created the acoustic speech data entirely through rules on the acoustic correlates of the various speech utterances. Second, diphone synthesis systems controlled only F0 and duration (and possibly intensity). Third, unit selection synthesis used phoneme string, duration and F0 to preserve the features of the recorded speech. This paper was presented in a short overview of the prosody rule given that it has been successfully employed to express a number of emotions.

Wee Ser et al. [11] proposed a novel hybrid scheme that combined the Probabilistic Neural Network (PNN) and the Gaussian Mixture Model (GMM), for identifying emotions from speech signals. Their experimental results show that the proposed scheme is able to achieve a much higher recognition accuracy compared to that obtained by using PNN or UBM-GMM alone.

4 Experiment and Results

There are 4 sequential steps in the experiment with Thai interjection utterances: (1) data recording from movies, (2) pre-processing of each utterance, (3) prosody parameters extraction, and (4) prosody parameters analysis. They are described as follows.

Data Recording. The speech emotion database used in this research is extracted from Thai utterances in movie recordings. Especially, interjection utterances that are in Thai utterances movies by Thai actors and actresses. There are three types of movies that consist of 3 cartoons, 45 Thai movies and 4 international movies. All utterances comprise expressions spoken by male and female actors. The original utterances were manually selected only 91 interjection utterances using Adobe Audition 1.5 software with 44100 sample rates and 16 bit stereo resolution.

Preprocessing. All 352 utterances were adjusted to the amplitude of 3 dB peak level and noise was reduced.

Prosody Parameters Extraction. After noise reduction in Adobe Audition 1.5 software, all utterances were measured using Praat software [5]. Several prosody parameters such as fundamental frequency (F0), intensity and duration were calculated for each utterance. All utterances were classified into 3 types of utterances such as one syllable, two syllables and all syllables.

Prosody Parameters Analysis. We analyze 3 prosody parameters such as fundamental frequency, intensity and duration from Thai emotional utterances. These data can further be subjected to statistical analysis.

Fundamental frequency (F0). Fundamental frequency is classified into 3 parameters consisting of average F0, range F0, and standard deviation F0. The results using these approaches are described below:

Average F0 analysis: 2 syllables of utterances are significantly different from one and all syllables. The comparison of each emotion shows that anger is the highest and neutral is the lowest average F0 based on one syllable. Unpleasant is the highest and pleasant is the lowest based on two syllables. The result of all syllables is the same as one syllable. For all emotions, anger, happiness, fear, surprise, and pleasant of one syllable are higher than other emotions. Sadness and unpleasant of all syllables are higher than other emotions. The precedence order of average F0 for all syllables of utterances is anger > fear > happiness > unpleasant > sadness > surprise > pleasant > neutral.

F0 range analysis: The comparison of each emotion shows that fear is the largest and neutral is the smallest based on one syllable. In the two syllables, happiness is the largest and surprise is the smallest. In all syllables, happiness is the largest and neutral is the smallest. The precedence order of F0 range for all syllables of utterances is happiness > pleasant > fear > sadness > anger > unpleasant > surprise > neutral.

F0 S.D. analysis: The comparison of each emotion shows that fear is the highest and neutral is the lowest, which is similar to F0 range in one syllable. In the two syllables, unpleasant is the highest and surprise is the lowest. In all syllables, fear is the largest and neutral is the smallest. The precedence order of F0 S.D. for all syllables of utterances is fear > happiness > pleasant > anger/sadness > unpleasant > surprise > neutral.

Intensity: This paper identified an average intensity. The results show that anger is the highest and happiness is the lowest. The precedence order of average of intensity for all types of utterances is anger > fear > sadness > pleasant > surprise > unpleasant > neutral > happiness.

Duration: This paper identified syllable duration. The results show that neutral is slowest and anger is fastest. The precedence order of average syllable duration for all types of utterances is neutral > pleasant > happiness > sadness > fear > unpleasant > surprise > anger.

5 Comparison with other Languages

This section presents a comparison of each parameter with previous other languages research.

5.1 F0 Analysis

There are several research studies investigated in F0 as an average of F0 and F0 range. This section presents the comparison of the F0 parameters in this research with other languages such as English, Italian, French, Spanish and Arabic. For the comparison between languages, this research classifies average F0 into 6 levels from the lowest (level 0) to the highest (level 6), F0 ranges into 5 levels from the smallest (level 0) to the largest (level 5) and F0 S.D. into 5 levels from the lowest (level 0) to the highest (level 5) respectively. These research studies are described below.

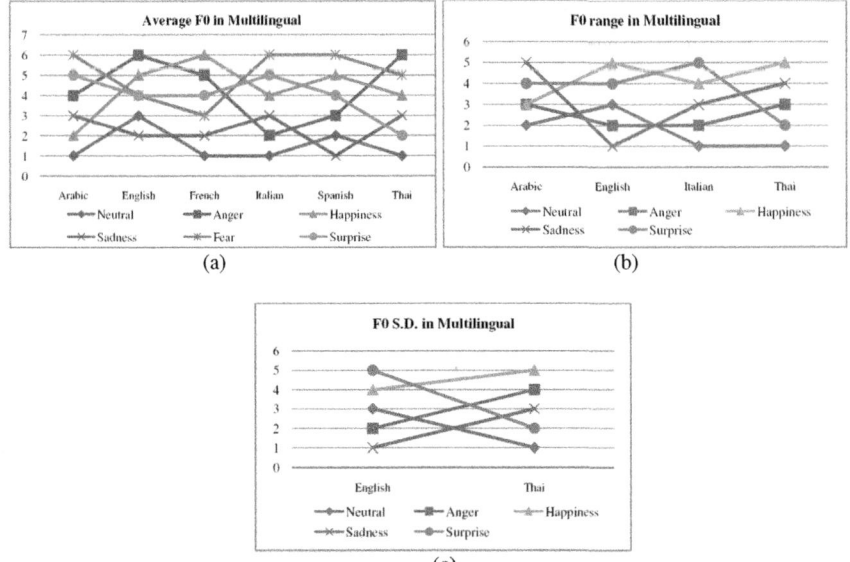

Fig. 1. (a) Average F0 in multilingual, (b) F0 range in multilingual, (c) F0 S.D in multilingual

Drioli et al. [17] investigated the average F0 and F0 range in Italian language. The average F0 for fear is the highest and is the same for Spanish and Arabic. The F0 range for surprise is the largest and is different from other languages. The comparison results of Drioli et al. with other languages are shown in figs 1 (a) and (b).

Boula de Mareüil et al. [18] investigated an average F0 for English French and Spanish languages. Anger is the highest in English, happiness is the highest in French and fear is the highest in Spanish. The comparison results of Boula de Mareüil et al. with other languages are shown in fig 1 (a).

Dakkak et al. [9] developed an automated tool for emotional Arabic synthesis. The average F0 for fear is the highest. Sadness is the largest F0 range of their paper. The comparison results of Dakkak et al. with other languages are shown in figs 1 (a) and (b).

Cecilia Ovesdotter Alm et al. [19] investigated the F0 range and F0 S.D. The F0 range of happiness and the F0 S.D. of surprise is the highest in English. The comparison results with other languages are shown in figs 1 (b) and (c).

Our experiment in Thai shows that anger is the highest average F0 and happiness is the largest F0 range and F0 S.D. The comparison results with other languages are shown in figs 1 (a), (b) and (c).

5.2 Intensity Analysis

In intensity analysis, this research classifies average intensity into 6 levels from the lowest (level 0) to the highest (level 6) for the comparison between languages. The results are described below.

Drioli et al. [17] presented an average for intensity of the Italian language. The comparison results of Italian language and Thai language are shown in fig 2 (a). Anger is the highest in Italian. The results are shown in fig 2 (a).

Boula de Mareüil et al. [18] investigated an average intensity in English. The result shows that surprise is the highest. The results are shown in fig 2 (a).

In our experiment in Thai, anger is the highest and is the same as Drioli et al. [17] and happiness is the lowest. The results are shown in fig 2 (a).

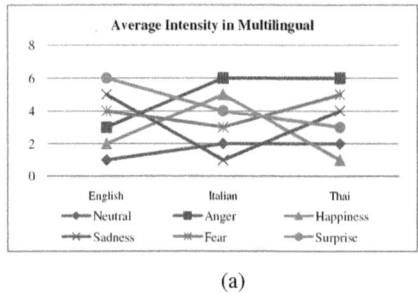 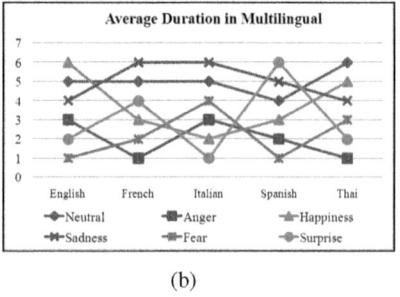

(a)   (b)

Fig. 2. (a) Average intensity in multilingual, (b) Average duration in multilingual

5.3 Duration Analysis

In duration analysis, this research classifies duration into 6 levels from the slowest (level 0) to the fastest (level 6) for the comparison between languages. The results are described below.

Drioli et al. [17] presented average values for duration of the Italian language. Sadness is the slowest and surprise is the fastest. The comparison results are shown in fig 2 (b).

Boula de Mareüil et al. [18] investigated average of duration in English, French and Spanish. Happiness, sadness and surprise are slowest in English, French and Spanish respectively. The comparison results are shown in fig 2 (b).

In our experiment in Thai, neutral is the slowest. Anger is the fastest. The comparison results are shown in fig 2 (b).

6 Conclusion and Discussion

This research investigated and classified prosodic parameters such as average F0, F0 range, standard deviation of F0, intensity and duration into 8 emotions. The results are described in several features analysis. Firstly, in F0 analysis, fear is the highest average of F0 in Arabic, Italian and Spanish. Happiness is the largest F0 range in English and Thai. Surprise and happiness F0 S.D. are the highest in English and Thai respectively. Secondly, in intensity analysis, anger is the highest in Italian and Thai. Surprise is highest in English. Lastly, in duration analysis, sadness is slowest in French and Italian and other languages are different in happiness, surprise, and neutral. The comparison results show that there are significant differences of prosodic features for each emotion in each language. Therefore, prosodic features of one language cannot be used with other languages in speech recognition and speech synthesis processing. Especially, the quality of text-to-speech synthesis relies heavily on the prosodic analysis of each language.

References

1. Tumtavitikul, A., Thitikannara, K.: Thai Intonation of Thai emotional speech. In: Proceedings of the 11th Australian International Conference on Speech Science & Technology, New Zealand, December 6-8 (2006)
2. Luksaneeyanawin, S., Intonation in Thai. Unpublished Doctoral Dissertation, University of Edinburgh (1983)
3. Cahn, J.: From Sad to Glad: Emotional Computer Voices. In: Proceedings of Speech Tech. 1988, Voice Input/Output Applications Conference and Exhibition, New York City, pp. 35–37 (1988)
4. Chuenwattanapranithi, S., Xu, Y., Thipakorn, B., Maneewongvatana, S.: Encoding emotions in speech with the size code. A perceptual investigation. Phonetica 65, 210–230 (2008)
5. Boersma, P.: Praat, a system for doing phonetics by computer. Glot. International 5(9-10), 341–345 (2001)

6. Burkhardt, F., Audibert, N., Malatesta, L., Türk, O., Arslan, L., Auberge, V.: Emotional Prosody – Does Culture Make A Difference? Speech Prosody, Dresden, Germany (2006)
7. Wutiwiwatchai, C., Furui, S.: Thai speech processing technology: a review. Speech Communication 49(1), 8–27 (2007)
8. Yimngam, S., Premchaisawadi, W., Kreesuradej, W.: Thai Emotion Words Analysis. In: The Eighth International Symposium on Natural Language Processing, SNLP (2009)
9. Dakkak, O., Ghneim, N., Abou Zliekha, M., Moubayed, S.: Emotion Inclusion in an Arabic Text-to-Speech. In:13th European Signal Processing Conference (2005)
10. Schröder, M.: Emotional Speech Synthesis - A Review. In: Proc. Eurospeech 2001, Aalborg, vol. 1, pp. 561–564 (2001)
11. Ser, W., Cen, L., Yu, Z.L.: A Hybrid PNN-GMM classification scheme for speech emotion recognition. In: ICPR 2008, pp. 1–4 (2008)
12. Ekman, P.: Basic emotions. In: Dalgleish, T., Power, M.J. (eds.) Handbook of Cognition & Emotion, pp. 301–320. John Wiley, New York (1999)
13. Schlosberg, H.: A scale for the judgement of facial expressions. Journal of Experimental Psychology 29, 497–510 (1941)
14. Mittrapiyanurak. P., Hansakunbuntheung, C., Tesprasit, V., Sornlertlamvanich, V.: Issues in Thai Text-to-Speech Synthesis: The NECTEC Approach. NECTEC, 483–495 (June 2000)
15. Tesprasit. V., Charoenpornsawat. P., Sornlertlamvanich. V.: A Context-Sensitive Homograph Disambiguation in Thai Text-to-Speech Synthesis. In: Proceedings of HLT-NAACL 2003 Short Papers, Edmonton, May-June 2003, pp. 103–103 (2003)
16. Cahn, J.E.: Generating expression in synthesized speech. Technical Report, MIT, Media Technology Laboratory, MA, USA (1990)
17. Drioli, C., Tisato, G., Cosi, P., Tesser, F.: Emotions and Voice Quality: Experiments with Sinusoidal Modeling. In: Proceedings of Voqual 2003, Voice Quality: Functions, Analysis and Synthesis, ISCA (2003)
18. Boula de Mareüil, P., Célérier, P., Toen, J.: Generation of Emotions by a Morphing Technique in English, French and Spanish. In: Proc. Speech Prosody, pp. 187–190 (2002)
19. Alm, C.O., Sproat, R.: Perceptions of Emotions in Expressive Storytelling. InterSpeech, 533–536 (2005)

Repurposing Social Tagging Data for Extraction of Domain-Level Concepts

Sandeep Purao[1], Veda C. Storey[2], Vijayan Sugumaran[3], Jordi Conesa[4],
Julià Minguillón[4], and Joan Casas[4]

[1] College of Information Sciences & Technology, The Pennsylvania State University,
University Park State College, PA 16802
spurao@ist.psu.edu
[2] Department of Computer Information Systems, J. Mack Robinson College of Business
Georgia State University, Box 4015 Atlanta, GA 30302
vstorey@cis.gsu.edu
[3] School of Business Administration, Oakland University
Rochester, MI 48309
sugumara@oakland.edu
[4] Estudis d'Informatica i Multimedia, Universitat Oberta de Catalunya,
Rambla del Poblenou, 156, E-08018, Barcelona, Spain
{jconesac,jminguillona,jcasasrom}@uoc.edu

Abstract. The World Wide Web, the world's largest resource for information, has evolved from organizing information using controlled, top-down taxonomies to a bottom up approach that emphasizes assigning meaning to data via mechanisms such as the Social Web (Web 2.0). Tagging adds meta-data, (weak semantics) to the content available on the web. This research investigates the potential for repurposing this layer of meta-data. We propose a multi-phase approach that exploits user-defined tags to identify and extract domain-level concepts. We operationalize this approach and assess its feasibility by application to a publicly available tag repository. The paper describes insights gained from implementing and applying the heuristics contained in the approach, as well as challenges and implications of repurposing tags for extraction of domain-level concepts.

1 Introduction

The World Wide Web, the world's most valuable information resource, has continued to evolve over time. It is no longer just a place to store information. Increasingly, it presents a platform for web citizens to communicate, collaborate and share content. This combination of information storing and collaboration is changing how we work, as well as how we carry out specific tasks such as information seeking [1]. A significant contributor to this change is the contribution of tags, or weak meta-data, by users. The result is a layer of social tagging data that contains significant potential. This research explores how to take advantage of this potential by repurposing social tagging data for extraction of domain-level concepts.

The objective of this research is to develop, implement and evaluate an approach to extract concepts that are implicitly represented by tags contributed by web citizens. The weak semantics in this meta-data layer, must be cleaned, structured, and

R. Muñoz et al. (Eds.): NLDB 2011, LNCS 6716, pp. 185–192, 2011.
© Springer-Verlag Berlin Heidelberg 2011

aggregated before extracting and identifying domain-level concepts. This paper develops a set of heuristics and procedures, applies them to tags extracted from a publicly available tagging site, and analyzes the results. The contribution of this research is to demonstrate the feasibility of extracting domain-level concepts from tags. It also shows that it is possible to repurpose content generated by the social web to contribute semantics to information contained in the World Wide Web.

2 Related Work

Tagging is the act of attaching a label that captures an interpretation of the underlying content. It can be a label that evokes what the content is about (e.g. green products), a label that captures an impression or action from the user (e.g. important, to read), or a combination of the two (e.g. important green products). Most tags fall into the first category, and evoke the underlying content [2]. As a result, they provide a short-hand for identifying key aspects of, or elements contained in, the domain. Each tag captures weak semantics about the underlying resources and contains the potential (with aggregation), to identify and extract important concepts in a domain [3].

The collection of tags may be viewed as 'wisdom of the crowd.' It reflects the independent and diverse opinions of a group of individuals [4]. The tags in this folksonomy [5] are often related to each other because they may be contributed by a user to tag related resources, or because they are contributed by different users to tag the same resource, or because different users may tag the same resource many times [6, 7]. The patterns that emerge from these tagging practices provide, not only a way to organize information for the users [8], but also a layer of weak semantics [9] that can be aggregated and linked to identify and extract domain-level concepts.

3 Identifying Domain-Level Constructs

The approach consists of multiple phases, outlined in Figure 1. This paper focuses on the first two phases: data cleansing and identification of concepts and connectedness. We describe the operationalization of heuristics and the results from investigating the potential feasibility of the approach.

The present work extends our prior work [10] by adding data cleansing and improving the identification of concepts connectedness. Further information about the heuristics that underlie these phases is available in [10].

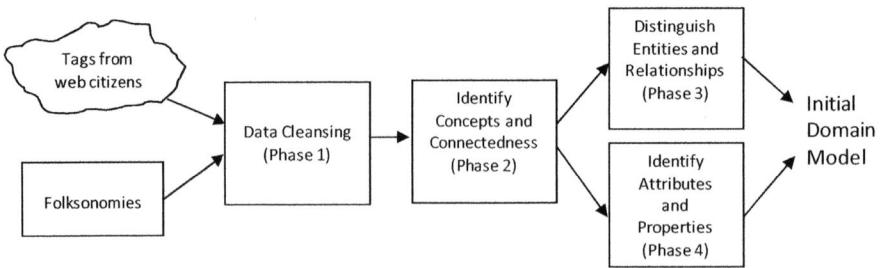

Fig. 1. A Multi-Phase Approach to Identify Domain-level Constructs

3.1   Phase 1: Data Cleansing

Before invoking any heuristics, the approach requires that the data obtained from a tagging web site be cleansed. This first phase, data cleansing, involves detecting and correcting corrupt or inaccurate records [11, 12] obtained from the social tagging web site. Often used within the context of data mining, the term 'data cleansing' refers to identifying incomplete, incorrect, inaccurate or irrelevant parts of the data in order to correct these errors by replacing, modifying or deleting these pieces of data.

The data cleansing phase is important because most tagging websites do not constrain nor validate the tags contributed by users. The result is a considerable amount of incorrect, inaccurate and sometimes, duplicated tags. For example, the use of punctuation in tags is common and often employed to compensate for the inability of delicious.com to deal with multi-word keywords. Tags such as *coolwebsites*, *cool_web*, *cool-site*, *coolsite* are, therefore, used. In addition, several tags represent variations of the same word (e.g., *bank* and *banking*) or different tags for the same concept, but in different languages (e.g. the concept *environment* represented by the tags *meio_ambiente, meioambiente, medio_ambiente, medi_ambient*).

The data cleansing phase, therefore, uses multiple techniques. Some of the operations performed in this phase for cleansing include:

* combining singular and plural forms of the tags (e.g. *auctions* and *auction*).
* converting all tags to lowercase (e.g. *static* and *Static*).
* replacing special characters, such as [, ; , .] (e.g. *web2.0* and *web2_0*).
* converting different conjugations of verbs to their infinitive (e.g. *awaiting* or *awaits* to *await*).
* expanding abbreviations (e.g. *sw* to *SouthWest*).
* dealing with synonyms (*achieved* to *attain*).
* deleting bookmarks that do not contribute any tags.
* deleting tags from users who only tagged few resources: e.g. discarding tags from users who tagged less than x resources, with x as a configurable variable.
* applying transformations manually created for each domain, which may include common mis-spellings and abbreviations (e.g. expand *diy* for "*do it yourself*").

Examples of these transformations include e-commerce -> Ecommerce (removing the hyphen), Auktionen -> auction (mapping words in different languages), Banking -> bank (removing the gerund), Financial -> finance (removing 'ial,' an adjective-forming suffix), supervise -> oversee (mapping words with similar meaning), fig->figure (by expanding abbreviations), ocw->open course ware (by expanding manually detected abbreviations) among others. Although the actual process is automated and can succeed in reducing the number of tags, much work from natural language processing can be applied, such as the use of *soundex* or similar algorithms to detect tags wrongly written (e.g. *auction* vs *aucktion*) or using bigrams to identify automatically the language of tags. Figure 2 outlines the outcomes that originated with a query on corporate sustainability.

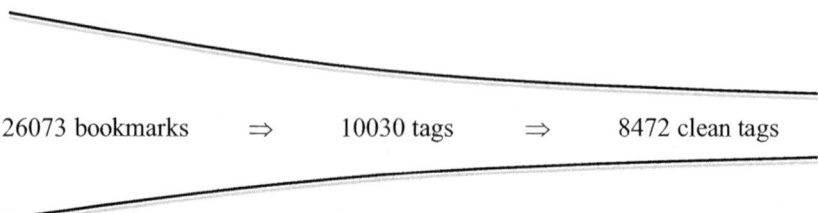

26073 bookmarks   ⇒   10030 tags   ⇒   8472 clean tags

Fig. 2. Results based on retrieval and cleansing of tags for Corporate Sustainability

The figure shows that the query retrieved 26,073 bookmarks. For each bookmark we collected the bookmarked resource, the user who made the bookmark, and all the tags the user used to tag the bookmark. This produced 10,030 tags. The data cleansing trimmed these to 8,472 tags, reducing the number by over 15%. This is relevant because the cleansing allows higher quality results.

3.2   Phase 2: Identifying Concepts and Connectedness

This phase detects the most relevant concepts of the domain and their connectedness. It is supported by five heuristics to assess the probability that a Tag is a *Candidate* domain-level construct and to identify pairwise *Candidate* connections. The heuristics progressively increase the probability that a tag represents a domain-level construct.

Heuristic 1: Number of Users using a tag. This heuristic leverages the number of users who use a tag. It follows the rationale that the larger the number of users who use a tag, the greater the agreement among them that the tag represents a meta-level concept.

Heuristic 2: Number of Resources tagged with a Tag. This heuristic suggests that the number of resources tagged with a given tag indicates the importance of that tag for the domain. It increases the probability that the candidate tag is a domain-level construct.

Heuristic 3: Frequency of Tag Use. This heuristic examines the frequency of tags and uses the rationale that the greater the frequency, the more important the tag. This heuristic examines only resources generated from the previous heuristic, effectively bounding the search, and results in a scoping decision. The heuristics, in effect, move from users to tags to resources, in each step, increasing the confidence in the assessment that a Tag is a domain-level construct.

Heuristic 4: Centrality of Tag. This heuristic leverages the centrality of tags interpreting their size and position in the Tag Cloud. The centrality is an indicator of the importance of the tag to further augment the probability that a candidate tag is a domain-level construct. This paper does not operationalize this heuristic.

Heuristic 5: Connectedness of Tags. The connectedness of the tags is assessed by measuring co-occurrence frequencies. Co-occurrence may indicate that tags are synonyms, or they represent related constructs that should eventually be bridged with a relationship, or they represent related constructs (e.g. one is an entity, the other is an attribute). This heuristic only considers candidate tags; that is, tags identified by previous heuristics as candidates for being domain-level constructs. The connectedness of tags has two facets:

- 5A: *Occurrence*: two tags are connected when they are used to tag the same resource. For example, the tags *Travel* and *Flights*, identified after executing the query "Air Travel", co-occur 31 times in the first 100 results returned.

- 5B: *Tendency*: two tags are connected when a significant amount of users tend to use the two tags to tag the same resource. In order to calculate this heuristic we used Principal Component Analysis (PCA) [13]. The PCA is conducted with the following parameters: using maximum likelihood as the method for extraction of the components, a Varimax rotation and only fields with weight >= 0.3 are taken into consideration.

The PCA analysis identifies a set of factors that conceptually can be seen as clusters whose tags represent a point of view from which the domain of discourse can be seen. The tags of each cluster may be semantically connected because they deal with the same semantic concept or sub-domain. Therefore, the tags of a given cluster may be used to infer tags potentially relevant for the user. (If a user decides that one of the tags of a cluster is relevant then all its tags are potentially relevant). For example, after applying the PCA analysis to the results obtained from the query "online auction" some of the obtained factors (or clusters) are: 1) craft, handmade, art, design, gifts and diy (related to crafting stuff), 2) daily, technology, gadgets, woot, electronics (related to technology stuff) and 3) paypal, money, finance and bank (related to the payment).

The result of these heuristics is a set of tags that are likely candidates for domain-level constructs, and potential connections among these constructs.

4   An Application

The approach has been applied to tags obtained from del.icio.us. A user can input a term or a phrase for a domain to retrieve documents and tags. The prototype carries out the procedures and applies the heuristics in a semi-automated manner. The result, concepts and connectedness among concepts, is visually represented as local graphs, as shown in Figure 4. At this time, a significant component of the experiments is the PCA analysis that emphasizes the connectedness of tags. For the experiment reported, the PCA has been conducted with the following parameters: using maximum likelihood as the method for extraction of the components; a Varimax rotation and only fields with weight >= 0.3 are considered. Because the PCA is conducted at the last step, the number of tags is relatively low. For example, in the cases described in this paper, the PCA analysis involved approximately 100 tags each.

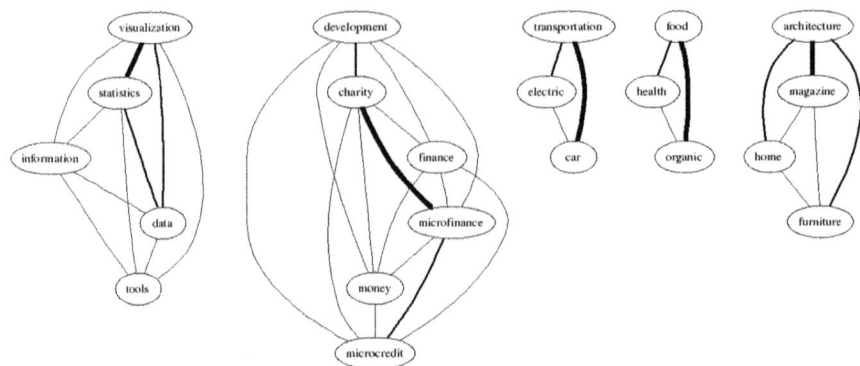

Fig. 3. Local graphs for the term Corporate Sustainability (Partial) obtained from PCA analysis

From the PCA analysis factors (aka components) are obtained. Each is composed of a set of tags. Although PCA is not a clustering technique, the set of tags in each component can be considered as a cluster, whose tags represent the annotation tendencies. For example, in the case of corporate sustainability, we can interpret one of the factors as an indication that a significant number of people tend to use the tags *movie, film* and *documentary* together. The tags within each cluster may, therefore, be considered semantically connected (see Figure 3). Here, the thickness of the line indicates the correlation between the connected tags. Table 1 shows the first factors (clusters) obtained from the PCA analysis for the "Corporate Sustainability" query.

An examination of the example of "Corporate Sustainability" shows that each component describes a different domain. It also shows that the first component is not related to the domain. It comes from the bookmarks of one website that is tangential to the domain. The analysis shows that a first-occurrence procedure can bias the results. The fifth component shows a similar bias. It would be useful to detect such tangential sites during the data cleansing process. In that case, the system could ask the users whether to take into account such sites.

Table 1. Tags contained in the main factors /clusters for "corporate sustainability"

Corporate Sustainability	
1st	visualization + statistics + information + data + tools
2nd	Development + charity + finance + microfinance + money + microcredit
3rd	Transportation + electric + car
4th	Food + health + organic
5th	Architecture + magazine + home +furniture
6th	trends + future + strategy + consulting
7th	Inspiration + flash + portfolio + webdesign
8th	architecture + collaboration + opensource
9th	Movie + film + documentary

Regardless of the insights gained from this experimentation, the local graphs generated for each component were then merged. The merged graph shows several interesting sub-domains that would be difficult to predict based on a layman's understanding of the terms 'corporate sustainability.' Even expert understanding of the term 'corporate sustainability' is unlikely to discover the varied sub-domains that this overall graph represents. Consider, for example, the sub-domains that are anchored towards micro-finance (second local graph of the figure) and the architecture (towards the right of the figure). The approach was also applied to other domains with the following outcomes: Online Auctions (102 domain-level concepts from 4,635 resources); Air Travel (150 concepts from 11,760 resources); and Emergency response (225 concepts from 1,664 resources). The implications are discussed further in the next section.

5   Conclusions

This research describes and validates an approach to repurpose tags from social tagging sites to extract domain-level concepts. The process and experiments reported demonstrate the feasibility of the approach. The experiments also point to interesting insights in two areas: domain restrictions as a way to advance the realization of results; and the possibility of additional heuristics. The key contribution of this research is the operationalization of the approach that may be broadly described as "designing with a crowd" to identify domain-level concepts for conceptual modeling and their connectedness. In particular, the PCA analysis provided useful insights to help in the identification of concepts in a domain. The results show that the selection (or rejection) of tags cannot be automatic; rather, it requires some feedback from a user. The results also highlight challenges such as the generic nature and ambiguity of tags that necessitate cleansing, and the need to aggregate tags that represent the same underlying concept. In addition, the components generated from the PCA analysis (such as *microfinance* and *architecture*) point to the possibility that the proposed approach has the potential to greatly exceed the limitations of naïve or even expert perspectives on domain models by accounting for a much wider array of concepts that can be part of the representation of a domain.

Future research will focus on additional experiments, improving the data cleansing process to increase the efficacy of discarding irrelevant tags and merging similar ones, developing techniques to extract more insights from the tags, and using fuzzy variables. Research is also needed to refine the heuristics, implement the remaining phases, and conduct empirical analyses.

Acknowledgments. This work has been supported by Sogang Business School's World Class University Program (R31-20002) funded by Korea Research Foundation, the Georgia State University, the IN3 institute and the Spanish Ministry MICINN (TIN2008-00444).

References

1. Hendler, J., Golbeck, J.: Metcalfe's Law Applies to Web 2.0 and the Semantic Web. Journal of Web Semantics 6, 14–20 (2008)
2. Furnas, G.W., Landauer, T.K., Gomez, L.M., Dumais, S.T.: The vocabulary problem in human-system communication. Communications of the ACM 30, 964–971 (1987)
3. Weikum, G., Kasneci, G., Ramanath, M., Suchanek, F.: Database and information-retrieval methods for knowledge discovery. Communications of the ACM 52, 56–64 (2009)
4. Sunstein, C.: Infotopia: How Many Minds Produce Knowledge. Oxford University Press, Oxford (2006)
5. Angeletou, S., Sabou, M., Specia, L., Motta, E.: Bridging the gap between folksonomies and the semantic web: An experience report. In: Workshop: Bridging the Gap between Semantic Web and Web 2.0 at 4th ESWC (2007)
6. Wu, X., Zhang, L., Yu, Y.: Exploring social annotations for the semantic web. In: Proceedings of the 15th International Conference on World Wide Web, pp. 417–426 (2006)
7. Campbell, D.G.: A phenomenological framework for the relationship between the semantic web and user-centered tagging systems. In: Proceedings of the 17th ASIS&T SIG/CR Classification Research Workshop, Austin, TX, USA (2006)
8. Kipp, M.E.I., Campbell, D.G.: Patterns and inconsistencies in collaborative tagging systems: An examination of tagging practices. In: Proceedings of the ASIST, Austin, TX, USA, pp. 1–18 (2006)
9. Dye, J.: Folksonomy: A game of high-tech (and high-stakes) tag. EContent 29, 38–43 (2006)
10. Sugumaran, V., Purao, S., Storey, V., Conesa, J.: On-Demand Extraction of Domain Concepts and Relationships from Social Tagging Websites. In: Hopfe, C.J., Rezgui, Y., Métais, E., Preece, A., Li, H. (eds.) NLDB 2010. LNCS, vol. 6177, pp. 224–232. Springer, Heidelberg (2010)
11. Maletic, J.I., Marcus, A.: Data cleansing: Beyond integrity analysis. In: Proceedings of the Conference on Information Quality, pp. 200–209 (2000)
12. Tan, K.W., Han, H., Elmasri, R.: Web data cleansing and preparation for ontology extraction using WordNet. In: Tan, K.W., Han, H., Elmasri, R. (eds.) Proceedings of the WISE, pp. 11–18. IEEE, Los Alamitos (2000)
13. Ding, C., He, X.: K-means clustering via principal component analysis. In: Proceedings of the Twenty-First International Conference on Machine Learning (2004)

Ontology-Guided Approach to Feature-Based Opinion Mining

Isidro Peñalver-Martínez[1], Rafael Valencia-García[1],
and Francisco García-Sánchez[2]

[1] Dpto. Informatica y Sistemas, Facultad de Informatica, Universidad de Murcia,
30100, Espinardo (Murcia), Spain
ipmartinez1@gmail.com, valencia@um.es
[2] Dpto. Informatica, Escuela Tecnica Superior de Ingenieria, Universidad de Valencia,
46100, Burjassot (Valencia), Spain
Francisco.Garcia-Sanchez@uv.es

Abstract. The boom of the Social Web has had a tremendous impact on a number of different research topics. In particular, the possibility to extract various kinds of added-value, informational elements from users' opinions has attracted researchers from the information retrieval and computational linguistics fields. However, current approaches to so-called opinion mining suffer from a series of drawbacks. In this paper we propose an innovative methodology for opinion mining that brings together traditional natural language processing techniques with sentimental analysis processes and Semantic Web technologies. The main goals of this methodology is to improve feature-based opinion mining by employing ontologies in the selection of features and to provide a new method for sentimental analysis based on vector analysis. The preliminary experimental results seem promising as compared against the traditional approaches.

Keywords: Opinion mining, Ontology, Sentimental Analysis, Feature extraction, Polarity identification.

1 Introduction

With the Social Web, the number of online reviews in which people freely express their opinions on a whole variety of topics is constantly increasingly. Opinion mining is a recent subdiscipline at the crossroads of information retrieval and computational linguistics which is concerned not with the topic a text is about, but with the opinion it expresses [7]. It determines critics' opinions about a given product, book review, etc. which are expressed in online forums, blogs or comments. Existing techniques that involve checking the similarity between a text and a seed list of words is not enough.

Ontologies can be used to structure information. In this work, an ontology is seen as "a formal and explicit specification of a shared conceptualization" [16]. Ontologies provide a formal, structured knowledge representation, with the advantage of being reusable and shareable. They also provide a common vocabulary

R. Muñoz et al. (Eds.): NLDB 2011, LNCS 6716, pp. 193–200, 2011.
© Springer-Verlag Berlin Heidelberg 2011

for a domain and define, with different levels of formality, the meaning of the terms attributes and the relations between them.

The objective of this work is to describe the architecture of an approach for feature-based opinion mining based on the identification of features using a domain ontology. Once the features have been identified, new polarity identification and opinion mining approaches are used to get an efficient sentiment classification. The proposed approach has been validated in the movie reviews domain. The rest of the paper is organized as follows: related work is expounded in Section 2; the proposed method is explained in Section 3; a validation of this ontology-guided approach in the movie reviews domain is performed in Section 4; and finally, conclusions and future work are put forward in Section 5.

2   Related Work

Hatzivassiloglous and McKeown worked in the semantic orientation (polarity) of adjectives [9]. Since then, several techniques have been employed for opinion mining tasks: probabilistic measures of word association [17], techniques that use information about lexical relations [11], and techniques that use lexical databases [1]. They all use the data coming from the reviews of either automobiles, bank, movies, or travel destinations. These classification mechanisms are useful and improve the effectiveness of a sentiment classification but cannot determine what the opinion holder liked and disliked on each particular feature.

There are also studies that determine not only the polarity of a text (positive or negative), but also indicate the level of polarity as high/medium/low positive/negative [6,15,3].

Mukras R. J. [13] and Michael G. [8] exploit the data coming from movie reviews, customer feedback review and product review. These experiments show that machine learning techniques do not obtain satisfactory performing results for sentiment classification. In 2004, Hu y Liu [10] focused on obtaining user's opinion about features of a product. These features were obtained using mining association between words.

More recently, there have been two important works concerning ontologies and feature-based opinion mining. The first work is the study of Zhou and Chaovalit in [19]. The other work is the study of Zhao and Li in [18]. The aim of these works is to calculate the polarity by taking into consideration the features of a concept.

Our work focuses on obtaining an alternative sentimental analysis method based on ontologies to improve the results obtained in the works [19,18].

3   Ontology Based Opinion Mining

The main goals of the approach proposed in this manuscript are to improve feature-based opinion mining by employing ontologies in the selection of features and to provide a new method for sentimental analysis based on vector analysis.

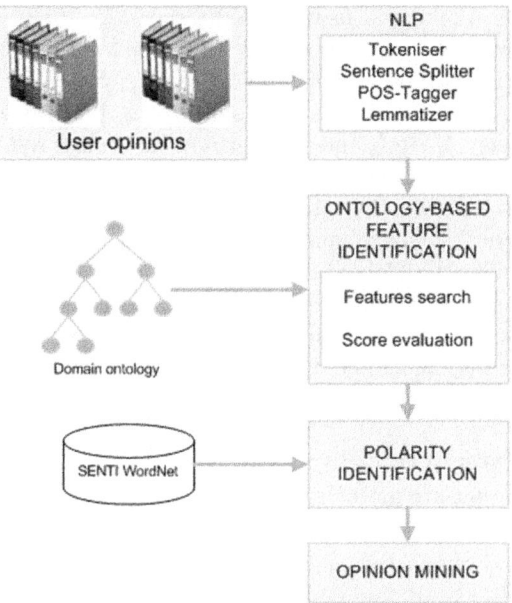

Fig. 1. Proposed system architecture

This approach is composed of four main modules (see Fig. 1): the Natural Language Processing (NLP) module, the ontology-based feature identification module, the polarity identification module and the opinion mining module. These components are described in detail below.

3.1 NLP Module

The main objective of this module is to obtain the morphologic and syntactic structure of each sentence. For this, a set of NLP tools including a sentence detection component, tokenizer, POS taggers, lemmatizers and syntactic parsers has been developed using the GATE framework (http://gate.ac.uk/).

For example, the morphological analysis that is carried out by the NLP module for the sentence *"Return of the Jedi is the latest episode of Star Wars but it's the one I liked. I love the soundtrack"* obtains the following lemmatized words accompanied by their grammatical categories:

```
Return_of_the_Jedi_NNP be_VBZ the_DT late_JJS episode_NN of_IN
Stars_War_NNP but_CC it_PRP be_VBZ the_DT one_NN I_PRP like_VBD.
I_PRP love_VBP the_DT soundtrack_NN.
```

3.2 Ontology-Based Feature Identification Module

This module receives both a corpus of opinions and the domain ontology as input. Let us suppose that the domain ontology represents the movies domain. The ontology has been populated using The Internet Movie Database (IMDb) data available on www.imdb.com. The features identified within the sentence given in the previous section will be { "Return of the Jedi", "Star Wars", "Soundtrack"}. More concretely, "Return of the Jedi" and "Stars War" are identified as individuals of the class "Movie", and "Soundtrack" is a class related to "Movie". At this point, thanks to the semantic structure of ontologies, the opinions that are concerned with the classes that are directly related to the class "Movie", such as "Actor", "Producer", "Genre" and their individuals would be taken into account in order to calculate the global polarity.

Traditional feature identification methods give equal importance to all the features identified in the text such us the following works [3,19,18,12]. In our approach, a score is calculated for each feature on the basis of the following assumptions:

1. Not all the features related to the same class have the same relevance.
2. The features that are more often cited by users in their opinions will be more relevant.
3. The polarity of the last paragraph coincides roughly with the global polarity of the text [12]. Start and end with positive or negative views enhances the positive or negative polarity.

Once the features have been identified in the opinions, the score of the features in each user opinion is calculated. For each retrieved feature the total score, called $Tscore$ (see equation 1), is calculated as the arithmetic mean of the score of the feature in each user opinion.

$$Tscore(f) = \sum_{i=1}^{n} \frac{score(f, userop_i)}{n} \tag{1}$$

In the formula to calculate the score of a given feature in a user opinion, the position of the linguistic expression that represents the feature within the text is taken into consideration. For this, the text is divided into three equal parts: (1) the beginning, (2) the middle, and (3) the end of the opinion. This score is defined in equation 2.

$$score(f, userop_i) = z_1 * |O_1| + z_2 * |O_2| + z_3 * |O_3| \tag{2}$$

where $|O_i|$ is the number of ocurrences of the feature f in the part of the text i of the $userop_i$ opinion, and z_i is a parameter that represents the importance of the occurrence of the feature in this part of the text. The segmentation of the texts is done by dividing them into three equal parts based on the number of words in the texts. In this work, the values showed in equation 3 have been used.

$$z_1 = 0,3 \quad z_2 = 0,2 \quad z_3 = 0,5 \tag{3}$$

3.3   Polarity Identification Module

The polarity of the features is calculated using SentiWordNet 3.0 [4]. This tool allows to obtain the values of the positive, negative and neutral nouns, adjectives and verbs that are located next to the linguistic expressions that represent a given feature in the opinion. The neutral sense is calculated as follows: $ScoreObj = 1 - ScorePos - ScoreNeg$. For this, the most common sense in SentiWordNet is used. Then, for each value of the polarity senses (positive, negative and neutral) of each feature a vector is defined as shown in equation 4.

$$V(f) = Tscore(f) * (a * ScorePos, a * ScoreNeg, ScoreObj) \qquad (4)$$

where a indicates whether a negative clause has been found in the sentence. If so, the positive and negative coordinates are multiplied by -1, otherwise the a parameter is set to 1.

3.4   Opinion Mining Module

In this section, an opinion mining method based on vector analysis for feature sentiment classification is described. As shown in the previous section, features are expressed by three coordinates, that is, each feature is determined by its euclidean vector (V) in R^3. A euclidean vector is given by two points, the origin point and the target point. Since the point of origin is always $(0, 0, 0)$, the expression of the position vector is reduced to express the target point. Therefore, a position vector is expressed by $V = (x, y, z)$.

The polarity of a feature f in the user opinions set, $Polarity(f)$, is calculated as follows (equation 5):

$$Polarity(f) = \begin{cases} +x \text{ (positive)} & \text{if } inside(P; P_P) = true \\ -y \text{ (negative)} & \text{if } inside(P; P_N) = true \\ 0 \quad \text{(neutral)} & \text{if } inside(P; P_O) = true \end{cases} \qquad (5)$$

where:

- $P = (x, y, z)$, is the target point of $V(f)$.
- P_P is the geometric pyramid whose volume is composed of all target points of position vectors with positive direction.
- P_N is the geometric pyramid whose volume is composed of all target points of position vectors with negative direction.
- P_o is the geometric pyramid whose volume is composed of all target points of position vectors with neutral direction.
- $inside(P; P_X)$ is a function that returns $true$ if the point P is inside the geometric pyramid P_X. Otherwise, it returns $false$.

In order to implement $inside(P; P_X)$ it is necessary to know if a point $P = (x, y, z)$ is an interior point of P_P, P_N or P_o. The elements of topology on sets of points definition is used [2]:

- *Inside Point*: be S a subset of R^n and P is a point of Rn. We suppose that $P \in S$. Then P is called interior point of S if there is an open n-ball centred at P, contained in S.
- *N-ball*: be P a point of R^n and R is a positive number given. The set of all points X in R^n such that $\|X - P\| < R$ is called open n-ball of radius R and centre P. Denote this set by $B(P)$ or $B(P; R)$.

On the other hand, all the features are assigned an score in accordance with the $V(f)$ value. Then, the polarity of all user opinion (the global sentiment analysis), called *Polarity*, is determined by the position vector resulting from the summation of all the (scored) position vectors for every identified feature (see equation 6):

$$Polarity = \sum_{i=1}^{n} \boldsymbol{V}(f_i) \tag{6}$$

The resulting *Polarity* value will be an Euclidean vector with three coordinates (x, y, z). Be P the target point of the resultant position vector, $P = (x, y, z)$, then (see equation 7):

$$Polarity = \begin{cases} +x \ (positive) & \text{if } inside(P; P_P) = true \\ -y \ (negative) & \text{if } inside(P; P_N) = true \\ 0 \quad (neutral) & \text{if } inside(P; P_O) = true \end{cases} \tag{7}$$

4   Use Case Scenario. Movie Reviews

In this section, the experimental results obtained by the proposed method in the movie reviews domain are presented.

The Web Ontology Language (OWL) is used to specify the MO ontology. Several other ontologies that are available within the Linked Data cloud are considered and integrated to highly couple the MO ontology with the Linked Data cloud, and so take advantage of synergy effects. A total number of 100 different movies have been inserted in the ontology as individuals for experimental purposes. The corpus of the experiment contains 15,323 words and comprises 64 opinions (31 negative and 33 positive). This corpus has been extracted from the corpus used in [14].

In the experiment, we study the results obtained by labeling manually as baseline in contrast to the results of our approach, and approaches [19] and [18]. The results for accuracy in features identification (see table 1) and sentimental analysis (see table 2) seem promising.

Table 1. Feature Identification Results

	Baseline	Our method	Zhou and Chaovalit	Zhao and Li
Features	15	14	13	11
Accuracy		93.3%	86.6%	73.3%

Table 2. Results for Sentiment Analysis

	Baseline	Our method	Zhou and Chaovalit	Zhao and Li
Pos	33	28	24	28
		Accuracy: 84.8%	Accuracy: 72.7%	Accuracy: 78.7%
Neg	31	27	23	25
		Accuracy: 87.1%	Accuracy: 74.1%	Accuracy: 80.6%

5 Discussion and Conclusion

Opinion mining is a challenging problem. It is concerned with analyzing the opinions that appear in a series of texts and are given by users towards different features, and determine whether these opinions are positive, negative or neutral and how strong they are [18]. In this paper, we focus on defining a new feature-based opinion mining method. It provides a complete set of new features such as (i) an ontology-based mechanism for identifying features, (ii) a method for assigning a score to each feature based on the number of occurrences and their relative position in each user's opinion, and (iii) a new approach for opinion mining based on calculations based on vectors analysis.

The proposed approach has been validated in the movie reviews domain and the results seem promising. A larger validation of the system is planned to be done by applying the system to opinions from the product domain and by using statistical methods for analysing the results obtained.

Feature-based opinion mining from product reviews is a difficult task, due to both the high semantic variability of opinion expression, as well as the diversity of characteristics and sub-characteristics describing the products and the multitude of opinion words used to depict them [5].

We are also currently working on upgrading this system and validating it into product reviews domain in Spanish language.

Acknowledgments. This work has been partially supported by the Spanish Ministry for Science and Education through project TIN2010-18650.

References

1. Andreevskaia, A., Bergler, S.: Mining WordNet for a fuzzy sentiment: Sentiment tag extraction from WordNet glosses. In: Proceedings of the European Chapter of the Association for Computational Linguistics, EACL (2006)
2. Apostol, T.M.: Mathematical Analysis. AddisonWesley, London (1974)
3. Baccianella, S., Esuli, A., Sebastiani, F.: Multi-facet rating of product reviews. In: Boughanem, M., Berrut, C., Mothe, J., Soule-Dupuy, C. (eds.) ECIR 2009. LNCS, vol. 5478, pp. 461–472. Springer, Heidelberg (2009)
4. Baccianella, S., Esuli, A., Sebastiani, F.: Sentiwordnet 3.0: An enhanced lexical resource for sentiment analysis and opinion mining. In: Calzolari, N., Choukri, K., Maegaard, B., Mariani, J., Odijk, J., Piperidis, S., Rosner, M., Tapias, D. (eds.) LREC. European Language Resources Association (2010)

5. Balahur, A., Montoyo, A.: Semantic approaches to fine and coarse-grained feature-based opinion mining. In: Horacek, H., Métais, E., Muñoz, R., Wolska, M. (eds.) NLDB 2009. LNCS, vol. 5723, pp. 142–153. Springer, Heidelberg (2010)
6. Dave, S.L.K., Pennock, D.M.: Mining the peanut gallery: Opinion extraction and semantic classification of product reviews (2003)
7. Esuli, A., Sebastiani, F.: Sentiwordnet: A publicly available lexical resource for opinion mining. In: Proceedings of the 5th Conference on Language Resources and Evaluation (LREC 2006), pp. 417–422 (2006)
8. Gamon, M.: Sentiment classification on customer feedback data: noisy data, large feature vectors, and the role of linguistic analysis. In: Proceedings of the International Conference on Computational Linguistics, COLING (2004)
9. Hatzivassiloglou, V., McKeown, K.R.: Predicting the semantic orientation of adjectives. In: Proceedings of the Eighth Conference on European Chapter of the Association for Computational Linguistics, EACL 1997, pp. 174–181. Association for Computational Linguistics, Stroudsburg (1997)
10. Hu, M., Liu, B.: Mining and summarizing customer reviews. In: Proceedings of the tenth ACM SIGKDD International Conference on Knowledge Discovery and Data Mining. KDD 2004, pp. 168–177. ACM, New York (2004)
11. Kim, S.-M., Hovy, E.: Determining the sentiment of opinions. In: Proceedings of the International Conference on Computational Linguistics, COLING (2004)
12. Moreno Ortiz, A., Pineda Castillo, F., Hidalgo García, R.: Análisis de valoraciones de usuario de hoteles con sentitext: un sistema de analisis de sentimiento independiente del dominio. Procesamiento del Lenguaje Natural 45, 31–40 (2010)
13. Mukras, R.: A comparison of machine learning techniques applied to sentiment classification. Masters thesis University of Sussex Falmer Brighton (2004)
14. Pang, B., Lee, L.: A sentimental education: Sentiment analysis using subjectivity summarization based on minimum cuts. In: Proceedings of the ACL, pp. 271–278 (2004)
15. Shimada, K., Endo, T.: Seeing several stars: A rating inference task for a document containing several evaluation criteria. In: Washio, T., Suzuki, E., Ting, K.M., Inokuchi, A. (eds.) PAKDD 2008. LNCS (LNAI), vol. 5012, pp. 1006–1014. Springer, Heidelberg (2008)
16. Studer, R., Benjamins, V.R., Fensel, D.: Knowledge engineering: Principles and methods. Data Knowl. Eng. 25(1-2), 161–197 (1998)
17. Turney, P.D., Littman, M.L.: Measuring praise and criticism: Inference of semantic orientation from association. ACM Trans. Inf. Syst. 21(4), 315–346 (2003)
18. Zhao, L., Li, C.: Ontology based opinion mining for movie reviews. In: Karagiannis, D., Jin, Z. (eds.) KSEM 2009. LNCS, vol. 5914, pp. 204–214. Springer, Heidelberg (2009)
19. Zhou, L., Chaovalit, P.: Ontology-supported polarity mining. JASIST 59(1), 98–110 (2008)

A Natural Language Interface for Data Warehouse Question Answering

Nicolas Kuchmann-Beauger[1,2] and Marie-Aude Aufaure[2]

[1] SAP France, 157/159 rue Anatole France, 92309 Levallois-Perret Cedex, France
nicolas.kuchmann-beauger@ecp.fr
[2] Ecole Centrale Paris, MAS laboratory, 92290 Chatenay-Malabry, France
marie-aude.aufaure@ecp.fr

Abstract. Business Intelligence (BI) aims at providing methods and tools that lead to quick decisions from trusted data. Such advanced tools require some technical knowledge on how to formulate the queries. We propose a natural language (NL) interface for a Data Warehouse based Question Answering system. This system allows users to query with questions expressed in natural language. The proposed system is fully automated, resulting low Total Cost of Ownership. We aim at demonstrating the importance of identifying already existing semantics and using Text Mining techniques on the Web to move toward the users's need.

Keywords: Question Answering, Natural Language Interface, Data Warehouses.

1   Introduction

Business Intelligence (BI) aims at providing a better understanding of the business environment to make decision. In this perspective, software companies have developed applications and tools that ease this understanding through data warehousing and data visualization which leads to better decision-making.

To access these data, all databases (DB) have an interface from which any user may write a query. This query is usually a technical query expressed in a specific language (e.g. MDX or SQL).

An issue is that such language is not *natural* to a non-expert user. In order to address this, several tools have been proposed such as SAP BusinessObjects Explorer [9]. One notice that once a user has entered a query, they modify it through classical operation (roll up, drill down). Indeed, building the best user interface for data warehouses is a subject that raises lots of interests.

The question on how to best display an answer for a real user's question belongs to the Information Retrieval (IR) domain. Systems that handle questions expressed in Natural Language (NL) are called Question Answering (Q&A) systems. Traditional IR systems allow users to express their queries using keywords or using a restricted syntax.

The basic approach in IR consists in identifying keywords or known entities in the user's need expression, and in linking these entities to objects defined in the data model. We have implemented this approach, and we propose an original

R. Muñoz et al. (Eds.): NLDB 2011, LNCS 6716, pp. 201–208, 2011.
© Springer-Verlag Berlin Heidelberg 2011

Table 1. Extract of questions used in the evaluation of the "String Matching" component

domain	question
eFashion	Which store has not sold which product category?
eFashion	How many orders have been generated in New York?
eFashion	Where customers have not bought which product?
eFashion	What was the most profitable product category?
eFashion	What product quantity was ordered in Chicago by quarter and what was the average for all quarters?
eFashion	What revenue did we achieve in Florida?

method to get better results in this context. A thesaurus is used to rewrite the user query when no answer has been found.

In the following, *question* is used for a well-formed question expressed in NL, while *query* stands for a query that the system can render (e.g. expressed in SQL). The expression *user's query* stands for *question*, or a user question that is not a well-formed question but composed of keywords (e.g. "Sales revenue France 2008" is a *user's query* that refers to an abstract *question*, one of whose occurrence is "What is the sales revenue in France in 2008?").

Our experiments show that the keyword-based aprroach is not sufficient, and that the same query may be expressed with different words. We evaluated the relevance of the keyword-based approach from a set of hundreds of complex business questions (see an extract table 1). The toy dataset to which the questions belongs is related to business sales and orders ("eFashion" dataset). The questions could all be answered by the system. Among these questions, only 52% contain at least one object recognized in the component (which does not mean that the analysis of the component is correct).

2   Related Work

Q&A is one of the first applications of Artificial Intelligence, and the first proposals were based on structured data [7,15].

These systems deal with databases interfaces, and more specifically natural language database interfaces [2]. [11] is an example of natural language interface to a XML database. [5] proposes such a system that links keywords present in the user's question into a SQL-statement. The data are split into several tables that have to be joined to get the requested information. Usually, a grammar dedicated to NL interpretation is needed [8] which is a limitation, because the NL processing is tailored to the database itself.

[13] aims at determining whether a question can be answered by the system, and if so it asserts that the answer corresponds to the correct interpretation. In the field of enterprise Q&A, one generally prefer retrieving pieces of answers, even if they are not the exact interpretation. [10] is a system that can be used with any database, however it requires resources in addition to the database.

Recently, [6] proposed a Q&A system dedicated to enterprise questions but also to general domain questions.

Building an ontology [14,3,16] is often used to support Q&A, especially in the semantic Web community. Ontologies are often used in the Question Rewriting component [4], [1]. The ontology is then used to translate the user's queries into a technical query. Accessing a database through an ontology in not new: e.g. [12] presents how an ontology can be used in order to get data from a relational database.

We leverage traditional Q&A techniques with the specificities of an enterprise IR system. In this perspective, a user-specific ontology is being automatically built and enriched, this will be further discussed below.

3   Architecture

In this paper we are interested in the first component, aiming at describing the semantic analysis of the user's query, as described in the figure 1. The Named

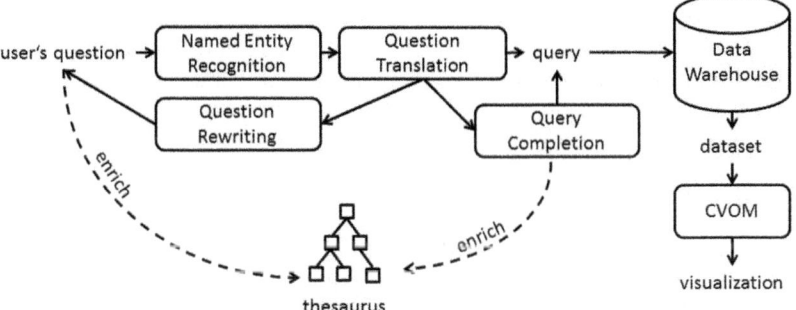

Fig. 1. Analysis of the user's question

Entity Reconizer (NER) aims at identiying *known objects* of the data model, that comprise dimensions, measures and values of dimensions. The Question Translation component build a technical query from a user's question. The Question Rewriting component rewrites the query if no answer can be retrieved. The Query Completion component is used when queries are not expressive enough (e.g. when no dimension is comprised in the query). The CVOM[1] component provides a visualization to a dataset. In this paper, we focus on three components: the Question Tranlation component, the Question Rewriting component and the thesaurus-based Q&A.

4   Question Translation

Identified semantic units from the user's question are translated into a machine-readable query (technical query), which is a graph composed of objects and

---

[1] SAP BusinessObjects Common Visual Object Modeler.

constraints described in the data warehouse model. For example, the user's question "What are the sales revenue in France in 2008" will be translated into the following technical query:

```
AGGREGATE MEASURE 'Sales Revenue' AGAINST guessed DIMENSION
FILTERING ON 'France' BEING 'Country'  AND '2008' BEING 'Year'
```

There is here a direct link between the entities expressed in the user's query and the objects modeled in the data warehouse. If the query is empty (no object can be recognized) the Question Rewriting component provides new question formulations from which the Question Translation component is called. If the query is not empty but does not return any answer, the Query Completion component is called. We exploit scores from a search service[2] to infer semantic closeness between terms or expressions from the user's query and labels. Scores are the number of pages that contains the expression.

Let consider the question "What are the admission fees in Asia?". No entities are identified in this question. One of the rewritten questions is "What are the admissions in Asia?" (one statement in the thesaurus is that *admission fees* and *admissions* are synonyms, and Admissions is an object defined in the data model. In this example, "Asia" is not defined in the data model, and no row in the data contains this token. The following section shows how it is possible to deal with such cases.

5   Thesaurus

Our current work implements automatically built thesaurus from structured data. Thesaurus building is done in three-steps: 1) building of a minimal thesaurus, 2) validating this minimal thesaurus and 3) adding terms to be validated in the thesaurus.

5.1   Minimal Thesaurus

The minimal thesaurus is built from objects defined in the data model. They are represented in the thesaurus by terms as well as values of dimensions that are represented as terms sharing the *is-a* semantic relationship with the concerned dimensions. For each terms, WordNet is used to add relationships (e.g. synonymy relationship) and new terms (e.g. synonyms). The sense disambiguation is the main focus of the thesaurus validation. Our heuristic consists in choosing the first sense, that corresponds to the most common sense.

The validation of the thesaurus is performed as the user queries the system. The algorithm is based on the identification of terms in the user's query which are semantically linked with other terms in the thesaurus.

The figure 2 presents terms of the minimal thesaurus automatically built from the *eFashion* dataset. The top concept is labelled "TOP". The first-level edges are *is-a* relationships, while other relationships are synonymy relationships.

---

[2] See Yahoo BOSS at http://developer.yahoo.com/search/boss/.

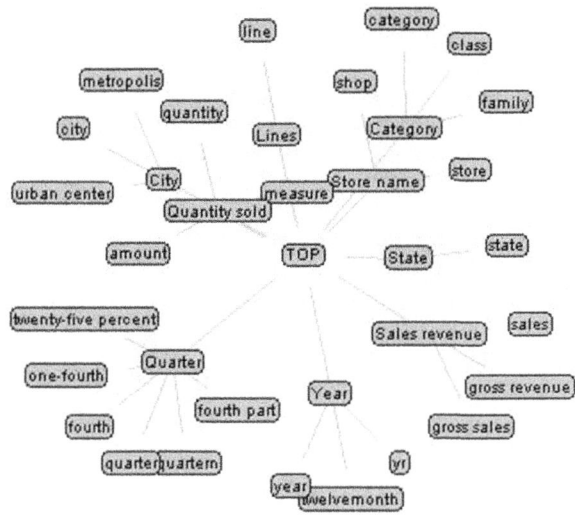

Fig. 2. Minimal thesaurus

5.2 Learning Approach

To address the problem of unknown terms (such as "Asia" in the user's question "What are the admission fees in Asia?"), we try to categorize them using known terms as labels. Our problem is similar to classification problems. The algorithm 1 presents how new terms are added to the minimal thesaurus. In our example, if we remove the known entity (*admission fees*) and the stop words (defined in a specific dictionary), the remaining is "Asia". The algorithm returned two related entities, "Malaysia" and "Japan", both values of the dimension *Admission Country*. The algorithm ranks these entities according to the frequency of the related dimensions, and adds them to the query as filters.

New instances are also added when the user queries the system. We use patterns to get new instances from existing ones, similar to those defined in the Ontology Design Pattern (ODP) portal[3].

6 Query Reformulation

Query Reformulation aims at rewriting queries so that the user's need is better satisfied. Two common ways of rewriting a query is expanding or refining it. Other kinds of reformulation replace terms by other ones sharing a specific relationship. This process is iterated, until a valid technical query is available.

In our present work, we only use the synonymy relationship among terms to reformulate user's queries. Word-level N-grams are identified in the user's query, and matched to the thesaurus. The corresponding terms are listed, and

---

[3] Go to http://ontologydesignpatterns.org/wiki/Main_Page for more details.

Algorithm 1. Query completion

Require: question$\in L\star$
1: question←remove_known_entities(question)
2: question←remove_stop_words(quesiton)
3: **for** word **in** question.words **do**
4:     abstracts←web_search(word)
5:     **for** abstract **in** abstracts **do**
6:         **for** token **in** abstract.words **do**
7:             **if** is_known(token) **then**
8:                 add_related_entity(get_entity(token))
9:             **end if**
10:        **end for**
11:    **end for**
12: **end for**
13: rank(related_entities)

semantically related terms are retrieved trom the thesaurus using the SPARQL[4] query language. An ordred list of rewritten queries are then processed by the system, and corresponding answers are presented to the user. The order is defined by the semantic closeness of retrieved terms.

7   Evaluation

We consider a movie-related dataset (containing data about movies with associated business data such as the corresponding admission fees and budget). One typical query would be "What are the admission fees in *this* country". Let analyze the user's query "Compare the admission fees in Europe and in America". The traditional keyword-based method cannot retrieve any information for two reasons: 1) the modeled fact is not called "admission fees", 2) "Europe" and "America" are not defined. The first concern is resolved by the reformulating component using the minimal ontology. The second concern need a deeper analysis because the terms "Europe" and "America" are not found in the thesaurus. Some of the patterns used to validate "Europe" from the user's terms is "Europe is composed of $\star$" and "$\star$ compose Europe". Table 2 shows the results of the example. The entity type COUNTRY gets the best score, which means that the terms must be of the COUNTRY type. Once the new terms are discovered, the algorithm selects known terms that best represent the user's need. In this example, the retrieved terms of type COUNTRY are: United States, Belarus, Portugal, Slovakia, Greece, Spain, Ireland, England, Northern Ireland, Luxembourg, Switzerland, Italy, Great Britain, France, Moldova, Liechtenstein, Denmark, Netherlands, the Netherlands, Britain, US, Finland, UK, Austria, Wales, United Kingdom, Andorra, Scotland, America, Hungary, Czech Republic, Slovak Republic, Belgium, Poland, Romania, Bulgaria, Russia, Norway, Germany, Sweden. One notice that

---

[4] See http://www.w3.org/TR/rdf-sparql-query/ for more details.

Table 2. Entities identified in a Web corpus dedicated to ontology enrichment

Entity type	found entities	distinct found entities
PLACE_OTHER	3	3
TIME_PERIOD	3	2
ORGANIZATION	19	9
PLACE_REGION	117	17
COMPANY	7	6
COUNTRY	106	38
PERSON	13	12
DATE	9	9
CITY	4	4

this output is not perfect ("United States"). In our example, Norway, Germany and Sweden were already known, so the rewritten query is composed of these three filters.

8   Conclusion and Future Work

We have proposed a Q&A system based on structured data enabling users formulating their queries in NL. The system is composed of a string matching component performing keyword-based Q&A and of a learning thesaurus. This thesaurus is automatically built from the data model, and is enriched through users' queries. It is used to rewrite the queries.

One improvement will be turning the thesaurus into an ontology, wich will provide a reasoning feature. The ontology could be used for other purpose than question rewriting, like gathering data from unstructured sources to validate provided answers. Improvements when formulating search patterns (e.g. using lemmas) are also planned.

References

1. An, Y., Borgida, A., Mylopoulos, J.: Inferring complex semantic mappings between relational tables and ontologies from simple correspondences. In: Chung, S. (ed.) OTM 2005. LNCS, vol. 3761, pp. 1152–1169. Springer, Heidelberg (2005)
2. Androutsopoulos, L.: Natural language interfaces to databases - an introduction. Journal of Natural Language Engineering 1, 29–81 (1995)
3. Ben Mustapha, N., Zghal, H.B., Aufaure, M.A., ben Ghezala, H.: Semantic search using modular ontology learning and case-based reasoning. In: EDBT 2010: Proceedings of the 2010 EDBT/ICDT Workshops, pp. 1–12. ACM, New York (2010)
4. Bizer, C.: D2r map - a database to rdf mapping language (2003)
5. Chaudhuri, S., Das, G.: Keyword querying and ranking in databases. In: Proc. VLDB Endow., vol. 2(2), pp. 1658–1659 (2009)
6. Ferrucci, D.A., Brown, E.W., Chu-Carroll, J., Fan, J., Gondek, D., Kalyanpur, A., Lally, A., Murdock, J.W., Nyberg, E., Prager, J.M., Schlaefer, N., Welty, C.A.: Building watson: An overview of the deepqa project. AI Magazine 31(3), 59–79 (2010)

7. Green, B., Wolf, A., Chomsky, C., Laughery, K.: Baseball: an automatic question answerer, pp. 545–549 (1986)
8. Hendrix, G.G., Sacerdoti, E.D., Sagalowicz, D., Slocum, J.: Developing a natural language interface to complex data. ACM Trans. Database Syst. 3(2), 105–147 (1978)
9. Hilgefort, I.: Inside SAP BusinessObjects Explorer. SAP Press, Bonn (2010)
10. Knowles, S., Mitrovic, S.T.: A natural language database interface for sql-tutor (1999)
11. Li, Y., Yang, H., Jagadish, H.V.: Nalix: an interactive natural language interface for querying xml. In: SIGMOD 2005: Proceedings of the 2005 ACM SIGMOD International Conference on Management of Data, pp. 900–902. ACM, New York (2005)
12. Poggi, A., Lembo, D., Calvanese, D., Giacomo, G.D., Lenzerini, M., Rosati, R.: Linking data to ontologies. J. Data Semantics 10, 133–173 (2008)
13. Popescu, A.M., Armanasu, A., Etzioni, O., Ko, D., Yates, A.: Modern natural language interfaces to databases: composing statistical parsing with semantic tractability. In: Proceedings of the 20th International Conference on Computational Linguistics, COLING 2004. Association for Computational Linguistics, Stroudsburg (2004), http://dx.doi.org/10.3115/1220355.1220376
14. Serhatli, M., Alpaslan, F.N.: An ontology based question answering system on software test document domain. World Academy of Science, Engineering and Technology (2009)
15. Woods, W.A.: Progress in natural language understanding: an application to lunar geology. In: AFIPS 1973: Proceedings of the National Computer Conference and exposition, June 4-8, pp. 441–450. ACM, New York (1973)
16. Zghal, H.B., Aufaure, M.A., Mustapha, N.B.: A model-driven approach of ontological components for on-line semantic web information retrieval. J. Web Eng. 6(4), 309–336 (2007)

Graph-Based Bilingual Sentence Alignment from Large Scale Web Pages

Yihe Zhu, Haofen Wang, Xixiu Ouyang, and Yong Yu

Apex Data & Knowledge Management Lab
Shanghai Jiao Tong University
{yhzhu,whfcarter,oyxx,yyu}@apex.sjtu.edu.cn

Abstract. Sentence alignment is an enabling technology which extracts mass of bilingual corpora automatically from the vast and ever-growing Web pages. In this paper, we propose a novel graph-based sentence alignment approach. Compared with the existing approaches, ours is more resistant to the noise and structure diversity of Web pages by considering sentence structural features. We formulate sentence alignment to be a matching problem between nodes (bilingual sentences) of a bipartite graph. The maximum-weighted bipartite graph matching algorithm is first applied to sentence alignment for global optimal matching. Moreover, sentence merging and aligned sentence pattern detection are used to deal with the many-to-many matching issue and the low probability of aligned sentences with few mutual translated words issue respectively. We achieve good precision over 85% and recall over 80% on manually annotated data and 1 million aligned sentence pairs with over 82% accuracy are extracted from 0.8 million bilingual pages.

1 Introduction

With the development of the World Wide Web, more and more multilingual Web sites appear. Meanwhile, a lot of bilingual pairs of sentences are found on Web pages. We call a pair of Web pages *Bilingual Parallel Page Pair* if one page is the translated version of the other or most of their contents are mutual translated. On the other hand, a Web page is called *Bilingual Web Page* if its content contains several sentences that can be aligned to some other sentences written in another language on the same page. Facing the ever growing bilingual pages or page pairs on the Web, people have begun to realize that it is possible to leverage these rich resources to build very large bilingual corpora automatically. This task has two main steps. First, we need to collect Bilingual Web Pages and Bilingual Parallel Page Pairs in an automatic way. Some developed systems such as PTMiner [3], STRAND [9], and WPDE [11] have paid much attention to dealing with research issues involved in this step. Once we have collected abundant bilingual resources on the Web, the next step is to extract bilingual aligned sentences from a Web page or a page pair with high accuracy.

Sentence Alignment is the enabling technology for the latter step, which is the focus of the paper. However, existing approaches [5,4] which work well on bilingual parallel corpora with articles and books cannot be directly applied to sentence alignment on Web pages or Web page pairs. This is due to the fact that sentences in articles or books

R. Muñoz et al. (Eds.): NLDB 2011, LNCS 6716, pp. 209–216, 2011.
© Springer-Verlag Berlin Heidelberg 2011

contain pure textual contents while sentences in Web pages are surrounded by HTML tags or other structural information for display. Thus, sentences are always aligned in a sequential order in such corpora but can be cross aligned in Web pages. More recent research work [10,11,6] focuses on sentence alignment against Web pages by additionally considering more specific features. They also reformulate the alignment task as a binary classification problem to check whether the input sentence pair is aligned or not. Nevertheless, there still remain some unsolved issues: 1) classification tends to find sub-optimal solutions of aligning sentences; 2) most approaches have not considered the many-to-many alignment problem; 3) some implicit aligned sentences with few mutual translated words can only be detected with the help of discovering aligned sentence hidden patterns.

In this paper, we propose a novel graph-based sentence alignment approach to tackle the above issues. We construct a bipartite graph for the Web Page where we represent sentences as nodes and organize sentences in different languages on two sides. The weight of an edge connecting two potentially aligned sentences is given by the classification result. The maximum-weighted bipartite graph matching is first applied to sentence alignment for global optimal matching. Moreover, we use some heuristic rules to merge sentences and update the bipartite graph with new nodes and edges accordingly, which deals with many-to-one matching issue and aligned sentence pattern recognition.

2   Related Work

Some researchers treat sentence alignment as the classification task. They have used different kinds of classifiers to extract bilingual aligned sentences from Web pages. Munteanu and Marcu [8] used a maximum entropy classifier for this task. Zhang et al. [11] used multiple features to identify aligned sentences through a k-nearest-neighbor classifier. Regarding unsupervised learning approaches, Chuang and Yeh [4] proposed an unsupervised EM approach to align sentences. Gale and Church [5] developed a dynamic programming algorithm to find the maximum likelihood of sentence alignment. Recently Braune and Fraser [2] present here our two-step clustering approach to sentence alignment.

Both Martínez et al. [7] and Zhen et al. [12] formalized sentence alignment to a bipartite matching problem. Martínez et al. [7] desgined the TasC algorithm which aligns each source sentence to the target with the most similarity. They also solved the possible conflicts when a sentence matches some already aligned target sentences. Zhen et al. [12] developed the BAM system which aligns sentence pairs according to their translational similarity in a descending order. They also added constraints to make sure translators maintain the order of original texts in their counterparts. However, these approaches did not use maximum-weighted bipartite matching for global optimization.

Among the state of the art, different types of features have been proposed for sentence alignment. Resnik and Smith [9] used structural features and proposed a new content-based measure of translational equivalence to improve the classification performance. Shi et al. [10] presented a DOM tree alignment model for mining parallel sentences on a Web page. Zhen et al. [12] used the sentence pair location information to improve the alignment accuracy. Chuang et al. [4] exploited the statistically ordered matching of punctuation marks to achieve high-accuracy sentence alignment. More recently, Jiang et al. [6] pointed out that bilingual resources in many Web pages share similar patterns and proposed an adaptive pattern-based mining method with over 80% accuracy and recall for sentence alignment.

3   Extracting Bilingual Aligned Sentences from Web Pages

3.1   Feature Extraction

Sentence Features

- *DOM Path.* A Web page can be parsed into a DOM tree with HTML tags as the intermediate nodes and contents as the leaves. The DOM path is denoted by the identifier of the leaves (i.e. sentences).
- *DOM Depth.* It indicates the length of the DOM path of a sentence.
- *Sentence Position.* In a Web page, we record the sequence number of a sentence as its sentence position feature.
- *Sentence Surrounding Environment.* We count the number of sentences prior to a given sentence as well as the number of sentences following that sentence written in the same language in a Web page.
- *Leading Term Type.* As an empirical evidence, if a sentence starts with any leading terms with three types (i.e., ordinal number, letter, frequent term, and others), it indicates that the sentence might have the translated counterpart. For instance, "1. I see." started by an ordinal number '1', "a) It's so hot outside." begins with a letter 'a', and "Ques: How do you feel?" with "Ques" used as the leading term of a lot of sentences in Q&A.
- *End of Sentence Type.* We segment sentences according to three types: punctuation marks, carefully selected HTML tags, and language changing at the sentence boundary.

Sentence Pair Features

- **Translational Similarity Features**
  - *Count and Proportion of Matched Term Pairs.* A matched term pair refers to a pair of mutual translated terms in their corresponding sentences.
  - *Sentence Length Ratio.* We calculate the ratio of the length of the shorter sentence to that of the longer one as another feature.
  - *Translational Entropy.* Even though two bilingual sentence pairs have the same count of matched term pairs, the position distribution of matched term pairs in each pair might differ a lot. We treat the position difference of the next two matched term pairs as an event and calculate its probability. Then a normalized entropy is computed to characterize the distribution. According to this feature, we can tell a bilingual sentence pair with matched term pairs of uniform distribution is more likely to be aligned than that of biased distribution.

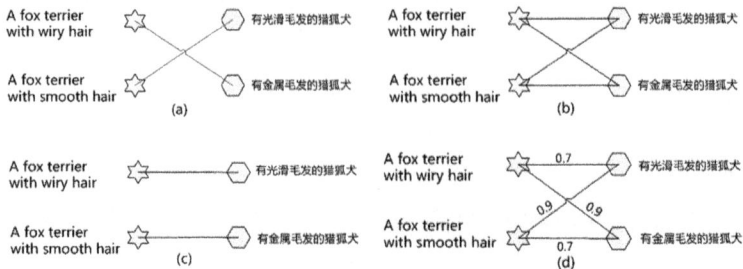

Fig. 1. Bipartite Graphs with Different Configurations

- **Structural Features**
  - *Sentence Position Distance.* It captures the different between sentence positions of the two sentences in their respective Web pages.
  - *Environment Distance.* It counts the difference between sentence surrounding environment of two sentences.
  - *DOM Edit Distance.* It calculates the edit distance between DOM paths of two sentences considering the replacing and inserting costs.
- **Miscellaneous Features**
  - *Source Type.* We distinguish the source type of two sentences (i.e., either a Bilingual Web Page or a Bilingual Parallel Page Pair) to assign different weights of other features during the training stage for classification.
  - *Leading Term Type Pair.* Since the leading term type represent the beginning of the sentence, we take the leading term type pair of two sentences as another sentence pair feature to measure the commonality between two bilingual sentences.
  - *End of Sentence Type Pair.* Similar to leading term type pair, the end of sentence type pair is also included in our sentence pair features.

3.2   Graph-Based Sentence Alignment

Maximum-Weighted Bipartite Graph Matching. Once we get initial alignment results from classification, a bipartite graph is constructed. We organize sentences in two languages on different sides and edges are added between two bilingual sentences if the classifier thinks the sentence pair should be aligned. Formally speaking, a bipartite graph G is in the form of $(V_1 \cup V_2, E)$ where V_1 and V_2 represents sentences in two different languages, and $E = \{(v_i, v_j), v_i \in V_1, v_j \in V_2\}$ in which each (v_i, v_j) is a sentence pair classified to be bilingual. Then we perform maximum bipartite graph matching (MBGM) to find the global optimal sentence alignments. Furthermore, we extend the bipartite graph G to G' by additionally assigning edge weights. The weight of an edge shows the log value of the probability of the sentence pair to be aligned and is denoted as $w_{i,j}$ where $(v_i \in V_1, v_j \in V_2)$. Hence, we adopt maximum-weighted bipartite graph matching (MWBGM) to G' and regard sentence pairs as bilingually aligned if their matching scores exceed the empirical threshold which will be discussed in the experiments section. The objective of MWBGM is to maximize the multiplication of probabilities of the finally aligned sentence pairs. The process of MWBGM

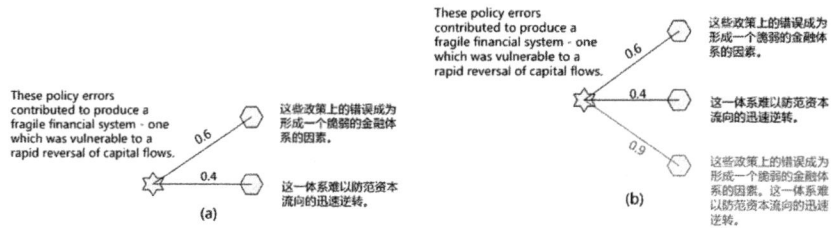

Fig. 2. An Example of Many-to-One Matching

is to estimate the maximum likelihood which can be simply reduced to maximize the sum of those log probabilities of the final alignments. The well-known state of the art MWBGM runs in $O(|V| \times |E| + |V|^2 log |V|)$ [1] while the time complexity of ours is $O(max(|V_1|, |V_2|)^3)$ where $|V|$, $|V_1|$ and $|V_2|$ are vertex count, and $|E|$ is the edge count. Figure 1 illustrates bipartite graphs with different configurations given four sentences (i.e., two English sentences and two Chinese sentences). Figure 1(a) shows the bipartite graph with the correct alignments, Figure 1(b) shows the one based on classification results but without weights on edges, Figure 1(c) shows an example which performs a global optimal matching using MBGM but involves incorrect results, and Figure 1(d) shows the graph with weighted edge, which leads to the correct results.

Dealing with Many-to-One and Many-to-Many Matching. A sentence might be segmented to several sentences while the bilingual counterpart remains to be one, which leads to a many-to-one matching issue during sentence alignment. Figure 2(a) shows a many-to-one matching example where the English sentence is not aligned to any Chinese sentence separately but their merging sentence. Neither MBGM nor MWBGM can deal with the issue because they find at most one alignment for a sentence. An intuitive way is to re-formulate graph-based sentence alignment into a net flow maximum flow minimum cost (MFMC) problem. Note that maximum weighted bipartite graph matching is a special case of MFMC. While MFMC could relax the limitation of the matching number for each sentence, it suffers from high time complexity and it is still hard to decide how many matches a sentence shall have for sentence alignment. In this paper, we propose a simple but effective approach to solve the above issue. For any two sentences $v_{i1}, v_{i2} \in V_1$ that connect to the same sentence $v_j \in V_2$ on the other side, if $j = \arg\max_k w_{i,k}, v_k \in V_2$ satisfies, a new vertex representing the merged sentence v_i is added on the bipartite graph. Figure 2(b) illustrates the newly added vertex by merging two Chinese sentences and the new edge whose weight is also given by the classifier in red. Similarly, the same merging strategy can be used for adding merged vertices in V_2. In this way, the more general many-to-many matching issue can be solved by iteratively merging two vertices on each side on the graph.

Aligned Sentence Pattern Detection. While we propose sophisticated sentence pair features for classification to output the probability as the weight of an edge in the bipartite graph, we still surfer from the problem that some bilingual sentence pairs which

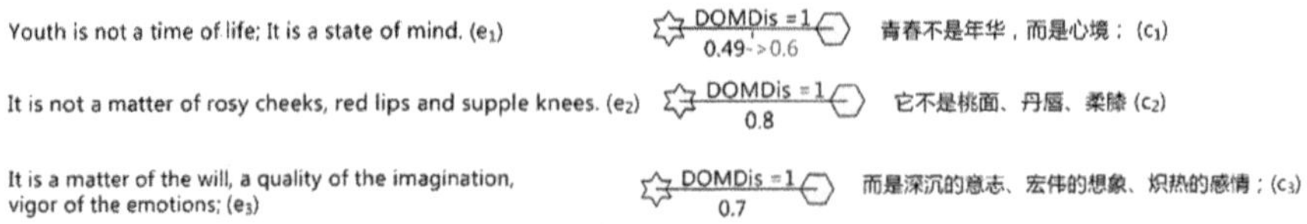

Youth is not a time of life; It is a state of mind. (e_1)　　DOMDis =1　　青春不是年华，而是心境；(c_1)
0.49->0.6

It is not a matter of rosy cheeks, red lips and supple knees. (e_2)　　DOMDis =1　　它不是桃面、丹唇、柔膝 (c_2)
0.8

It is a matter of the will, a quality of the imagination,
vigor of the emotions; (e_3)　　DOMDis =1　　而是深沉的意志、宏伟的想象、炽热的感情；(c_3)
0.7

Fig. 3. An Example of Aligned Sentence Pattern

should be aligned get low probabilities. On the other hand, Jiang et al. [6] have shown that aligned sentences in Web pages will probably have hidden patterns like frequent feature combinations, which can be used to solve the issue. Figure 3 shows an example of the necessity of discovering aligned sentence patterns. As shown in the figure, both DOM edit distances and position distances of all three example sentence pairs are the same (i.e., 1). Since $(e2, c2)$ and $(e3, c3)$ are classified to be aligned (with high probabilities), the above combination of DOM edit distance and position distance (shared by both alignments) is an aligned sentence pattern. As the pattern is also shared by $(e1, c1)$ with low probability, it is reasonable to increase the edge weight accordingly. A lightweight pattern mining strategy is used in this paper. For a set of bilingual sentence pairs $\{(e_i, c_i)\}$, they have a common pattern (a feature combination) if each e_i chooses c_i as its best aligned sentence (i.e., $c_i = \arg\max_k w_{e_i,k}$, and each sentence pair has the same value for each feature involved in the pattern. Then for each discovered pattern, we will increase the probability of each sentence pair whose alignment probability is lower than the average one of all pairs sharing the pattern. The increasing step is half of the difference between the current probability and the average probability.

4　Experiments

4.1　Data Collection and Training Details

To collect bilingual Web Pages, we crawled a list of Web sites from Google Web Directory. For each Web site, we first tentatively crawled a few pages to estimate the site's language distribution. If a certain proportion of pages have contents in bilingual languages, we treat the site as a bilingual one and will download all its pages. In this way, we collected 3,137 candidate Web sites which may contain bilingual sentence pairs, which results in 801,409 pages in total. Regarding Bilingual Parallel Page Pairs, we collected 22,962 pairs from 500 multi-lingual sites (e.g., `http://www.gov.hk/sc/residents/` and `http://www.gov.hk/en/residents/`).

In order to train a classifier, we randomly select 200 Bilingual Web Pages and 50 Bilingual Page Pairs as the training set. Another 200 pages and 50 page pairs are used for testing. We annotated aligned bilingual sentence pairs as positive examples and randomly selected other pairs of the similar number as negative examples. After feature extraction, we used LIBSVM [1] for the classification purpose.

---

[1] `http://www.csie.ntu.edu.tw/~cjlin/libsvm`

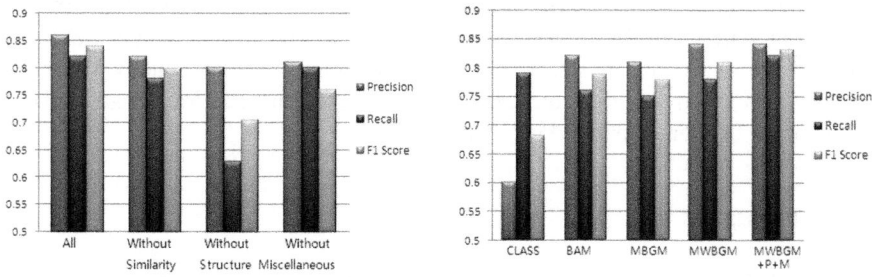

Fig. 4. Performance Comparison a) using Different Features b) under Different Methods

4.2   Sentence Alignment Using Different Features

We combine different features to see their impacts on the performance (i.e., precision, recall, and F1 score) of sentence alignment on the manually labeled data set. Figure 4(a) shows the performance variance when using different feature combinations. *All* means that we use all three types of sentence pair features, *Without Similarity* represents that we exclude those translational similarity features and re-train a new classifier for performance measurement. Similarly, *Without Structure* and *Without Miscellaneous* indicate that we do not consider structural features and others in the feature set respectively. As shown in the figure, all types of features have positive effects to improve the alignment performance. In particular, structural features play the most important role with respect to the recall and F1 score.

4.3   Performance Comparison under Different Methods

We compared different methods on their precisions, recalls, and F1 scores for sentence alignment using all proposed sentence pair features and based on the same training data. CLASS [7] is the first baseline for comparison which just applied a classifier to find aligned bilingual sentence pairs. BAM [12] is another method to compare which performs sentence alignment on a bipartite graph with edge weights given by a classifier. Figure 4(b) shows the performance results of our approach variants (i.e., MBGM, MW-BGM, and MWBGM+P+M) as well as the two baselines. Note that MWBGM+P+M denotes the maximum-weighted bipartite graph matching with aligned sentence pattern detection and many-to-many matching considered. From the figure, we find CLASS gets very poor precision and the lowest score for F1, BAM obtain 82% for precision and 76% for recall, MBGM has the similar performance as BAM, MWBGM performs consistently better than the above three, and MWBGM+P+M achieves the best performance by increasing the recall (by 4%) and thus the F1 score accordingly.

4.4   Evaluation on the Large Scale Web Page Collection

In order to make the experimental results more convincing, we test our approach on the large scale Web page collection (i.e., 801,409 Bilingual Web Pages and 22,962 Bilingual Parallel Page Pairs in total). After removing duplications, we get 1,094,455 unique bilingual sentence pairs from the Bilingual Web Pages and 50,298 from the Bilingual Parallel Page Pairs. Since it is hard to estimate the recall of sentence alignment properly for a so large data set, we only report the precision by manually judging 500 randomly selected pairs. we achieve 0.856 precision on Bilingual Web Pages and 0.824 precision on Bilingual Parallel Page Pairs.

5   Conclusion

This paper introduces a novel graph-based sentence alignment approach on Web pages. We proposed some novel sentence pair features for classification. Moreover, we were the first to apply the maximum-weighted bipartite graph matching to find the global optimal alignments of sentences. Experimental results show the effectiveness of our approach on both manually labeled data and large scale Web pages.

References

1. Ahuja, R., Magnanti, T., Orlin, J.: Network flows: theory, algorithms, and applications (1993)
2. Braune, F., Fraser, A.: Improved unsupervised sentence alignment for symmetrical and asymmetrical parallel corpora. In: Proceedings of the 23rd International Conference on Computational Linguistics: Posters, pp. 81–89. Association for Computational Linguistics (2010)
3. Chen, J., Nie, J.: Parallel web text mining for cross-language IR. In: Proceedings of RIAO 2000: Content-Based Multimedia Information Access, vol. 1, pp. 62–78 (2000)
4. Chuang, T., Yeh, K.: Aligning parallel bilingual corpora statistically with punctuation criteria. Computational Linguistics and Chinese Language Processing 10(1), 95–122 (2005)
5. Gale, W., Church, K.: A program for aligning sentences in bilingual corpora. Computational Linguistics 19(1), 75–102 (1993)
6. Jiang, L., Yang, S., Zhou, M., Liu, X., Zhu, Q.: Mining bilingual data from the web with adaptively learnt patterns. In: Proceedings of the Joint Conference of the 47th Annual Meeting of the ACL, pp. 870–878. Association for Computational Linguistics (2009)
7. Martínez, R., Abaitua, J., Casillas, A.: Bitext correspondences through rich mark-up. In: Proceedings of the 17th International Conference on Computational Linguistics, vol. 2, pp. 812–818. Association for Computational Linguistics (1998)
8. Munteanu, D.S., Marcu, D.: Improving machine translation performance by exploiting non-parallel corpora. Computational Linguistics 31, 477–504 (2005)
9. Resnik, P., Smith, N.: The web as a parallel corpus (2002)
10. Shi, L., Niu, C., Zhou, M., Gao, J.: A DOM tree alignment model for mining parallel data from the web. In: Proceedings of the 21st International Conference on Computational Linguistics, pp. 489–496. Association for Computational Linguistics (2006)
11. Zhang, Y., Wu, K., Gao, J., Vines, P.: Automatic Acquisition of Chinese–English Parallel Corpus from the Web. Advances in Information Retrieval, 420–431 (2006)
12. Zhen, L., Sheng, L.: Aligning Bilingual Corpora Using Sentences Location Information. 哈尔滨工业大学信息检索研究室论文集 2, 40 (2004)

On Specifying Requirements Using a Semantically Controlled Representation

Imran Sarwar Bajwa[1] and M. Asif Naeem[2]

[1] School of Computer Science, University of Birmingham
B152TT Birmingham, United Kindom
[2] Department of Computer Science, University of Auckland
Auckland, New Zealand
i.s.bajwa@cs.bham.ac.uk, mnae006@aucklanduni.ac.nz

Abstract. Requirements are typically specified in natural languages (NL) such as English and then analyzed by analysts and developers to generate formal software design/model. However, English is ambiguous and the requirements specified in English can result in erroneous and absurd software designs. We propose a semantically controlled representation based on SBVR for specifying requirements. The SBVR based controlled representation can not only result in accurate and consistent software models but also machine process able because SBVR has pure mathematical foundation. We also introduce a java based implementation of the presented approach that is a proof of concept.

Keywords: Requirement Specifications, SBVR, Natural Language Processing.

1 Introduction

Since early days of software engineering, the use of a natural language (NL) to specify requirements [1] has been a challenge due to inherent semantic ambiguities of NL [2]. In this paper, we propose the use of a semantically controlled representation [3] for specifying requirements. The used representation is based on OMG's recently introduced standard, Semantic of Business Vocabulary and Rules (SBVR) [4]. A SBVR based representation can be beneficial in manifold aspects: SBVR is easy to understand for humans as it is close to NL, SBVR is easy to machine process as SBVR has a pure mathematical foundation [4].

The presented approach works as the user captures the requirements from users in simple English and inputs a piece of English specification of requirements. NL2SBVR approach [5] processes English text, extracts SBVR vocabulary (such as concepts, fact types, etc) and then uses SBVR vocabulary to construct SBVR business rules. SBVR Conceptual Formalization [4] and Semantic Formulation [4] are applied on SBVR vocabulary to generate SBVR rules. Finally, the SBVR Structured English [4] notation is applied to form a complete SBVR rules.

R. Muñoz et al. (Eds.): NLDB 2011, LNCS 6716, pp. 217–220, 2011.
© Springer-Verlag Berlin Heidelberg 2011

2   NL to SBVR Approach

The NL to OCL is a modular NL-based approach that generates SBVR based semantically controlled representation from English text in following two steps:

2.1   Processing NL Text

Following is the detail of processing of English text:

Lexical Processing. In lexical processing, the input English text is read and the sentence margins are identified to split sentences. Each sentence is stored separately and passed to parts-of-speech tagger to identify the basic POS tags for input text. For POS tagging, the Stanford POS tagger v3.0 [6] is used that can identify 44 POS tags.

Syntactic and Semantic Interpretation. In this syntactic processing, various parts of a sentence are identified [7] and in semantic interpretation phase [8], the basic SBVR vocabulary is extracted using following mapping rules:

- All proper nouns are mapped to the individual concepts
- All common nouns are mapped object types.
- All auxiliary verbs and action verbs are mapped to verb concepts.
- The adjectives and possessive nouns are mapped to the characteristics.
- All cardinal numbers are mapped to quantification.

2.2   Generating SBVR Rule

In this phase, a SBVR business rule is constructed from the SBVR vocabulary generated in previous phase. SBVR rule is generated in two phases as following:

Applying Semantic Formulation. A set of semantic formulations such as logical formulations, modal formulations, quantifications are applied to each fact type to construct a SBVR rule.

Applying Structured English Notation. The last step in generation of a SBVR is application of the Structured English notation in SBVR 1.0 document, Annex C [4].

3   A Case Study

A small case study is discussed from the domain of robot systems that was originally presented by Callan [15] and [16]. The problem statement for the case study is as follows:

> *"An assembly unit consists of a user, a belt, a vision system, a robot with two arms, and a tray for assembly. The user puts two kinds of parts, dish and cup, onto the belt. The belt conveys the parts towards the vision system. Whenever a part enters the sensor zone, the vision system senses it and informs the belt to stop immediately. The vision system then recognizes the type of part and informs the robot, so that the robot can pick it up from the belt. The robot picks up the part, and the belt moves again. An assembly is complete when a dish and cup are placed on the tray separately by the arms of the robot."*

The problem statement of the case study was given as input (NL specification) to the tool. The tool parses and semantically interprets English text and extracts the SBVR vocabulary from the case study as shown in table 1:

Table 1. SBVR vocabulary generated from English text

Category Count Details	#	Details
Object Types /General Concepts	13	assembly_unit, user, belt, vision_system, robot, tray, dish, cup, parts, senso_zone, senses, arms, parts
Verb Concepts	10	consists, puts, conveys, enters, informs, stop, recognizes, pick, moves, placed
Individual Concepts	00	-
Characteristics	01	Type
Quantifications	10	1 Each, 7 at least *n*, 2 exactly *n*
Unary Fact Types	03	vision_system *senses*; belt *stops;* belt *moves;* assembly *is_complete*
Associative Fact Types	08	belt *conveys* parts, parts *enter* sensor_zone, vision_system *informs* belt, vision_system *recognizes* part, vision_system *informs* robot, robot *picks_from* belt, robot *picks* part, dish and cup *placed_on* tray
Partitive fact Types	01	assembly_unit *consists of* user, ….
Categ. Fact Types	01	user *puts* two kinds of parts

Here, types and arms are characteristics but wrongly classified as object types. Similarly, a verb concept senses is wrongly characterized as noun concept. Here, the fact types are further processed to generate SBVR rule types. There were one structural constraint and six behavioural constraints as shown in table 2:

Table 2. SBVR Rule representation of structural and behavioural constraints

Category	#	Details
Structural Requirements	01	It is possibility that each assembly unit *consists* of at least one user, at least one belt, at least one vision system, at least one robot with exactly two arms, and at least one tray for assembly.
Behavioural Requirements	06	It is permitted that the user *puts* exactly two kinds of parts, dish and cup, onto the belt.
		It is permitted that the belt *conveys* the parts towards the vision system.
		It is obligatory that at least one part *enters* the sensor zone, implies the vision system senses and *informs* the belt to *stop* immediately.
		It is obligatory that the vision system *recognizes* the *type* of part and *informs* the robot, so that it is obligatory that the robot *can pick* from the belt.
		It is obligatory that the robot *picks* the part, and the belt *moves* again.
		It is obligatory that an assembly *is* complete when at least one dish and cup *are placed* on the tray separately by the arms of the robot.

A matrix representing software constraints accuracy (recall and precision) test (%) for structural and behavioural constraints has been constructed.

Table 3. Evaluatin results of NL to OCL

Example	N_{sample}	$N_{correct}$	$N_{incorrect}$	$N_{missing}$	Rec%	Prec%
Structural	16	15	1	0	93.75	93.75
Behavioural	27	24	2	1	88.88	92.30
Total	43	39	2	2	73.3	93.02

The results of this initial performance evaluation are very encouraging and support both the approach adopted and in this paper and the potential of this technology in general.

4 Conclusion

The primary objective of the paper was to address the ambiguous nature of natural languages (such as English) and generate a controlled representation of English so that the accuracy of machine processing can be improved. To address this challenge we have presented a NL based automated approach to parse English software requirement specifications and generated a controlled representation using SBVR. Automated object oriented analysis of SBVR based software requirements specifications using NL2SBVR provides a higher accuracy as compared to other available NL-based tools.

The future work is to extract the object-oriented information from SBVR specification of software requirements such as classes, instances and their respective attributes, operations, associations, aggregations, and generalizations. Automated extraction of such information can be helpful in automated conceptual modelling of natural language software requirement specification.

References

1. Denger, C., Berry, D.M., Kamsties, E.: Higher Quality Requirements Specifications through Natural Language Patterns. In: Proceedings of IEEE International Conference on Software-Science, Technology & Engineering (SWSTE 2003), pp. 80–85 (2003)
2. Ormandjieva, O., Hussain, I., Kosseim, L.: Toward A Text Classification System for the Quality Assessment of Software Requirements written in Natural Language. In: 4th International Workshop on Software Quality Assurance (SOQUA 2007), pp. 39–45 (2007)
3. Kuhn, T., Controlled English for Knowledge Representation. Doctoral Thesis. Faculty of Economics, Business Administration and Information Technology of the University of Zürich (2010)
4. OMG. Semantics of Business vocabulary and Rules. (SBVR) Standard v.1.0. Object Management Group, (2008), http://www.omg.org/spec/SBVR/1.0/
5. Bajwa, I.S., Lee, M.G., Bordbar, B.: SBVR Business Rules Generation from Natural Language Specification. In: AAAI Spring Symposium 2011, San Francisco, USA, pp. 2–8 (2011)
6. Toutanova, K., Manning, C.D.: Enriching the Knowledge Sources Used in a Maximum Entropy Part-of-Speech Tagger. In: The Joint SIGDAT Conference on Empirical Methods in Natural Language Processing and Very Large Corpora, Hong Kong, pp. 63–70 (2000)
7. Bajwa, I.S., Samad, A., Mumtaz, S.: Object Oriented Software modeling Using NLP based Knowledge Extraction. European Journal of Scientific Research 35(01), 22–33 (2009)
8. Bajwa, I.S., Choudhary, M.A.: A Rule Based System for Speech Language Context Understanding. Journal of Donghua University (English Edition) 23(6), 39–42 (2006)

An Unsupervised Method to Improve Spanish Stemmer

Antonio Fernández[1], Josval Díaz[1], Yoan Gutiérrez[1],
and Rafael Muñoz[2]

[1] Departamento de Informática
Universidad de Matanzas.
Autopista a Varadero, Matanzas Cuba
{antonio.fernandez,josval.diaz,yoan.gutierrez}@umcc.cu
[2] Departamento de Lenguaje y Sistemas Informáticos
Universidad de Alicante, España
rafael@dlsi.ua.es

Abstract. We evaluate the effectiveness of using our edit distances algorithm to improving an unsupervised language-independent stemming method. The main idea is to create morphological families through the automatic words grouping using our distance. Based on that grouping, we make a stemming process. The capacity of the edit distance algorithm in the task of words clustering and the ability of our method to generate the correct stem for Spanish was evaluated. A good result (98% precision) for the morphological families' creation and also a remarkable 99.85% of correct stemming was obtained.

Keywords: Stemming, words clustering, edit distance, morphological families.

1 Introduction

Perhaps the first developed linguistic stemming was the Lovins [1] algorithm. He uses varied lists of exceptions to remove different endings. Paice-Husk's stemmer [2] was another rules based algorithm. Other famous is Porter's stemmer [3]. It is a lineal sequential algorithm considered one of the better. The principal drawback of linguistic stemmer is his language-dependent nature. Before Porter's approach, the successor variety stemmer [4] was developed, the problem with this algorithm is the cut-off method used. In 1993 Krovetz uses frequency of English derivational endings as the basic for his stemmer. Smirnov reported in [5] that they are: Corpus-based [6], Context Sensitive [7] and the N-gram stemming [8]. They do not need a prior linguistic knowledge; instead they use unsupervised training. In spite of the increment in the stemming study for language other than English [9], [10], however the majority of the stemmers have been done for English. We differed in some aspects from other words: the way of morphological characteristic acquisition, our method for pre-process the input corpus is not case sensitive and has no problem with special character; in our distance some etymological characteristics are taken into account to improve the morphological families' formation. Also a charters penalization is introduced; it concedes vulnerability to orthography. This paper show a new algorithm based on an Extended Edits Distance (EDx) for automatic grouping words; we named it

R. Muñoz et al. (Eds.): NLDB 2011, LNCS 6716, pp. 221–224, 2011.
© Springer-Verlag Berlin Heidelberg 2011

Morphological Families (MF). Our proposal does not need neither external resource, nor knowledge about the language we are performing the stemming for.

2 Grouping MFs Using Extended Edit Distance

The algorithm is created utilizing the matrix generated for the ED using the same way adopted in [11]. As a result, we get a character string with the edit operations, where every O character will be part of the common subsequence and the concatenation of all them represents the Longest Common Subsequence (LCS). We obtain the LCS, EDx, and a different attributes for similarity determination. In addition, we offer the position where the modifications have come true, also the kind they are. We also give penalty costs that depend on which characters are transformed. More detail of the EDx is presented in [12].

$$EDx = \sqrt[8]{\frac{\sum_{i=0}^{l-1} V_{(O_i)} * \left(P_{(c1_j)}, P_{(c1_k)} \right) \cdot (2L+1)^{i-1}}{\sum_{i=0}^{L-1} (2R_{max}-1) \cdot (2L+1)^{i-1}}} \tag{1}$$

where: V - Transformations accomplished on the words (O, I, D, S). O- Not operations at all, I- Insertion, D- Deletion, S- Substitution. We formalize V as a vector: $((0,0):O)$, $((1,0):I)$, $((0,1):D)$, $((1,1):S)$. $j=j+1$ If O_i is different from (I), $j=j$ otherwise. $k=k+1$ If O_i is different from (D), $k=k$ otherwise. c1 and c2- The examined words, $c1_j$- The j-th character of the word c1, $c2_k$- The k-th character of the word c2. P- An assigned weight for each character, $P_{(c1_j)}$ - Character weight in c1j, $P_{(c2_k)}$- Character weight in $c2_k$. L- Edit operation length, l- The biggest word length of the language. O_i- Operation at (i) position.

The equation (1) resulted from simples hypotheses about similarity between two words: Two words are similar if their differing letters occur outermost possible of the root and these differences are less than a threshold. The algorithm consists of the following steps: First of all, a list of unique words is obtained from a corpus of the language we are performing the stemming for. For our experiment we used 28390 words from the Spanish terms list of MiKTeX 2.6 Dictionary [13]. The algorithm works as follow:

Step 1-Organize the list of words alphabetically. Step 2-The first word of the list is taken as a pivot and will be compared, utilizing the EDx, with every other word until a range R. Step 3-The resulting list must be sorted taking into consideration the comparison. So, words closer to the pivot will be at the very beginning of the list. Step 4-A cut-off point is applied in order to separate the possible MFs. A word will be part of the MF only if it is true the first comparison EDx(x, yo) ≤ 0.1. If not, the MF will be formed only by the pivot. Step 5-They will be selected, in order to belong to the MF, all the words which the comparisons obey ΔEDx=EDx(x, yi+1) - EDx(x,yi) ≤ E, where E is a threshold experimentally determined. Step 6-The obtained MF is registered and eliminated from the original list before repeating the process, that will be continue until the list will end up.

3 Experiments

Thanks to the lexicon of MFs, made with our EDx, the algorithm for the stemming can be reduced to search for the common initial subsequence (CIS) of this last word. CIS will be the minimal necessary stem to conflate the different words in a MF. Two experiments were conducted; one to evaluate the capability of the EDx to conflate the correct words. The second was to verify the precision of the stemming process.

We utilized the lexicon of 28390 to conflating Spanish words with the EDx and a significant sample (384 words) was calculated for the described population. We decide to verify 356 MF instead of 384 words. There are 10628 words in the 356 MF selected. They were obtained at random trying to select a representative group for each alphabet letter. The verifications come true interrogating the index generated by a computer program designed with regard to this matter and contrasted with the sample of 356 MF, elaborated by a group of experts. In order to evaluate our proposal we adapted the well-know recall and precision measure as follow.

$$\text{precision} = \frac{\text{CPF}}{\text{CPF+FP}} \quad (2); \quad \text{recall} = \frac{\text{CPF}}{\text{CPF+FN}} \quad (3); \quad \text{accuracy} = \frac{\text{CPF}}{\text{CPF+FN+FP}} \quad (5)$$

where: CPF: identified words that belong to a MF; FP: identified words that not belong to a MF; FN: not identified words that belong to a MF.

$$\text{Fm}\beta = \frac{(\beta^2+1)*\text{precision}*\text{recall}}{\beta^2*\text{precision}+\text{recall}} \quad (4)$$

The results for the MF creation with our algorithm were: precision=0.985240289, recall=1, accuracy=0.990417723 and Fm$_\beta$=0.98416281

In another experiment the results for the stem's determination was checked. Each stem is obtained at random from the index and is verified and compared with the hand-made list. We get errors only in 16 of 356 MF. Errors were provoked by the inclusion of several wrong words in the MF; there were not FP at all in the 356 MF. This is caused for the selected static cut-off point E=0.08. We decide to do the same experiment with Porter's stemmer and we get the following results:

Table 1. Comparison with porter's algorithm

Stemmer	recall	precision	accuracy
Our	1	0,98	0,99
Porter	0,53	0,99	0,53

4 Conclusions and Future Works

Using the EDx, for the MF's creation, a substantial 98% of precision and a maximum recall is obtained, this mean a good capability of our algorithm to conflate words. The usefulness of the MF to develop a good stemmer was demonstrated with the experiments. A good cut-off point's selection guaranties a more accurate stem. Precisely, this is the main drawback of the algorithm. The FP and FN introduced in the MFs where due to the static cut-off point used. Nevertheless, it was not necessary to define the language's affixes, neither his behavior, nor to construct the rules of affixes elimination. All these steps get substituted for the process of grouping the MF and then the CIS selection. The experiment of stemming only show 0.15% wrongly estimated stems. In the future we expect to create a dynamic cut-off point for truncating the MFs.

References

1. Lovins, J.B.: Development of a stemming algorithm. Mechanical Translation and Computational Linguistic 11(1-2), 22–31 (1968)
2. Paice, C.D.: Another Stemmer. ACM SIGRIR Forum 24 (3), 56–61 (1990)
3. Porter, M.: An Algorithm for suffix stripping. Program 14(3), 130–137 (1980)
4. Hafer, M., Weiss, S.: Word segmentation by letter succesor varieties. Information Storage and Retrieval 10, 371–385 (1974)
5. Smirnov, I.: Overview of Stemming Algorithms. DePaul University (2008)
6. Jinxi, X., Bruce, C.: Corpus-based stemming using co-ocurrence of word variants. ACM Transactions on Information Systems 16, 61–81 (1998)
7. Peng, F., Ahmed, N., Li, X., Lu, Y.: Context sensitive stemming for web search. In: Proceedings of the 30th Annual International ACM SIGIR Conference on Research and Development in Information Retrieval, pp. 639–646 (2007)
8. James, M., Paul, M.: Single N-gram stemming. In: Proceedings of the 16th Annual International ACM SIGIR Conference on Research and Development in Information Retrieval, pp. 415–416 (2003)
9. Popovic, M.F., Willet, P.: The effectiveness of stemming for natural language accesess to Slovene textual data. Journal of American Society for Information Science 3(5), 384–390 (1992)
10. Braschler, M., Schäuble, P.: Experiments with the Eurospider REtrieval System for CLEF 2000. In: Peters, C. (ed.) CLEF 2000. LNCS, vol. 2069, pp. 140–148. Springer, Heidelberg (2001)
11. Levenshtein, V.I.: Binary codes capable of correcting deletions, insertions, and reversals. Soviet Physics - Doklady 10, 707–710 (1966)
12. Fernández Orquín, A., Díaz, J., Fundora, A., Muñoz, R.: Un algoritmo para la extracción de características lexicográficas en la comparación de palabras. In: IV Convención Científica Internacional CIUM, Matanzas, Cuba (2009)
13. Knuth, D.E.: MiKTeX 2.6 (May 28, 2007), http://www.miktex.org

Using Semantic Classes as Document Keywords[*]

Rubén Izquierdo[1], Armando Suárez[1], and German Rigau[2]

[1] GPLSI Group, University of Alicante, Spain
{ruben,armando}@dlsi.ua.es
[2] IXA NLP Group, EHU, Donostia, Spain
german.rigau@ehu.es

Abstract. Keyphrases are mainly words that capture the main topics of a document. We think that semantic classes can be used as keyphrases for a text. We have developed a semantic class–based WSD system that can tag the words of a text with their semantic class. A method is developed to compare the semantic classes of the words of a text with the correct ones based on statistical measures. We find that the evaluation of semantic classes considered as keyphrases is very close to 100% in most cases.

1 Introduction

Keyphrases are mainly words that capture the main topics of a document. Keyphrases can be single words or compound words or multiwords. They can provide a high level view of the content of a document. For example, they can be used by a reader to know if some document is relevant for him or not. In this sense, the authors of a lot of scientific papers are required to include some keyphrases describing the paper. These topics can be used also for classifying the paper.

There are more applications of keyphrases. For example in Document Summarization, keyphrase represent the meaning of a whole document in a few topics, providing a very concise summary. Also in clustering and Information Retrieval, keyphrases have an important role. In general, keyphrases provide a powerful way for representing the meaning and content of a document in a bunch of topics. There are a lot of documents that have no keyphrases pre–assigned, and doing it manually is a hard task. Hence, the development of automatic techniques for perform keyphrase identification is a very interesting research area. There are two sub-types of this task: keyphrase extraction (selecting some relevant phrases from a document) and keyphrase assignment (assign a predefined list of categories as keyphrases).

The work in [6] presents one of the first developed systems for keyphrase extraction. It was implemented by means of a genetic algorithm. This algorithm

[*] This paper has been supported by the European Union under the project KYOTO (FP7 ICT-211423), the Valencian Region Government under projectss PROMETEO/2009/119 and ACOMP/2011/001 and the Spanish Government under the project TEXT MESS 2.0 (TIN2009-13391-C04-01) and KNOW2 (TIN2009-14715-C04-01).

R. Muñoz et al. (Eds.): NLDB 2011, LNCS 6716, pp. 225–229, 2011.
© Springer-Verlag Berlin Heidelberg 2011

adjusts a set of heuristic rules to extract the correct keyphrases. [7] developed the KEA system, a machine learning system for keyphrase extraction. Recently, the interest in keyphrase extraction has reemerged, leading to the development of several new systems and techniques. Moreover, in last SemEval conference[1], a specific task was proposed in order to evaluate the performance of keyphrase extraction systems. The task was called *Automatic Keyphrase Extraction from Scientific Articles*.

2 Using Semantic Classes as *Keyphrases*: Evaluation

Our idea is to use semantic classes as keyphrases for representing the main topics of a document. We think than the semantic classes of the words can be informative about the topics of a text. This effect could be due to the semantic classes provide a high semantic level of abstraction, not considering individual words but general concepts. To obtain these semantic classes, we employ a semantic class based *Word Sense Disambiguation* (from now on WSD) system that we have developed to tag each noun and verb with its proper semantic class.

We have developed a WSD based on machine learning and semantic classes. Our system uses an implementation of a Support Vector Machine algorithm to train the classifiers using SemCor [4] Corpus for acquiring examples. The system uses set of traditional features for representing these examples, expanded with two semantic features. More details about the WSD system can be found in [2].

Semantic classes are concepts that subsume sub–concepts and words with a related meaning and common characteristics. We use two semantic class sets: the first one automatically extracted from WordNet, the Basic Level Concepts, and the second one, SuperSense, was created manually by the developers of WordNet. On the one hand, **Basic Level Concepts**[2] (BLC) [1] are small sets of meanings representing the whole nominal and verbal part of WN. **SuperSenses** (SS) are based on open syntactic categories (nouns, verbs, adjectives and adverbs) and logical groupings, such as person, phenomenon, feeling, location, etc.

Our method of keyphrase extraction consists of considering the semantic classes of the words within a document, and select the most frequent ones as keyphrases for the document. One important criterion is the selection of what words are considered to calculate the overall set of semantic classes. Semantic classes of domain general words are maybe not informative and can introduce noise, so they must not be taken into account. To remove this general words, we use a filter, implemented by means of an statistical measure traditionally used in the field of Information Retrieval: the TF-IDF measure [3]. To apply this formula we need a big corpus belonging to a general domain, and The British National Corpus[3] (BNC) is chosen[4].

---

[1] http://semeval2.fbk.eu

[2] http://adimen.si.ehu.es/web/BLC

[3] http://www.natcorp.ox.ac.uk

[4] The BNC is a 100 million word collection of samples of written and spoken language from a wide range of sources, designed to represent a wide cross-section of current English.

To evaluate our approach we use the corpora from the WSD All Words tasks from SensEval-2[5], SensEval-3[6] and SemEval-1[7]. We use these corpora due to the texts contained on them are supposed to be domain specific and show a good coherence to perform keyphrase extraction and topic detection. Moreover, existing corpora for keyphrase extraction are not well designed, and are mainly focused on keyphrase assignment, and are not designed for keyphrase extraction. Specifically, each test corpus (SensEval-2, SensEval-3 and SemEval-1) consists of three independent documents.

We weigh each word in a document according to the TF–IDF formula. The formula is calculated combining two values: the frequency of a word within a document (SenEval of SemEval document), and the number of BNC files in which the word appears. So, words that are related with the specific domain of the document (words with a high document frequency) and are not general domain words (words contained on a low number of BNC files), will have a high TF–IDF value. Once we have all words of a document weighted, we sort them according to this weight. We propose two experiments for calculating the semantic classes of a document: (1) consider the semantic classes of all words within the document, and (2) consider the semantic classes of the 99% words with higher weight. As explained before, in the second experiment we expect to remove general domain words.

3   Results

We have three ranks of semantic classes: the guessed by our system (G), the correct rank (C), and the most frequent rank (M). We compare pairs of ranks by means of the Spearman's rank correlation[5]. This is a non-parametric measure of statistical dependence between two variables. In other words, it gives a measure of the similarity or two sorted ranks, and this fits very well with our case.

We first show the evaluation using **BLC–20**[8] **semantic classes**. In table 1 the values for the Spearman's rank correlation are shown, including the experiment considering all words (no filtering), and the experiment considering only the 99% words with higher weight (TF–IDF filtering). We can see that in the most cases, the values are very close to 1, what indicates identical ranks. Moreover, in a lot of cases, our rank (G) is more similar to the correct (C) than the most frequent rank (M). The behaviour of the system in the three corpora is similar, obtaining higher results in SensEval-3 corpus. We can also see that the results considering TF–IDF filtering are not always higher that without filtering. This is maybe due to the filtering process is not well tuned, and some general domain words removed by the filter contain actually relevant semantic information about the document. Anyway, the results are very close to 1, indicating

---

[5] http://86.188.143.199/senseval2

[6] http://www.senseval.org/senseval3

[7] http://nlp.cs.swarthmore.edu/semeval

[8] BLC–20 is a kind of BLC where each BLC concept must subsume at least 20 sub-concepts.

Table 1. Spearman values for BLC–20

Corpus	File	No filtering			TF–IDF filtering		
		G–C	G–M	M–C	G–C	G–M	M–C
SV2	d00	0.9974	0.9961	0.9906	0.9948	0.9894	0.9773
	d01	0.9906	0.9901	0.9697	0.9761	0.9661	0.9930
	d02	0.9899	0.9983	0.9905	0.9835	0.9979	0.9828
SV3	d000	0.9981	0.9995	0.9978	0.9969	0.9987	0.9962
	d001	0.9885	0.9962	0.9897	0.9786	0.9966	0.9637
	d002	0.9977	0.9998	0.9981	0.9966	0.9991	0.9971
SEM1	d00	0.9515	0.9893	0.9103	0.9253	0.9940	0.8593
	d01	0.9514	0.9995	0.9514	0.9040	1	0.9040
	d02	0.9721	0.9444	0.9690	0.9560	0.9995	0.9679

Table 2. Spearman's values for SuperSense

Corpus	File	No filtering			TF–IDF filtering		
		G–C	G–M	M–C	G–C	G–M	M–C
SV2	d00	0.7994	0.7325	0.6150	0.8464	0.6091	0.3275
	d01	0.7069	0.6151	-0.2917	0.7309	0.6029	-0.3221
	d02	0.7875	0.9754	0.7911	0.5182	0.9231	0.4818
SV3	d000	0.8029	0.9201	0.8431	0.8765	0.9456	0.9235
	d001	-0.4923	0.9701	-0.2352	0.5099	0.9868	0.5099
	d002	0.8820	0.9802	0.8323	0.9657	0.9816	0.9397
SEM1	d00	0.4524	0.6426	0.5	0.4405	0.9818	0.5
	d01	0.9182	0.5412	0.9318	0.9030	0.9955	0.9212
	d02	0.5955	-0.1448	0.9126	0.3333	0.7394	0.9273

that we can use our BLC–20 semantic class WSD system to detect and extract relevant concepts representing the main topics of a document. In other words, we can use our semantic class WSD system as keyphrase extractor.

Similarly, we perform the same experiment considering the **SuperSense**-based WSD system. The results of the Spearman's rank correlation are shown in table 2. In this case the results are considerably lower, although SuperSense semantic classes have a higher level of abstraction and the average polysemy is lower, and the results could be expected to be higher. This can indicate that the BLC concepts work very well as keyphrases. The words within a document seem to maintain a certain semantic coherence considering BLC classes, whereas this coherence is not hold in the case of SuperSense, maybe due to this classes are too coarse–grained.

4   Conclusions

In this paper we propose the use of a semantic class–based WSD system in order to perform keyphrase extraction. To do this, we use different kinds of semantic classes as keyphrases, expecting semantic classes to represent the topic information of a document. We represent the topic information of a document starting from the semantic classes assigned automatically by the WSD system to the words within the document. Then we use an statistical measure, the Spearman's rank correlation, to evaluate our system, and compare the semantic classes as keyphrases with the correct ones.

In general, we obtain very good results with the Basic Level Concepts–based system, better than using Supersense. Therefore, the performance of the keyphrase extractor based on a WSD system not depends only in the abstraction level of the semantic classes used, but also in the discriminative power and the coherence of the set.

References

1. Izquierdo, R., Suarez, A., Rigau, G.: Exploring the automatic selection of basic level concepts.In: Angelova, G., et.al. (eds.) International Conference Recent Advances in Natural Language Processing, Borovets, Bulgaria, pp. 298–302 (2007)
2. Izquierdo, R., Suárez, A., Rigau, G.: An empirical study on class-based word sense disambiguation. In: Proceedings of the 12th Conference of the European Chapter of the Association for Computational Linguistics, Athens, Greece, EACL 2009, pp. 389–397. Association for Computational Linguistics, Stroudsburg (2009)
3. Jones, K.S.: A statistical interpretation of term specificity and its application in retrieval. Journal of Documentation 28, 11–21 (1972)
4. Miller, G., Leacock, C., Tengi, R., Bunker, R.: A Semantic Concordance. In: Proceedings of the ARPA Workshop on Human Language Technology (1993)
5. Spearman, C.: The proof and measurement of association between two things. The American Journal of Psychology 15(1), 72–101 (1904)
6. Turney, P.D.: Learning algorithms for keyphrase extraction. Inf. Retr. 2, 303–336 (2000), http://portal.acm.org/citation.cfm?id=593957.593993
7. Witten, I.H., Paynter, G., Frank, E., Gutwin, C., Nevill-Manning, C.G.: KEA: Practical Automatic Keyphrase Extraction. In: Proceedings of Digital Libraries 1999 (DL'99), pp. 254–255 (1999),
http://citeseerx.ist.psu.edu/viewdoc/summary?doi=10.1.1.55.3127

Automatic Term Identification by User Profile for Document Categorisation in Medline

Angelos Hliaoutakis and Euripides G.M. Petrakis

Intelligent Systems Laboratory, Electronic and Computer Engineering Dept.
Technical University of Crete (TUC), Chania, Crete, Greece
{angelos,euripides}@intelligence.tuc.gr
http://www.intelligence.tuc.gr

Abstract. We show how term extraction methods such as $AMTE_X$ and MMT_X can be used for the automatic categorisation of medical documents by user profile (novice users and experts). This is achieved by mapping document terms to external lexical resources such as WordNet, and MeSH (the medical thesaurus of NLM).

Keywords: term extraction, document indexing, Medline, MeSH, MMTx, AMTEx.

1 Introduction

Medical information systems such as MedLine[1] must be capable of providing dedicated, domain specific answers to experts or, simple, easy to comprehend answers to novice users respectively. MedLine documents are currently indexed by human experts by assigning to each one, a number (typically 10 to 12) of terms, deriving from the MeSH[2] (Medical Subject Headings) thesaurus. The automatic mapping of biomedical documents to UMLS Meta-thesaurus[3] term concepts has been undertaken by the U.S. National Library of Medicine (NLM) with the development of MMTx[4] (MetaMap Transfer tool). $AMTE_X$ [1] aims at improving the efficiency of MMT_X based on the extraction and mapping of document terms to the MeSH Thesaurus, rather than the full UMLS Meta-thesaurus mapping of MMT_X. It is therefore more selective resulting in more compact representations than MMT_X.

In this work, we show how MMT_X and $AMTE_X$ can be used for filtering medical information for targeted audiences such as experts and novice users. An obvious application of this filtering operation is retrieval of medical information by user profile.

---

[1] http://www.nlm.nih.gov/databases/databases_medline.html
[2] http://www.nlm.nih.gov/mesh
[3] http://www.nlm.nih.gov/research/umls
[4] http://mmtx.nlm.nih.gov

R. Muñoz et al. (Eds.): NLDB 2011, LNCS 6716, pp. 230–233, 2011.
© Springer-Verlag Berlin Heidelberg 2011

2   Document Categorisation by User Profile

MeSH terms are distinguished into i) general medical terms expressing known concepts (e.g., "pain", "headache") which are easily conceived by all users, ii) domain specific terms which are used mainly by experts, iii) general - non medical terms. Fig. 1 illustrates the respective categorisation of Medline documents and MeSH terms.

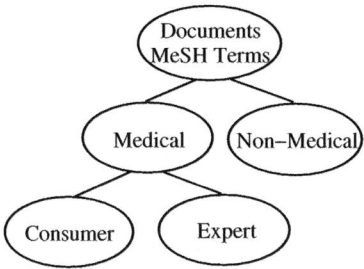

Fig. 1. Categorisation of Medline documents and MeSH terms

By combining information from WordNet[5] and MeSH the following three term vocabularies are constructed:

Vocabulary of General Terms (VGT): these are terms that belong to WordNet but not to MeSH:

$$VGT = (WordNet) - (MeSH)$$

It follows that VGT contains 105.675 general (WordNet) terms.

Vocabulary of Consumer Terms (VCT): these are terms that belong to both, WordNet and MeSH:

$$VCT = (WordNet) \cap (MeSH)$$

It follows that VCT contains 7,165 consumer (MeSH) terms.

Vocabulary of Expert Terms (VET): these are MeSH terms that do not belong to WordNet:

$$VET = (MeSH) - (Wordnet)$$

It follows that VET contains 16,719 consumer (MeSH) terms.

The more expert (or consumer) terms a document contains, the higher its probability to be a document suitable for experts (or consumers respectively). Document categorisation by user profile is realized by computing the percentage of expert (VET) and consumer (VCT) terms in a document term vector.

---

[5] http://wordnet.princeton.edu

For example, a document with VET% = 0.62 has 62% probability of being a document suitable for experts.

We design an information retrieval method capable of both i) ranking documents by similarity with a query, and ii) bringing documents matching a given user profile higher in the ranked list of similar documents. Documents are represented by term vectors [2] extracted by $AMTE_X$ or MMT_X respectively. As it is typical in information retrieval (IR), the similarity between a query and a document is computed by matching their term vectors according to VSM [2]. More specifically, the query is matched against all Medline documents and the returned list of documents is ranked by decreasing similarity. For ranking query results by user profile we distinguish between the following two cases:

Known user profile: The user identifies her/himself as an expert (or consumer) prior to issuing a query. The similarity score by VSM is multiplied by its percentage of VET (or VCT) terms that is, its probability of being a document for experts (or consumers respectively).

Unknown user profile: The system determines her/his profile from the query. If the query contains at least one expert term, the user is considered to be an expert (a consumer otherwise). Retrievals are then processed similar to the previous case.

3   Evaluation

The experimental results are obtained using OHSUMED, a standard TREC[6] collection of 348,566 medical document abstracts from Medline, published between 1988-1991. OHSUMED is commonly used in benchmark evaluations of IR applications. OHSUMED provides 64 queries and the relevant answer set (documents) for each query. The correct answers were compiled by the editors of OHSUMED and are also available from TREC. For the evaluations, we applied all 64 queries available.

To evaluate our document categorisation method we run a retrieval experiment on the OHSUMED dataset. The results were evaluated against all 64 TREC provided queries and answers; 15 out of the 64 queries contain no expert terms and are suitable for consumer users. The remaining queries are suitable for experts. The objective is to measure the ability of a method in retrieving information for consumer and expert users respectively. We run this experiment twice, once for experts and once for consumer users. A method is deemed successful if it retrieves documents suitable for the particular type of users under consideration.

The candidate methods are i) retrieval using vectors of the manually assigned MeSH terms ii) retrievals using the MMT_X extracted terms and iii) retrievals using terms extracted by $AMTE_X$. The performance of each method is represented by a precision/recall plot. Each point in such a plot is computed as the average

---

[6] http://trec.nist.gov/data/t9_filtering.html

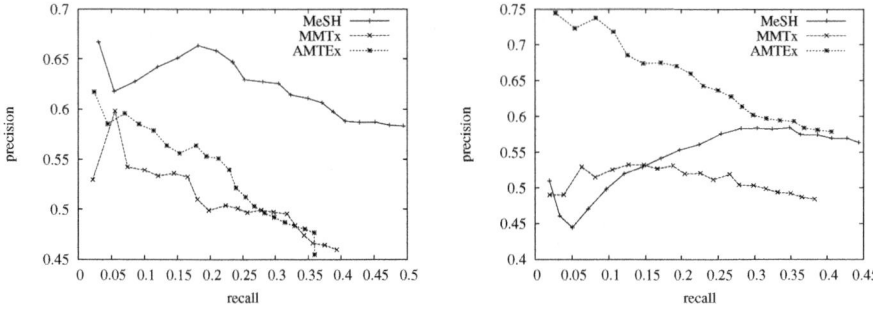

Fig. 2. Average precision/recall for the consumer (left) and expert users (right) retrieval task

precision/recall over all queries. Each method retrieves the best 20 answers for each TREC query so that, each plot contains exactly 20 points.

Fig. 2 (left) illustrates the relative performance of the three retrieval methods examined for the consumer retrieval task. Retrievals with the manually assigned MeSH terms performs better than any other method. This result reveals a tendency of the human indexers to assign simpler terms for the indexed documents. Both $AMTE_X$ and MMT_X perform similarly.

Fig. 2 (right) illustrates results for the retrieval experiment for expert users. $AMTE_X$ outperforms all other methods. This experiment demonstrates the selective ability of $AMTE_X$ towards extracting complex medical terms which can be found in the majority of Medline documents. It also reveals a weakness of manually assigning MeSH terms to documents, as the human indexers may be not familiar with the content complexity and specificity of Medline publications.

4   Conclusions

We introduce to the research community the problem of automatic categorisation of medical publications in Medline by user profile by investigating two common types of users of medical information (i.e., consumers and experts). Based on our experiments, we conclude that $AMTE_X$ selective term output is more effective than MMT_X (the state-of-the-art method of the U.S. NLM).

References

1. Hliaoutakis, A., Zervanou, K., Petrakis, E.: The $AMTE_X$ Approach in the Medical Document Indexing and Retrieval Application. Data and Knowledge Engineering 68(3), 380–392 (2009)
2. Baeza-Yates, R., Ribeiro-Neto, B.: Modern Information Retrieval. Addison Wesley, Longman (1999)

AZOM: A Persian Structured Text Summarizer

Azadeh Zamanifar and Omid Kashefi

School of Computer Engineering, Iran University of Science and Technology, Tehran, Iran
az_zamanifar@comp.iust.ac.ir,
kashefi@{ieee.org, iust.ac.ir}

Abstract. In this paper we propose a summarization approach, nicknamed AZOM, that combines statistical and conceptual property of text and in regards of document structure, extracts the summary of text. AZOM is also capable of summarizing unstructured documents. Proposed approach is localized for Persian language but easily can apply to other languages. The empirical results show comparatively superior results than common structured text summarizers, also than existing Persian text summarizers.

Keywords: Summarization, Persian, Fractal Theory, Statistic, Structure, Conceptual.

1 Introduction

Summarization is a brief and accurate representation of input text such that the output covers the most important concepts of the source in a condensed manner [1]. The summarization process could be extractive or abstractive. Extractive summaries contain sentences that are copied exactly from the source document [2]. In abstractive approaches, the aim is to derive the main concept of the source text, without necessarily copying its exact sentences [2]. The traditional text summarization methods use statistical properties of the text such as term frequency, sentence position, and cue terms. Researchers enhance the traditional statistical methods using multiple techniques such as extract the relation between words by lexical database in order to generate more coherent summaries [3], and applying rhetorical structure analysis [4]. In this paper we propose a summarization approach in regards of statistical, conceptual, and structural features of the document for Persian text.

2 Proposed Summarization Approach

Referring to our studies, there is not any notable Persian summarizers that consider the structure of the document along with conceptual properties. In this paper we present an automatic text summarizer considering statistical, conceptual, and structural features of the text. The proposed approach uses a lexical database in order to determine the relationship between words as conceptual feature of the text. Proposed approach consists three steps includes (1) preprocessing of the text, (2) text interpretation, and (3) summary generation that are described in follow.

R. Muñoz et al. (Eds.): NLDB 2011, LNCS 6716, pp. 234–237, 2011.
© Springer-Verlag Berlin Heidelberg 2011

2.1 Preprocessing

In order to reduce the dimensionality, the original text must be preprocessed [5]. At the first, the text is segmented in order to detect the boundary of the sentences [6]. Then the stop words are eliminated. After that, each word is lemmatized using the method proposed in [7] and inflected forms of words are unified. Now the whole text is converted to a uniform text in which the text processing can be applied.

2.2 Interpretation

In this step, whole document is scanned and the structure of the text such as chapters, sections, paragraphs and sentences are extracted by constructing the corresponding fractal tree of the document. If the document does not have any structure, nothing is done in this phase. Then, each word is looked up at Persian lexical in order to extract the relations (e.g. synonym, hypernym and hyponymy) between words.

2.3 Statistical Weighting

Static score of each term T_i can be calculated by modified version of entropy as Equation 1.

$$W(T_i) = 1 + log\left(\frac{tf_{ir}/f_i}{M}\right) \times tf_{ir}/f_i \tag{1}$$

Where tf_{ir} is the frequency of T_i in block r and f_i is the total frequency of T_i in the whole document; M is the number of blocks in the document. If the document does not have any structure, term frequency of each word in the document is considered.

2.3.1 Conceptual Weighting

After calculating the statistical weight of each term, we update the weight of each term in each block with the summation of the weights of the terms in its lexical chain according to the lexical relation type. If the relationship between terms is synonyms, the weights of both terms are updated by summation of the weights of their corresponding synonym terms. If the relationship between terms is not synonym and they are in the same block, the weight of each term sums up with 0.7 weight of other term; if they are not in the same block, the weight is increased by half of the weight of the other word. Equation 2 shows the conceptual weighting of terms.

$$W(T_i) = \sum_{T_j \in Document, \ Relation(T_j, T_i) = synonym} W(T_j) + 0.7 \times$$
$$\sum_{T_i, T_j \in B_k, \ Relation(T_j, T_i) \neq sysnonym} W(T_j) + 0.5 \times \tag{2}$$
$$\sum_{T_i \in B_k, \ T_j \in B_m, \ k \neq m, \ Relation(T_j, Ti) \neq synonym} W(T_j)$$

2.3.2 Structural Weighting

Next step is scoring each sentence of a block. The weight of each sentence is the summation of each word's score divided by total number of sentence words. The score of sentence k is computed as (Equation 3).

$$Score(S_k) = \frac{\sum_{T_i \in S_k} W(T_i)}{Num(T_i)} \tag{3}$$

Raw score of each block (i.e. $Score(B_r)$) is calculated as the summation of the score of its sentences divided by total number of its sentences. To normalize the score, raw score of each block is divided to the raw scores of its sibling blocks in corresponding fractal tree of the current text (Equation 4).

$$NormalizedScore(B_r) = \frac{Score(B_r)}{\sum_{B_s \in Sibling(B_r)} Score(B_s)} \qquad (4)$$

Therefore, in text interpretation step, terms are statistically weighted by entropy metric, terms weight are updated through conceptual potential of the text, sentences are weighted based on terms' weights, and document blocks are weighted regarding their sentences' weights and the structure of the text.

2.4   Summary Generation

We generate the summary based on fractal theory. To extract sentences from each block according to the importance of that block, the block that is more important have more sentences in summarized text. Therefore, the normalized score for each block is calculated according to Equation 6. Compression ratio is variable and can be adaptive to the user request.

$$NumOfSummarizedSentence_{B_r} = CompressionRatio \times NormalizedScore(B_r) \times$$
$$\sum_{B_s = B_r or B_s \in Sibling(B_r)} NumOfSentence(B_s) \qquad (6)$$

3   Evaluation

Usually, performance of a summarization technique is calculated by comparing the results with manual (intrinsic) extracted summary; but to our knowledge, since there is not any manual extracted summary corpus in Persian, we employ a novel strategy to evaluate the effectiveness of automatic summarization. We consider the abstract part of the scientific and scholarly papers as the ideal manual summarization. Abstract of the scientific and scholarly papers are written by educated authors who try to capsulate the summaries of all section of the document in abstract, so it is a good candidate for manual summary. We have used 100 different Persian scientific papers to construct the benchmark. We use the abstract of each paper as ideal summary, and the body of the paper (i.e. except the abstract, keywords, acknowledgment, and references sections) as the original text. We compare our result with fractal based method proposed by Yang and Wang [8] that is one of the few structured summarized approach with good result and our previous method [9] that is one of the few text summarization method with good result which is applied on Persian. We also compare the results with flat summary which is the same as our proposed method except the fact that whole document is considered as one block.

As it is shown in Table 1, if we compare the exact sentences of abstract and the output of our approach, the precision and recall would not be high. It is because of the fact that human abstract does not necessarily contain the exact sentences of the text. Therefore we calculated the similarity of extracted summaries with the abstract based on the method proposed in [10]. The results are shown in Table 2. As it is shown in Table 2 precision and recall significantly increase compared to Table 1. As it is shown in Table 1 and Table 2 the precision and result in score based summary- in which the importance factor of block is considered in order to determine the number of sentences from each block- is better than distributed summary, where equal number of sentences are extracted from each block, this is because of the fact that in most of the cases human extract more sentences form important block of the text.

Table 1. Comparative evaluation result with matching sentences

Compression Rate	Parameter	Our Approach	Fractal Yang	Flat Summary	Co-occurrence
Distributed Structured Summary	Precision	**0.64**	0.61	0.55	0.45
	Recall	**0.65**	0.57	0.53	0.48
Score based structured Summary	Precision	**0.73**	0.61	0.57	0.42
	Recall	**0.71**	0.60	0.55	0.4

Table 2. Comparative evaluation result with matching similarity

Compression Rate	Parameter	Our Approach	Fractal Yang	Flat Summary	Co-occurrence
Distributed Structured Summary	Precision	**0.73**	0.67	0.62	0.55
	Recall	**0.71**	0.68	0.59	0.58
Distributed Structured Summary	Precision	**0.81**	0.71	0.65	0.54
	Recall	**0.76**	0.70	0.66	0.50

References

1. Luhn, H.P.: The Automatic Creation of Literature Abstract. IBM Journal of Research and Development 2, 159–165 (1958)
2. Yatsko, V.A.: Special Features of the Communication Syntatical Structure of Summary Utterances. NTI 2, 1–5 (1993)
3. Barzilay, R., Elhadad, M.: Using Lexical Chains for Text Summarization. In: Proceedings of the ACL Workshop on Intelligent Scalable Text Summarization, Spain, pp. 10–17 (1997)
4. Mann, W.C., Thompson, S.A.: Rhetorical Structure Theory: A Theory of Text Organization (1987)
5. McCarty, L.T.: Deep Semantic Interpretations Of Legal Texts. In: Proceedings of the 11th International Conference on Artificial Intelligence and Law, USA, pp. 217–224 (2007)
6. Reynar, J.C., Ratnaparkhi, A., Maximum, A.: Entropy Approach to Identifying Sentence Boundaries. In: Proceedings of the 5th Conference on Applied Natural Language Processing, pp. 16–19 (1997)
7. Kashefi, O., Mohseni, N., Minaei, B.: Optimizing Document Similarity Detection in Persian Information Retrieval. Journal of Convergence Information Technology 5, 101–106 (2010)
8. Yang, C.C., Wang, F.L.: Hierarchical Summarization of Large Document. American Society or Information Science and Tecnology 10, 888–902 (2008)
9. Zamanifar, A., Minaei-Bidgoli, B., Sharifi, M.: A New Hybrid Farsi Text Summarization Technique Based on Term Co-Occurrence and Conceptual Property of the Text. In: 9th SNPD Conference, pp. 635–639. IEEE Computer Society, Thiland (2008)
10. Zamanifar, A., Minaei, B., Kashefi, O.: A New Technique for Detecting Similar Documents based on Term Co-occurrence and Conceptual Property of the Text. In: Int. Conf. on Digital Information Management, England, pp. 526–531 (2008)

POS Tagging in Amazighe Using Support Vector Machines and Conditional Random Fields

Mohamed Outahajala[1,4], Yassine Benajiba[2], Paolo Rosso[3], and Lahbib Zenkouar[4],

[1] Royal Institut for Amazighe Culture, Morocco
[2] Philips Research North America, Briacliff Manor, USA
[3] NLE Lab – EliRF, DSIC, Universidad Politécnica de Valencia, Spain
[4] Ecole Mohammadia d'Ingénieurs, Morocco
outahajala@ircam.ma, yassine.benajiba@philips.com,
prosso@dsic.upv.es, zenkouar@emi.ac.ma

Abstract. The aim of this paper is to present the first Amazighe POS tagger. Very few linguistic resources have been developed so far for Amazighe and we believe that the development of a POS tagger tool is the first step needed for automatic text processing. The used data have been manually collected and annotated. We have used state-of-art supervised machine learning approaches to build our POS-tagging models. The obtained accuracy achieved 92.58% and we have used the 10-fold technique to further validate our results.

Keywords: POS tagging, Amazighe language, supervised learning.

1 Introduction

The part-of-speech (POS) tagging task consists of disambiguating the category of each word in a sentence and tagging them with the adequate lexical category, i.e. part-of-speech. This enriches the text by providing a more abstract layer which in its turn endows other NLP tasks to better perform [1]. In the literature, proof is abound that the most effective approaches to build an automatic POS-tagger are based on supervised learning machines, i.e. relying on a manually annotated corpus and often other resources, such as dictionaries and word segmentation tools, to pre-process the text and extract features. In our approach we use sequence classification techniques based on two state-of-art machine learning approaches, namely: Support Vector Machines (SVMs) and Conditional Random Fields (CRFs), to build our automatic POS-tagger. We use a ~20k tokens manually annotated corpus [9] to train our models and a very cheap feature set consisting of lexical context and character n-grams to help boost the performance.

2 Related Work on POS Tagging

POS tagging has been well researched both for English and other European and non European languages. The very first POS taggers were mainly rule-based systems. Building such systems requires a huge manual effort in order to handcraft the rules

R. Muñoz et al. (Eds.): NLDB 2011, LNCS 6716, pp. 238–241, 2011.
© Springer-Verlag Berlin Heidelberg 2011

and to encode the linguistic knowledge which governs the order of their application. For instance, in 1970 Green and Robin [5] developed a system named TAGGIT containing about 3,000 rules and achieving an accuracy of 77%. Later on, machine learning based POS-tagging proved to be both less laborious and more effective than the rule based ones. In the literature, many machine learning methods have been successfully applied for POS tagging, namely: the Hidden Markov Models (HMMs) [4], the transformation-based error driven system [3], the decision trees [11], the maximum entropy model [10], SVMs [6], CRFs [7]. Results produced by statistical taggers obtain about 95%-97% of correctly tagged words. There are also, hybrid methods that use both knowledge based and statistical resources.

3 Amazighe Language

The Amazighe language is spoken in Morocco, Algeria, Tunisia, Libya, and Siwa (an Egyptian Oasis); it is also spoken by many other communities in parts of Niger and Mali. Due to its complex morphology as well as the use of the different dialects in its standardization (Tashlhit, Tarifit and Tamazight the three more used ones), the Amazighe language presents interesting challenges for NLP researchers which need to be taken into account.

Defining the adequate tag set is a core task in building an automatic POS tagger. It aims at defining a processable tag set which provides enough information to be used by the potential federate systems. In [8], a tag set containing 13 elements (verb, noun, adverb…etc.) was developed. For each element we define morpho-syntactic features and two common attributes: "wd" for "word" and "lem" for "lemma", whose values depend on the lexical item in question. The utilized tag set comprises 15 tags representing the major parts of speech in Amazighe plus 2 tags assigned to words containing two morphemes: N_P and S_P. This tag set is derived from the larger one presented in [8]. Gender, person and number information have not been included in the tag set and were considered as a separate investigation subject to be pursued in the future.

4 Experiments and Error Analysis

Our corpus consists of a list of texts extracted from a variety of sources [9], annotated using AnCoraPipe tool [2]. Annotation speed of this corpus was between 80 and 120 tokens/hour. Randomly chosen texts were revised by different annotators. On the basis of the revised texts inter-annotator agreement is 94.98%.Common remarks and corrections were generalized to the whole corpora in the second validation by a different annotator.

The training process has been carried out by YamCha[1], an SVM based toolkit. Also, we have used CRF++[2], an open source implementation of CRFs. In this paper, we explore two features sets. Both of them are based on the actual text and are very cheap to extract. In the first feature set, we use the surrounding words in a window of -/+2, and their POS tags. The second feature set, we add to the first feature set

---

[1] http://chasen.org/~taku/software/yamcha/
[2] http://crfpp.sourceforge.net/

character n-gram feature which consists of the last and first i character n-gram, with i spanning from 1 to 4.

We have taken 50 corpora size points, with a step of 25 sentences between each two points. An extract of 10 points of the results are summarized in Fig.1. They have been done using SVMs and CRFs with and without lexical features using 10-fold cross validation.

The POS tagger based on CRF++54 tool achieves an accuracy of 88.66% when using lexical features, so an improvement of 3.98% when using data without lexical features. The results of the tagger based on SVMs with lexical features gave an accuracy of 88.27%, so an improvement of 6.3% when using data without lexical features. With a data set of 1438 sentences and applied to 15 tags.

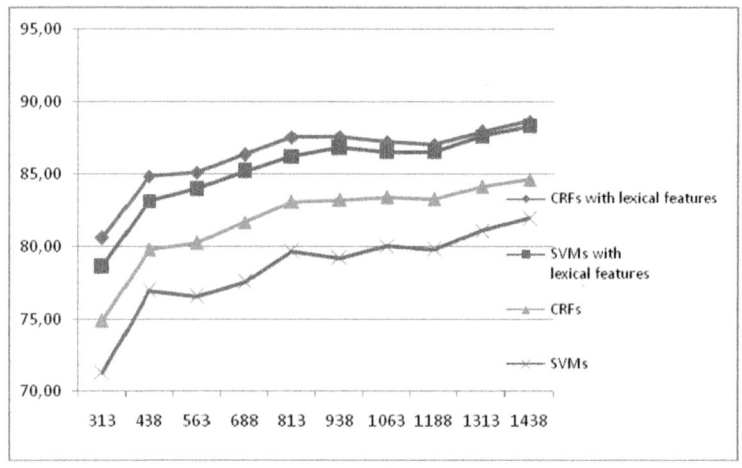

Fig. 1. Accuracy performance

We have run 10-fold cross validation over the corpus, i.e., training on 90% of the 1295 sentences and tagging the remaining 10%, with the experiment repeated 10 times, each time taking a different slice of the corpus.

By examining the confusion matrices of both SVMs and CRFs outputs, we found that adjectives are frequently tagged as nouns. This is due to the fact that adjectives may act as nouns. In line with this, many Amazighe linguists gave the name of quality nouns to adjectives. Error rate of pronouns is also high due to the large overlap between them and the determinants. Another common source of errors is verbs. The POS tagger based on CRFs tagged 4.5% of verbs as nouns and adjectives and 0.6% as prepositions, whereas the POS tagger based on SVMs tagged 6% of verbs as nouns and adjectives. Besides SVMs based POS tagger have better results in tagging kingship names, pronouns, determinants, prepositions when used together with pronouns, focalizers, particles and adverbs.

5 Conclusions and Further Work

In this paper we describe the morpho-syntactic features of the Amazighe language. We have addressed the design of a 15 tags tag set and two POS taggers based on SVMs and CRFs. The POS tagger based on CRFs achieves 88.66% whereas the POS tagger based on SVMs and using lexical features yields the accuracy of 88.27% based on a small corpus using 10-fold technique.

We are currently trying to improve the performance of the POS tagger by using additional features and more annotated data based on semi-supervised techniques and Active Learning. In addition, we are planning to approach base phrase chunking by hand labeling the already annotated corpus.

Acknowledgements. We would like to thank all IRCAM researchers for their valuable assistance. The work of the third author was funded by the MICINN research project TEXT-ENTERPRISE 2.0 TIN2009-13391-C04-03 (Plan I+D+i).

References

1. Benajiba, Y., Diab, M., Rosso, P.: Arabic Named Entity Recognition: A Feature-Driven Study. IEEE Transactions on Audio, Speech and Language Processing, Special Issue on Processing Morphologically Rich Languages, 15(5), 926–934 (2010), doi:10.1109/TASL.2009.2019927
2. Bertran, M., Borrega, O., Recasens, M., Soriano, B.: AnCoraPipe: A tool for multilevel annotation. Procesamiento del lenguaje Natural, Madrid,Spain, vol. (41) (2008)
3. Brill, E.: Transformation-Based Error-Driven Learning and Natural Language Processing: A Case Study in Part-of-Speech Tagging (1995)
4. Charniak, E.: Statistical Language Learning. MIT Press, Cambridge (1993)
5. Greene, B.B., Rubin, G.M.: Automatic Grammatical Tagging of English. Department of Linguistics. Brown University, Providence (1971)
6. Kudo, T., Matsumoto, Y.: Use of Support Vector Learning for Chunk Identification (2000)
7. Lafferty, J., McCallum, A., Pereira, F.: Conditional Random Fields: Probabilistic Models for Segmenting and Labeling Sequence Data. In: Proceedings of ICML 2001, pp. 282–289 (2001)
8. Outahajala, M., Zenkouar, L., Rosso, P., Martí, A.: Tagging Amazighe with AncoraPipe. In: Proc. Workshop on LR & HLT for Semitic Languages, 7th International Conference on Language Resources and Evaluation, LREC-2010, Malta, May 17-23, pp. 52–56 (2010)
9. Outahajala, M., Zenkouar, L., Rosso, P.: Building an annotated corpus for Amazighe. Will Appear in Proceedings of 4th International Conference on Amazigh and ICT, Rabat, Morocco (2011)
10. Ratnaparkhi, A.: A Maximum Entropy Model for Part-Of-Speech Tagging. In: Proceedings of EMNLP, Philadelphia, USA (1996)
11. Schmid, H.: Improvements in Part-of-Speech Tagging with an Application to German. In: Proceedings of the ACL SIGDAT-Workshop, pp. 13–26. Academic Publishers, Dordrecht (1999)

Towards Ontology-Driven End-User Composition of Personalized Mobile Services

Rune Sætre[1], Mohammad Ullah Khan[2], Erlend Stav[3],
Alfredo Perez Fernandez[1], Peter Herrmann[2], and Jon Atle Gulla[1]

[1] Computer and Information Science (IDI) NTNU
satre@idi.ntnu.no
[2] Telematics (ITEM) NTNU, 7491 Trondheim, Norway
[3] Information and Communication Technology (ICT)
SINTEF, 7031 Trondheim, Norway

Abstract. With mobile devices being an integral part of the daily life of millions of users having little or no ICT knowledge, mobile services are being developed to save them from difficult or tedious tasks without compromising their needs. Starting with a number of real life scenarios we have been working towards supporting end-users in managing their services in an efficient and user-friendly manner. We observe that these scenarios consist of sub-tasks that can be solved with collaborative service units. Therefore, a composition of such service units will serve the needs of the end-user for the complete scenario. We envisage that a visual formalism and tools can be developed to support these end-users in creating such service compositions. Moreover, methodologies and middleware can significantly reduce the complexity of developing composite services. This paper focuses on the role of ontologies within that context. Ontologies can assist the users in selecting appropriate services and setting composition parameters within a composition tool. For our prototype demonstration system we target the open source Android cellphone architecture supporting a number of different runtime platforms.

1   Introduction

The world of cell-phones and mobile WiFi-devices is changing rapidly. Modern mobile phones now have the same computational power as a desktop computer had only a few years ago. Many small programs and services are being made to harness this power in order to save the users from difficult or boring tasks. For example, the address book application on the phone remembers all the phone numbers for the user, so he or she can call someone just by typing (or saying) their name, or perhaps just by clicking on the correct portrait picture. We have studied several usage scenarios like "incoming call handling", "doctor's appointment" and "city-guiding". These scenarios consist of several sub-tasks that can be solved with various existing applications. We are developing a *composition tool* to allow non-experts (without programming experience) to compose their own services. The hypothesis is that a visual formalism and tools can be developed to support end-users in composing (their own) service collaborations. We

R. Muñoz et al. (Eds.): NLDB 2011, LNCS 6716, pp. 242–245, 2011.
© Springer-Verlag Berlin Heidelberg 2011

want to show that service composition can be supported by generic solutions in the form of methods and middleware that significantly reduce the complexity of developing these composite services. We have started implementing service- and domain-ontologies that allow reasoning about the user's intentions, and that can provide useful feedback in the compositional phase.

After a paper-prototyping phase to settle on an intuitive end-user composition notation and approach, and a comprehensive state-of-the-art survey, we are now moving into the runnable prototyping phase. In this paper we briefly provide the overall picture, while focusing on the usage of ontologies within the context of end-user service composition.

2   The City-Guide Scenario and the Ontology

We describe the City-guide scenario, which is one of the scenarios[1] that we are using to test our approach. We will first investigate the need for ontologies and then very briefly mention the concepts provided in the developed ontology to serve that purpose.

In this scenario, the composer creates an Android-based service which should assist a person (scientist or business man) in making a short guided tour in a not yet visited city within an afternoon. The available time is between the end of the conference or business meeting (which may be delayed), and the check-in time at the airport. The starting point is the meeting venue, and the final target is the airport. Means of transportations are walking, bus, train, taxi etc. The composer, still at home, wants to specify these constraints, together with a list of user profiles, stating what kind of sights are interesting to the user. For example, the *cultural* profile might include one visit to a church, like the *Nidaros Cathedral*, and one to a museum, e.g. the *Ringve Museum* in Trondheim.

To achieve that, the composer discovers the following service *building-blocks*[2]

1. Guided tours at the Nidaros Cathedral.
2. Guided tours at the Ringve Museum.
3. Bus time-tables and information about the necessary time to walk to and from the bus stops that are closest to the starting and the end point.
4. A call service for a taxi company.
5. A guiding map for walking from the present position to a certain point.
6. A scheduler for events, which provide alerts some time before the events.

Our scenarios are quite complex for an end-user composer, as it takes lot of effort to reflect the plan in the service composition. Also, we notice that the successful organization of events at different places seems to have a recurring pattern. It would be nice if we could relieve the composer from repetitive tasks such as transportation planning between composed events.

---

[1] More details on the other scenarios like "incoming call" or "doctor's appointment" are available in [2].

[2] A building block is a system abstraction used by a service composer. It can represent a component, a service or a system part in terms of a sequence of actions.

Ontologies can be of help here, since they define suitable terms which may guide the system to understand what it should do with the different building blocks. For example, they can be used to connect places of events with nearby places of transportation.

As the first step towards simplifying end-users' task in composing services with the intended support we have developed an ontology in OWL based on the presented scenario. This ontology will be integrated with the service composition tool (section 3) so that the service composer (an end-user) can benefit from it during the service composition. We foresee that the run time platform will also use the ontology in making automatic decisions; e.g., in the case of selecting alternative services.

The ontology consists of **classes (concepts)** representing different *entities* (OWL individuals). It contains entities like *Nidarosdomen (Nidaros Cathedral)* which is of subclass *Cathedral* in the class of *Church*. Other classes are *Museum*, *Station* and *Vehicle*. These concepts can be populated (semi-) automatically from a database of sights for the city/area, or from textual descriptions, using existing research, e.g., [1,3]. The classes also have properties, like location.

3   Methodology and Tools

We are developing a methodology that provides a step-by-step procedure both for the end-users creating service compositions and for the service developers creating individual services. This involves the usage of a composition tool that helps end-users to define the composition, using an end-user-friendly and intuitive notation Here we briefly introduce the contribution of the ontology within the methodology.

At design time, the composer creates application compositions by using the service building blocks. The composition is created in a scenario-like fashion; e.g., a particular block executes or a particular sequence of actions are performed only when a certain condition holds. Such conditions are dynamic in nature so that the evaluated result may depend on some information obtained at run time. Ontologies may be used to discover certain building blocks that help to solve a certain problem, in particular, to glue building blocks that should be executed in a specific sequence.

The composed scenarios are selected based on the conditions evaluated at run time. The run time platform should take care of the fact that the user's needs may change; e.g. the user may re-compose the application at run time. Ontologies can be used in identifying and selecting from alternative scenarios and services that serves the same composition and the same building block, respectively.

The methodology is supported by a meta-model representing the concepts, a set of end user-friendly notation and a composition tool that creates the composition taking into account. The tool is created using the Google Web Toolkit (GWT[3]), which is a development toolkit for building and optimizing complex

---

[3] http://code.google.com/webtoolkit/

browser-based applications. The composition tool works both on a PC and a mobile phone. We target end-users with limited or no ICT knowledge, but with the ability to use a PC or a mobile phone. Therefore, the composition tool also needs to be easy-to-use and easy-to-install. That triggered the choice of a composition tool that runs on a web browser. The tool development also takes advantage of tools like WindowBuilder[4] to work with GWT in addition to the automatic generation of editors from EMF meta-models.

4   Conclusion and Future Work

In this paper, we advocated the usage of *ontologies* within a methodology that supports end-user service compositions. End-users with minimum ICT knowledge needs intuitive, intelligent and easy-to-use notation and tools in managing their tasks. We argued that the use of ontologies can be of great help towards that direction. By analyzing different scenarios, we developed initial ontologies supporting different application domains, which are in the process of being integrated with an end-user service composition tool.

The usage of good ontologies can substantially make end-users' lives easy. Ontologies can help predicting what the user will type while composing a service. For example, there is only one *Church (or Cathedral)* with its name starting with the letter *N* in a particular city. These ontologies can be created either at compose-time for given cities, or for general-purpose (multi-city) services as soon as the composed service is started on the phone, and the GPS etc. is used to determine which city the user is in.

We are currently implementing the concepts presented in this paper. For our prototype demonstration system we target the open source Android cellphone architecture supporting a number of different runtime platforms. For example, we are currently developing end-user composition support for self-adaptive mobile applications running on the MUSIC platform. The implementation follows the incremental approach of updating the concepts and the implementation based on the feedback from user-testing on several different platforms/scenarios.

References

1. Gulla, J.A., Brasethvik, T.: A Hybrid Approach to Ontology Relationship Learning (chapter 9). In: Kapetanios, E., Sugumaran, V., Spiliopoulou, M. (eds.) NLDB 2008. LNCS, vol. 5039, pp. 79–90. Springer, Heidelberg (2008)
2. Shiaa, M.M., Vaskinn, J.E., Sanders, R.T.: Tool Chain for End-Users Service Composition. In: Proceedings of the 2010 VERDIKT Conference, p. 56 (November 2010)
3. Witte, R., Khamis, N., Rilling, J.: Flexible Ontology Population from Text: The OwlExporter. In: The Seventh International Conference on Language Resources and Evaluation (LREC 2010), pp. 3845–3850. ELRA, Valletta (2010)

---

[4] http://code.google.com/javadevtools/wbpro/

Style Analysis of Academic Writing

Thomas Scholz and Stefan Conrad

Heinrich-Heine-University Düsseldorf, Institute of Computer Science,
Universitätsstr. 1, D-40225 Düsseldorf, Germany
{scholz,conrad}@cs.uni-duesseldorf.de
http://dbs.cs.uni-duesseldorf.de

Abstract. This paper presents an approach which performs a Style Analysis of Academic Writing in terms of formal voice, readability and scientific language. Our intention is an analysis of academic writing style as a feedback for the authors and editors. The extracted features of a document collection are used to create Self-Organizing Maps which are the interim results to generate reports in our Full Automatic Paper Analysis System (Fapas). To evaluate this method, the system has to solve different tasks to verify the informative value of the generated maps and reports.

Keywords: Text Mining, Linguistic Style Analysis, User Interfaces and Visualization.

1   Motivation

The aim of this approach is a method which estimates style of writing. Today it is very important to share your ideas and developments with other people. Besides verbal presentations many new ideas are transferred by written texts. Therefore improving writing skills is very interesting for people who write publications or documentations. Furthermore, an automatic editorial department can be helpful for education institutes or companies who create or handle a great amount of written texts like publishers or press offices.

Also a qualitative style analysis can give hints for authorship identification and verification tasks and as well can be used to categorize texts of web pages. However, a major problem with this kind of application is the question: What is style in a text and how can it be evaluated? We design an approach which handles this problem in the specific domain of academic writing.

2   Related Work

In the field of analyzing style, some approaches [9,10] use lexical features to represent style in a text. These features are covering average lengths of words, token, sentences or word frequencies. By working with NLP techniques like Part-Of-Speech tagging, syntactic features can be extracted. Stamatatos et. al. [9] e.g. regard phrases like noun phrases, verbal phrases etc., while Stein and Meyer zu

R. Muñoz et al. (Eds.): NLDB 2011, LNCS 6716, pp. 246–249, 2011.
© Springer-Verlag Berlin Heidelberg 2011

Eissen [10] are looking for frequent POS-tags. A quite different way is the creation of big taxonomies to determine the effect of the written word. These function words are used by Koppel et. al. [1] to compute frequencies of their effects.

The most methods are trying to solve the authorship identification task [1,5,9], but some papers also identify the gender of the authors [6], their nationality [1], or deal with plagiarism detection [10]. But these approaches do quantify style and not qualify it (not even in specific domains). For text readability early approaches create indexes like [2] for the English language. These indexes are still used, e. g. in tools like the Readability Test Tool (RTT)[1].

3   Features for Style Analysis of Academic Writing

Now we present the feature set for our style analysis. Specialist literature about academic writing [3,7,8] is considered to design these features. To sum up, academic writing should be formal, concise and precise.

Feature Set Formal Voice (φ): The formal language is represented by three features. The System counts the usages of passive voice, subjective expressions and questions. E. g. domain experts concur that passive voice and little use of subjective words like "I", "my" or "our" are leading to a more formal voice.

Feature Set Readability (ρ): A text is easier to understand if its parts are well connected. So, the system identifies linking expressions like "furthermore" or "as a consequence". Also the average word count per sentence is computed and how much the authors tend to use nouns and noun phrases in their texts because if an author uses more long sentences and more nouns than verbs to explain processes, the sentences become harder to understand [8]. Likewise the system computes a staccato feature. This feature represents how much the author follows simple sentence structures [7].

Feature Set Language (λ): First of all, the system computes the proportion of academic words by identifying the most 200 words used in scientific texts (see [3] for the complete list). The second feature shows how many times the author uses hedging expression in contrast to statements. This shows how precise the announcements of the text are. To measure how precise the choice of words really is, the next feature counts the amount of synonyms and hyponyms of the words [5]. The last feature simply counts the amount of non stop words, because this is a another factor which expresses how concise the document is.

4   Implementation and Reports

Fapas generates Self-Organizing Maps [4] for a selected collection of analyzed documents. After the training process every document is mapped to its next output unit on the map. The red colour proportion of the unit stands for the formal style, green for the readability and blue for the language. The reports tell the user the differences between the style of a selected paper and area in detail.

---

[1] http://www.read-able.com/

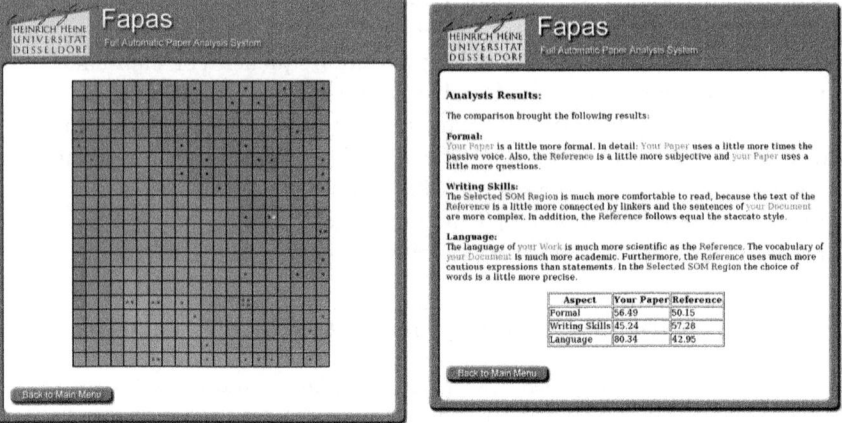

Fig. 1. A web interface shows the generated SOMs and reports

5 Evaluation

In the first task, 554 papers of 40 different authors were analyzed for authorship attribution. To use the SOMs for classification, and the test set is classified by a Nearest Neighbour Classifier on the output layer of the trained SOMs. Figure 2 (left) shows the accuracy results.

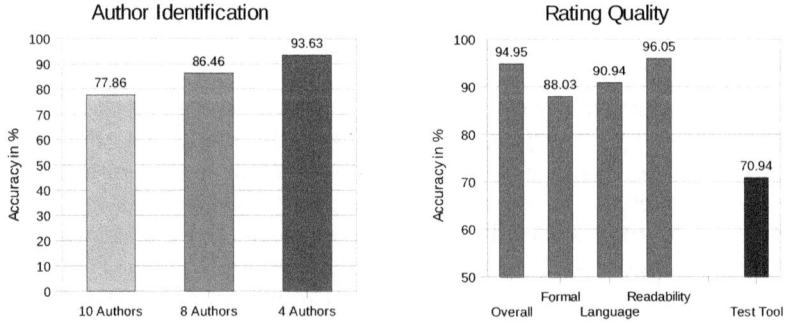

Fig. 2. Evaluations results: authorship attribution (left) and rating quality (right)

For the second task, we generate a corpus of 138 human rated documents. Then Fapas and the RTT have to make their review (in 1545 document-to-document comparisons). The RTT computes readability features like the Flesch Kincaid Readabilty Reading Ease [2], which we took for our comparison as the best index. The results are shown in figure 2 (right).

In the third task, the system has to classify documents as the right type and to rate the 3 feature aspects in the right way: novels have to get a higher in score

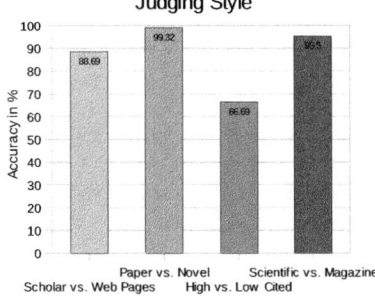

Category	Feature Set	Number
Google Scholar Article	$\varphi\rho\lambda$	800
Web Page		800
Paper	$\varphi\lambda$	365
Novel	ρ	365
High Cited	$\varphi\rho\lambda$	800
Low Cited		800
Scientific	λ	800
Magazine	$\varphi\rho$	800
Total		5530

Fig. 3. Classifying and rating style in different categories

in readability than scientific papers and so on. The strong feature aspects are shown in figure 3 (right). The left side shows the well performance.

6 Conclusion

The results suggest that our approach can effectively order, visualise and rate academic texts by style. Thus, the feedback can help to improve the user's writing abilities.

References

1. Argamon, S., Whitelaw, C., Chase, P., Hota, S.R., Garg, N., Levitan, S.: Stylistic text classification using functional lexical features: Research articles. J. Am. Soc. Inf. Sci. Technol. 58(6), 802–822 (2007)
2. Flesch, R.: A new readability yardstick. Journal of Applied Psychology 32, 221–233 (1948)
3. Gillett, A., Hammond, A., Martala, M.: Successful Academic Writing. Pearson, London (2009)
4. Kohonen, T., Schroeder, M.R., Huang, T.S. (eds.): Self-Organizing Maps. Springer-Verlag New York, Inc, Heidelberg (2001)
5. Koppel, M., Akiva, N., Dagan, I.: Feature instability as a criterion for selecting potential style markers: Special topic section on computational analysis of style. J. Am. Soc. Inf. Sci. Technol. 57(11), 1519–1525 (2006)
6. Mukherjee, A., Liu, B.: Improving gender classification of blog authors. In: EMNLP 2010, Morristown, NJ, USA, pp. 207–217 (2010)
7. Saran, A.: Sound - its importance in the short story. Writing Fiction (2009)
8. Skern, T.: Writing Scientific English. Facultas Wuv (2009)
9. Stamatatos, E., Fakotakis, N., Kokkinakis, G.: Computer-based authorship attribution without lexical measures. In: Computers and the Humanities, pp. 193–214 (2001)
10. Stein, B., Zu Eissen, S.M.: Intrinsic plagiarism analysis with meta learning. In: SGIR 2007 (2007)

Towards the Detection of Cross-Language Source Code Reuse

Enrique Flores, Alberto Barrón-Cedeño, Paolo Rosso, and Lidia Moreno

Dpto. de Sistemas Informáticos y Computación, Universidad Politécnica de Valencia
{eflores,lbarron,prosso,lmoreno}@dsic.upv.es

Abstract. Internet has made available huge amounts of information, also source code. Source code repositories and, in general, programming related websites, facilitate its reuse. In this work, we propose a simple approach to the detection of cross-language source code reuse, a nearly investigated problem. Our preliminary experiments, based on character n-grams comparison, show that considering different sections of the code (i.e., comments, code, reserved words, etc.), leads to different results. When considering three programming languages: C++, Java, and Python, the best result is obtained when comments are discarded and the entire source code is considered.

Keywords: Source code reuse, cross-language source code reuse analysis, plagiarism detection.

1 Introduction

In the digital era, massive amounts of information are available causing the material from other people to be exposed to reuse. Therefore, there is high interest in identifying whether a work has been reused.

As for documents in natural language, the amount of source code in Internet is huge, facilitating the reuse of all or part of previously implemented programs. People facing similar problems are frequently tempted to source code reuse and, if no reference to the original work is included, plagiarism.[1] As a counter measure different models for the automatic detection of source code reuse (in particular plagiarism) have been developed [2,3,6]. The challenge of cross-language reuse detection has been approached just recently [1].

Let L_1 and L_2 be two programming languages ($L_1 \neq L_2$), we define cross-language source code reuse as the translation of (part of) a source code $a \in L_1$ into $a' \in L_2$. As for texts written in natural language [5], detecting code reuse when a translation process occurred, is even more challenging; it is very likely that a' does not represent an exact translation of a because of implementation issues.

---

[1] Source code reuse is often allowed, thanks to licenses as those of Creative Commons (http://creativecommons.org/), but if the information related to the source is not included, plagiarism is being committed.

R. Muñoz et al. (Eds.): NLDB 2011, LNCS 6716, pp. 250–253, 2011.
© Springer-Verlag Berlin Heidelberg 2011

As far as we know the only approach that aims to detect cross-language source code reuse is that of [1]. Instead of processing source code, this approach compares intermediate language (RTL) produced by a compiler. The comparison is in fact monolingual and compiler dependent. Unfortunately, the corpus used is not available, making the direct comparison to this approach unfeasible.

This contribution represents a preliminary work attempting to detect source code reuse among three programming languages: C++, Java and Python on the basis of Natural Language Processing techniques.

2   Model

The following levels of edition in monolingual reuse are proposed by [2]:

0. No changes in source code.
1. Represents the changes in comments and indentation.
2. Includes level 1 plus changes in identifiers.
3. Groups level 2 and changes in declarations (i.e. declaring extra constants, changing the positions of declared variables, etc.).
4. Represents level 3 plus changes in program modules (i.e. merging procedures).
5. Comprises level 4 and changes in program statements (i.e. using *for* instead of *while*).
6. Represents level 5 and changes in control logic.

Our proposal aims to treat some of these levels, considering: (*i*) full code, i.e., source code and comments, for level 0; (*ii*) full code without comments (fc-without comments) for level 1; (*iii*) programming language reserved words only (fc-reserved words only) for levels 2 and 3. Additionally, three more exploratory experiments have been carried out: (*iv*) comments only (*v*) full code without reserved words (fc-without rw) and (*vi*) full code without comments and without reserved words (fc-wc-wrw).

The proposed model is divided in three steps: (*a*) Pre-processing: line breaks, tabs and spaces removal as well as case folding; (*b*) Features extraction: character n-grams extraction, weighting based on normalised tf; and (*c*) Comparison: cosine similarity estimation.

Once a' is compared to $a \in A$, a sorted list is generated that ranks the potential sources for the suspicious program a'. The top k pairs (a', a) in the ranked list are the most similar and, therefore, more likely to be reused.

3   Experiments

Our aim is to evaluate the proposed model to detect cross-language reuse between source codes. Our toy corpus is composed of a collection of programs including source code in C++, Java and Python (the programs are formerly part of a multi-agents system). For each language a collection of programs exist that maintains a correspondence to the programs in the other languages. The collections in C++ and Java have been partially reused. The cases Python→C++ represent real examples of cross-language reuse. The cases Python→Java represent simulated cases. Moreover, the cases C++−Java represent triangular reuse (having Python as pivot). Table 1 shows some statistics of the corpus.

Table 1. Statistics of the corpus used for the experiments

Language	Tokens	Avg. length of tokens	Types	Types per program	Programs
C++	1,318	3.46	144	28.8	5
Java	1,100	4.52	190	47.5	4
Python	10,503	3.24	671	167.75	4

We have tested our character n-grams model considering $n=\{1,\ldots,5\}$. Table 2 shows the average and standard deviation of the positions of that document a that is the source of a'. In the most of the experiments the best result is obtained when considering *full code* as well as *full code without comments* with the same values in both cases. The best results are obtained with $n = 3$. This is in fact the same cases as for text reuse detection [5].

Table 2. Results obtained with character 3-grams. The value represents the mean and standard deviation of the position of the source program in the ranked list.

Features	Java − C++	Python → C++	Python → Java
full code	1.00 ± 0.00	1.44 ± 0.83	1.62 ± 1.10
fc-without comments	1.00 ± 0.00	1.44 ± 0.83	1.62 ± 1.10
fc-reserved words only	1.56 ± 0.83	1.78 ± 1.02	1.75 ± 0.83
comments only	2.29 ± 1.57	2.83 ± 1.34	3.00 ± 0.67
fc-without rw	1.44 ± 0.83	1.78 ± 1.13	2.00 ± 1.32
fc-wc-wrw	1.44 ± 0.83	1.67 ± 0.94	1.44 ± 0.69

The comments in the source code has not had much impact, partly because the programmer has decided to rewrite the comment, write their own comments, or because they have not taken into account the comments when reusing the code. As a malicious programmer can modify the comments to introduce noise in the detection, it is better to ignore these sections of the program. The best results were obtained between the C++ and Java codes because they include common reused fragments from the Python implementations. Evidently, the syntax and vocabulary of C++ and Java is highly similar.

4 Conclusions and Future Work

This work is a preliminary attempt to detect cross-language source code reuse. The proposed approach is based on similarity computations at character n-grams level. The impact of comments, variable names, and reserved words of the different programming languages has been investigated. The best results are obtained when comments are ignored. This suggests that the comments can be safely discarded when aiming to determine the cross-language similarity between two programs. Presumably, the character 3-grams are able to represent programming style as in the case of documents written in natural language. No improvement was observed when weighting with tf-idf, but this could be due to the small corpus. Further experiments have to be carried out out on a larger corpus.

As future work, we identify the following avenues: (i) employing sliding windows [7] in order to compare blocks of codes, letting the location of similar fragments (for instance, a function at the beginning of a program could have been plagiarised and located at the end of another one); and (ii) applying cross-language alignment-based similarity analysis [4] that recently have given good results for texts (the necessary dictionary could be composed of reserved words).

Acknowledgments. This work has been developed with the support of the project TEXT-ENTERPRISE 2.0: Text comprehension techniques applied to the needs of the Enterprise 2.0 (MICINN, Spain TIN2009-13391-C04-03 (Plan I+D+i)).

References

1. Arwin, C., Tahaghoghi, S.M.M.: Plagiarism Detection across Programming Languages. In: Proceedings of the 29th Australasian Computer Science Conference, vol. 48, pp. 277–286 (2006)
2. Faidhi, J., Robinson, S.: An empirical approach for detecting program similarity and plagiarism within a university programming environment. Comput. Educ. 11, 11–19 (1987)
3. Jankowitz, H.T.: Detecting plagiarism in student pascal programs. The Computer Journal 31(1) (1988)
4. Pinto, D., Civera, J., Barrón-Cedeño, A., Juan, A., Rosso, P.: A statistical approach to crosslingual natural language tasks. Journal of Algorithms 64(1), 51–60 (2009)
5. Potthast, M., Barrón-Cedeño, A., Stein, B., Rosso, P.: Cross-Language Plagiarism Detection. Languages Resources and Evaluation. Special Issue on Plagiarism and Authorship Analysis 45(1) (2011)
6. Rosales, F., García, A., Rodríguez, S., Pedraza, J.L., Méndez, R., Nieto, M.M.: Detection of plagiarism in programming assignments. IEEE Transactions on Education 51(2), 174–183 (2008)
7. Stamatatos, E.: Intrinsic Plagiarism Detection Using Character n-gram Profiles. In: Proc. SEPLN 2009, Donostia, Spain, pp. 38–46 (2009)

Effectively Mining Wikipedia for Clustering Multilingual Documents

N. Kiran Kumar, G.S.K. Santosh, and Vasudeva Varma

International Institute of Information Technology, Hyderabad, India
{kirankumar.n,santosh.gsk}@research.iiit.ac.in, vv@iiit.ac.in

Abstract. This paper presents Multilingual Document Clustering (MDC) using Wikipedia on comparable corpora. Particularly, we utilized the cross lingual links, category, outlinks, Infobox information present in Wikipedia to enrich the document representation. We have used Bisecting k-means algorithm for clustering multilingual documents based on the document similarities. Experiments are conducted based on the usage of English and Hindi Wikipedia. We have considered English and Hindi Datasets provided by FIRE'10[1] for Ad-hoc Cross-Lingual document retrieval task on Indian languages. No language specific tools are used, which makes the proposed approach easily extendable for other languages. The system is evaluated using F-score and Purity measures and the results obtained are encouraging.

Keywords: Multilingual Document Clustering, Wikipedia, Document Representation.

1 Introduction

MDC is the grouping of text documents written in different languages, into semantically related groups. It plays a significant role in managing huge number of multlingual text documents present in the web. It has got applications in Cross Lingual Information Retrieval (CLIR) [1] systems, training of the parameters in statistical machine translation, among others. The text documents are represented as "bag of words" in traditional text clustering methods. The semantic information of the documents is not considered and this may result in forming false clusters. The most common way to solve this problem is to add semantic information to each document by enriching it's representation with an external knowledge or an ontology.

In this paper, we focus on using annotated multilingual Wikipedia structure (cross lingual links, outlinks, categories, etc.) in enriching the document representation. The content of a Wikipedia article is annotated by hyperlinks (references) to other articles and they denote the "outlinks" for that article.

---

[1] Forum for Information Retrieval and Extraction-http://www.isical.ac.in/~clia/

R. Muñoz et al. (Eds.): NLDB 2011, LNCS 6716, pp. 254–257, 2011.
© Springer-Verlag Berlin Heidelberg 2011

2 Related Work

MDC is normally applied on parallel [2] or comparable corpus [3]. In the case of the comparable corpora, the documents usually are news articles. In [4], the authors used existing knowledge structure EUROVOC thesaurus for measuring cross lingual document similarity. However the EUROVOC thesaurus supports only European languages. Steinberger *et al.* [5] proposed a method to extract language-independent text features using gazetteers and regular expressions besides thesaurus and classification systems. However, the gazetteers support only a limited set of languages. Hu *et al.* [6] has exploited Wikipedia concepts and categories for monolingual document clustering. In our previous work [7] we have implemented Centroid Similarity based MDC (CS-MDC) using Wikipedia. In this paper we have studied availing Wikipedia in enhancing the performace of MDC based on the Document Similarities (DS-MDC).

3 Proposed Approach

Each document (English or Hindi) in the dataset is represented with a Keyword vector. Three additional vectors namely Category vector, Outlink vector and Infobox vector are obtained by adding semantic information from Wikipedia using Keyword vector. Addition of semantic information to all the terms in the Keyword vector might lead to the distortion of original clustering. Hence, the information is added only for top-n terms based on their TFIDF scores, which also helps in reducing the dimensionality. Various experiments are conducted by varying the top-n value and clustering is performed based on Keyword vector alone. Best clusters are obtained for n=50%. For every term in the Keyword vector, either a Wikipedia article with exact title or a redirected article, if present, is fetched. From this article the outlink, category and Infobox terms are extracted to form the Outlink vector, Category vector and Infobox vector of that document. In all these vectors, the values are the TFIDF scores of those terms. The Keyword vector and the additional vectors are linearly combined for measuring the document similarity.

All the documents are mapped into English using Shabdanjali dictionary[2] and Wiki dictionary as implemented in [8] availing multilingual Wikipedia titles which are aligned using cross lingual links. We have also implemented Modified Levenshtein Edit Distance proposed in [7] to replace the purpose of Lemmatizers. Two different datasets (Dataset_$H_H E_E$ and Dataset_$H_E E_E$) are formed based on the usage of Wikipedia databases (English and Hindi). In Dataset_$H_H E_E$, additional vectors for Hindi documents are obtained from their respective Keyword vectors using Hindi Wikipedia. All these vectors are then mapped into English. In Dataset_$H_E E_E$, Hindi Keyword vectors are initially mapped into English, the additional vectors are then obtained using the English Wikipedia database. In both the datasets additional vectors for English documents are obtained from their respective Keyword vectors using English Wikipedia database. Clustering

---

[2] http://ltrc.iiit.ac.in/onlineServices/Dictionaries/Dict_Frame.html

is performed seperately on these two datasets based on their document similarities. We used Bisecting k-means algorithm for cluster formation and the results obtained are compared in Table 1. We choose the cosine distance to measure the similarity of two documents (d_i and d_j) which is defined as:

$$sim(d_i, d_j) = sim^{Keyword} + \alpha * sim^{Category} + \beta * sim^{Outlink} + \gamma * sim^{Infobox} \quad (1)$$

Here, sim(d_i,d_j) gives the cosine similarity of the documents d_i,d_j which is calculated as

$$sim = cos(v_i, v_j) = (v_i.v_j)/(|v_i| * |v_j|) \quad (2)$$

where v_i, $v_j \in$ {Keyword, Category, Outlink, Infobox} vectors of documents d_i and d_j respectively. The coefficients α, β and γ indicate the importance of Category vector, Outlink vector, and Infobox vector respectively.

Table 1. Clustering schemes based on different combinations of vectors

Document Representation	F-Score			Purity		
	CS-MDC	DS-MDC		CS-MDC	DS-MDC	
		H_EE_E	H_HE_E		H_EE_E	H_HE_E
keyword (baseline)	0.532	0.606		0.657	0.662	
keyword_Category	0.563	0.635	0.625	0.672	0.701	0.683
keyword_Outlink	**0.572**	**0.697**	0651	0.679	0.743	**0.705**
keyword_Infobox	0.544	0.612	0.609	0.661	0.687	0.673
Category_Outlink	0.351	0.527	0.453	0.434	0.601	0.501
Category_Infobox	0.243	0.420	0.391	0.380	0.507	0.483
Outlink_Infobox	0.248	0.438	0.383	0.405	0.543	0.492
keyword_Category_Outlink	0.567	0.689	**0.664**	**0.683**	**0.749**	0.697
keyword_Outlink_Infobox	0.570	0.653	0.647	0.678	0.726	0.690
keyword_Category_Infobox	0.551	0.622	0.620	0.665	0.699	0.684
Category_Outlink_Infobox	0.312	0.537	0.511	0.443	0.563	0.521
keyword_Category_Outlink_Infobox	0.569	0.681	0.658	0.682	0.739	0.689

4   Experimental Evaluation and Discussion

We have conducted experiments using the English and Hindi documents of FIRE 2010 dataset. There are 50 query topics represented in each of these languages. We used the topic-annotated 1563 documents of which, 650 are in English and 913 in Hindi for our experiments. Cluster quality is evaluated by F-score and Purity measures.

In our experiments, clustering based on Keyword vector is considered as the baseline. Various linear combinations of Keyword, Category, Outlink and Infobox vectors are examined in forming clusters. To determine the α value, experiments are conducted using Equation (1), by varying the α values from 0.0 to 1.0 with 0.1 increment (β and γ are set to 0). The α is set to the value for which best clusters are obtained. Similar experiments are repeated to determine β and γ values. In our experiments, it is found that setting $\alpha = 0.1$, $\beta = 0.4$ and $\gamma = 0.1$ yielded good results for Dataset_H_EE_E whereas $\alpha = 0.2$, $\beta = 0.1$ and $\gamma = 0.4$ achieved good results for Dataset_H_HE_E. From Table 1, it can be noticed that our experiments using Wikipedia have yielded better results than the baseline in both datasets. As the results obtained by DS-MDC are better than the results obtained by CS-MDC [7], it can be concluded that clustering based on document similarities perform better when compared to clustering based on centroid similarites.

We achieved better clustering results for Dataset $\_H_E E_E$ in both the measures when compared to Dataset $\_H_H E_E$. Using only the English Wikipedia database has proved to be beneficial for Dataset $\_H_E E_E$. This might be due the broader coverage of English Wikipedia compared to Hindi Wikipedia. The outlinks information has proved to perform better than categories followed by Infobox information. As the outlinks nearly overlap the context of an article, this might have improved the results better than the rest.

5 Conclusion and Future Work

We have performed MDC using Wikipedia. To evaluate the impact of Wikipedia in the cluster formation, we have experimented with English and Hindi Wikipedia databases. Bisecting k-means clustering algorithm is used for the cluster formation and our results showcases the effectiveness of Wikipedia in enhancing MDC performance. We are planning to extend our work by inspecting the role of Named Entities in enriching the document representation for better clustering the multilingual documents.

References

1. Pirkola, A., Hedlund, T., Keskustalo, H., Järvelin, K.: Dictionary-based cross-language information retrieval: Problems, methods, and research findings. Information Retrieval 4, 209–230 (2001)
2. Silva, J., Mexia, J., Coelho, C., Lopes, G.: A statistical approach for multilingual document clustering and topic extraction form clusters. Pliska Studia Mathematica Bulgarica, 207–228 (2004)
3. Romaric, B.M., Mathieu, B., Besançon, R., Fluhr, C.: Multilingual document clusters discovery. In: RIAO, pp. 1–10 (2004)
4. Steinberger, R., Pouliquen, B., Hagman, J.: Cross-lingual document similarity calculation using the multilingual thesaurus EUROVOC. In: Gelbukh, A. (ed.) CICLing 2002. LNCS, vol. 2276, pp. 415–424. Springer, Heidelberg (2002)
5. Steinberger, R., Pouliquen, B., Ignat, C.: Exploiting multilingual nomenclatures and language-independent text features as an interlingua for cross-lingual text analysis applications. In: Proc. of the 4th Slovenian Language Technology Conf. Information Society (2004)
6. Hu, X., Zhang, X., Lu, C., Park, E.K., Zhou, X.: Exploiting wikipedia as external knowledge for document clustering. In: SIGKDD, pp. 389–396. ACM, New York (2009)
7. Kumar, N.K., Santosh, G., Varma, V.: Multilingual document clustering using wikipedia as external knowledge. In: IRFC (2011)
8. Bharadwaj, G.R., Tandon, N., Varma, V.: An iterative approach to extract dictionaries from wikipedia for under-resourced languages. In: ICON (2010)

A Comparative Study of Classifier Combination Methods Applied to NLP Tasks

Fernando Enríquez, José A. Troyano, Fermín L. Cruz, and F. Javier Ortega

Departamento de Lenguajes y Sistemas Informáticos
Universidad de Sevilla
Av. Reina Mercedes s/n 41012, Sevilla, Spain
{fenros,troyano,fcruz,javierortega}@us.es

Abstract. There are many classification tools that can be used for various NLP tasks, although none of them can be considered the best of all since each one has a particular list of virtues and defects. The combination methods can serve both to maximize the strengths of the base classifiers and to reduce errors caused by their defects improving the results in terms of accuracy. Here is a comparative study on the most relevant methods that shows that combination seems to be a robust and reliable way of improving our results.

Keywords: Classifier Combination, Machine Learning.

1 Introduction

Precision and diversity are introduced by Hansen and Salamon [5] as the sufficient and necessary requirements for carrying out the combination of two or more classifier systems successfully. On the other hand Dietterich justifies the combination from three points of view, namely statistical, computational, and representational, making clear that it covers much better the search space allowing us to get closer to the optimal solution.

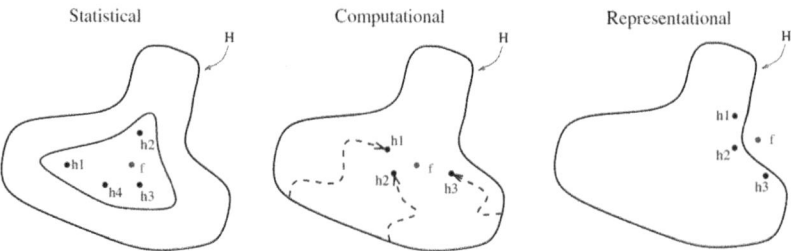

Fig. 1. Combination justification by Dietterich

In [7] the combination methods are organized in four levels based on where the combination is focused on considering the whole process. Therefore we can make

R. Muñoz et al. (Eds.): NLDB 2011, LNCS 6716, pp. 258–261, 2011.
© Springer-Verlag Berlin Heidelberg 2011

use of different collections of data (*data level*), different subsets of features used to represent the examples (*feature level*), different classifiers (*classifier level*) or different combining techniques (*combiner level*).

Still, not all existing combination methods are applicable to any set of classifiers. It is important to consider the type of information they produce as outputs. In [9] we find three possibilities: the 'abstract level' (the output is a single label or a subset of possible labels), the 'rank level' (all the tags or a subset of them are returned arranged in order of preference) and the 'measurement level' (the classifier assigns each tag a value indicative of the confidence you have in it.)

2   Combination and NLP

Since 1998, with the publication of papers like [4] and [2], many researchers developed their work using combination techniques to achieve better results on NLP tasks. Both articles were focused on POS tagging, and although many and varied works appeared afterwards, a comparative study considering a wider range of methods that could be useful to decide the most appropriate one for a particular scenery is still missing as far as we know.

After a literature review on a selection of seventy works that use some kind of combination for NLP tasks, we found the distribution of classifiers, combination methods and tasks shown in figure 2. Regarding classification and combination techniques we appreciate a very different allocation, as there is a greater balance in terms of frequency of use when we talk about classification algorithms although there are some methods that stand out slightly from the rest. In combination methods, however, voting methods and *stacking* account for most of the works, suggesting a possible lack of experimentation with other methods that could offer improvements in some NLP tasks or maybe some other characteristics that should be considered in some situations.

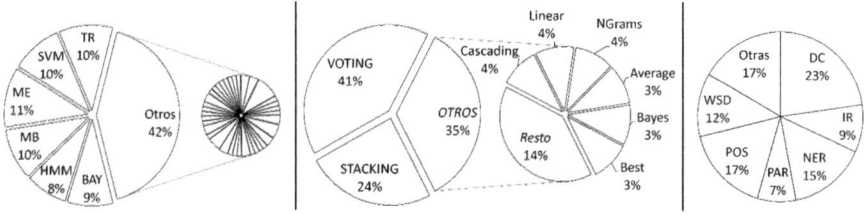

Fig. 2. Classifiers, combination methods and tasks in selected papers

With regard to the results presented in the papers, it is difficult to compare them because of the wide variety of classification and combination methods used, as well as the tasks and data, finding about fifty different datasets. Still we have prepared the table 1 which reflects the minimum, maximum and average improvements achieved in applying system combination to the NLP tasks faced by the different approaches.

Table 1. Summary of results obtained by selected papers

	min	max	mean
DC	0,01	8,10	2,02
NER	1,30	6,41	3,52
PAR	0,03	2,30	1,12
POS	-0,58	1,75	0,75
WSD	1,70	7,00	3,34

3   Comparative Study

In order to establish a more suitable setting to compare different combination methods, we performed experiments for a specific task and some specific individual base classifiers. We chose the POS task in which (thanks to the good performance of the base taggers) we can be sure that improvements obtained by the combination are not due to the low quality of the base classifiers. Three tools designed for this task have been used as base classifiers: TnT [1][1], TreeTagger [8][2] and MBT [3][3], adding a fourth classifier based on features and implemented using the software SVM^{light} [6][4]. The combination methods implemented are: Bayes (BAY), Behavior Knowledge Space (BKS), Stacked Generalization (SG), simple probabilistic combination (SPC), voting (VT) and bagging (BAG). It is also allowed to use the output of a method as input to another combination level, as if it were a base classifier, resulting in a cascading (CAS) scheme. In this case we tested with two combination levels, using a combination method to receive the outputs of other methods that work with the labels proposed by the base classifiers. All methods were evaluated using five very different corpus that differ from each other not only by their sizes but also in the language and the number of tags used in each case.

Table 2. Results obtained

CORPUS	Language	Classifiers				Combination						
		FV	MBT	TnT	TT	BAY	BKS	SG	SPC	VT	BAG	CAS
Brown	English	96,18	95,82	96,55	95,64	0,39	0,63	0,64	0,51	0,49	0,32	0,67
Floresta	Portuguese	96,52	95,81	97,02	96,66	0,55	0,72	0,78	0,60	0,63	0,36	0,71
Susanne	English	92,26	91,16	93,61	91,27	0,67	1,36	1,26	1,16	0,71	0,81	1,52
Talp	Spanish	94,59	94,80	95,82	95,62	0,96	1,08	1,10	1,10	0,76	0,75	1,18
Treebank	English	96,28	95,67	96,21	95,52	0,27	0,47	0,59	0,44	0,45	0,35	0,55
MEAN		95,17	94,65	95,84	94,94	0,57	0,85	0,87	0,76	0,61	0,52	0,93

---

[1] http://www.coli.uni-sb.de/thorsten/tnt
[2] http://www.ims.uni-stuttgart.de/Tools/DecisionTreeTagger.html
[3] http://ilk.uvt.nl/mbt/
[4] http://www.cs.cornell.edu/People/tj/svm_light/

Table 2 shows the results obtained by the classifiers and the improvements achieved by different combining methods. We can see that the improvements are significant in all cases being the stacking method the one with the best results obtained, showing it can be considered the best suited for dealing with different types of data because of its adaptability. Also cascading, with its two levels of combination, shows great robustness that should be taken into account. However there are also simpler methods that deserve our attention, like behavior knowledge space that achieved a noteworthy success and can be very useful in systems where speed is a critical requirement and not only the final accuracy.

References

1. Brants, T.: Tnt. a statistical part-of-speech tagger. In: In Proceedings of the 6th Applied NLP Conference ANLP 2000, pp. 224–231 (2000)
2. Brill, E., Wu, J.: Classifier combination for improved lexical disambiguation. In: Proceedings of the 17th International Conference on Computational Linguistics, pp. 191–195 (1998)
3. Daelemans, W., Zavrel, J., Berck, P., Gillis, S.: Mbt: A memorybased part of speech tagger-generator. In: Proceedings of the 4th Workshop on Very Large Corpora, pp. 14–27 (1996)
4. Halteren, H.V., Zavrel, J., Daelemans, W.: Improving data driven wordclass tagging by system combination. In: Proceedings of the 36th Annual Meeting of the Association for Computational Linguistics and 17th International Conference on Computational Linguistics, vol. 1, pp. 491–497 (1998)
5. Hansen, L., Salamon, P.: Neural network ensembles. IEEE Transactions on Pattern Analysis and Machine Intelligence 12(10), 993–1001 (1990)
6. Joachims, T.: Making large-Scale SVM Learning Practical, ch. 11. MIT Press, Cambridge (1999)
7. Kuncheva, L.I., Whitaker, C.J.: Measures of diversity in classifier ensembles and their relationship with the ensemble accuracy. Machine Learning 51, 181–207 (2003)
8. Schmid, H.: Probabilistic part-of-speech tagging using decision trees. In: Proceedings of the Conference on New Methods in Language Processing (1994)
9. Xu, L., Krzyzak, A., Suen, C.Y.: Methods of combining multiple classifiers and their application to handwriting recognition. IEEE Transactions on Systems, Man, and Cybernetics 22, 418–435 (1992)

TOES: A Taxonomy-Based Opinion Extraction System

Fermín L. Cruz, José A. Troyano, F. Javier Ortega, and Fernando Enríquez

University of Seville
Av. Reina Mercedes s/n 41012, Sevilla Spain
{fcruz,troyano,javierortega,fenros}@us.es

Abstract. Feature-based opinion extraction is a task related to opinion mining and information extraction which consists of automatically extracting feature-level representations of opinions from subjective texts. In the last years, some researchers have proposed domain-independent solutions to this task. Most of them identify the feature being reviewed by a set of words from the text. Rather than that, we propose a domain-adaptable opinion extraction system based on feature taxonomies (a semantic representation of the opinable parts and attributes of an object) which extracts feature-level opinions and maps them into the taxonomy. The opinions thus obtained can be easily aggregated for summarization and visualization. In order to increase precision and recall of the extraction system, we define a set of domain-specific resources which capture valuable knowledge about how people express opinions on each feature from the taxonomy for a given domain. These resources are automatically induced from a set of annotated documents. The modular design of our architecture allows building either domain-specific or domain-independent opinion extraction systems. According to some experimental results, using the domain-specific resources leads to far better precision and recall, at the expense of some manual effort.

1 Introduction

Sentiment analysis is a modern subdiscipline of NLP which deals with subjectivity, affects and opinions in texts (a good survey on this subject can be found in [5]). Within sentiment analysis, the *feature-based opinion extraction* is a task related to information extraction, which consists in extracting structured representations of opinions on features of some object from subjective texts [3,6,2]. For example, given the sentence *"The customer service is terrible"*, a negative opinion on feature *customer service* should be extracted. Some researchers have proposed several approaches to this task, often unsupervised, domain-independent ones. In most cases, they select a few words from the sentence representing the feature affected by the opinion (*opinion target* or *feature words*, depending on authors). This approach implies some problems. First, sometimes the same feature can be named in different ways. For example, *customer service* is also known as *helpline* or *help desk* in some contexts. So a further matching problem must

R. Muñoz et al. (Eds.): NLDB 2011, LNCS 6716, pp. 262–265, 2011.
© Springer-Verlag Berlin Heidelberg 2011

be solved in order to be able to aggregate opinions on the same feature. Besides, some features may include others; for example, someone looking for opinions about the *sound quality* of an audio system would be interested not only in those sentences explicitly referring to the sound quality (e.g., "The *sound quality* is superb", "Very clean, outstanding *sound*"), but also in sentences talking about some other related features (e.g., "The *low end* is clear and the *high* is twangy"). Dealing with these issues is important in order to properly aggregate the extracted opinions and exploit the whole amount of available information.

2   Our Approach

The main guidelines of our approach are (1) building a feature taxonomy for each new domain, so our system will extract opinions on those features and map them into the taxonomy, and (2) automatically generating domain-specific, feature-level resources which capture valuable knowledge about how people express opinions on each feature for a given domain. These resources lead to a higher quality opinion extraction, at the expense of a small manual effort to annotate some documents from the selected domain.

2.1   Feature Taxonomy

The feature taxonomy contains the set of features for which opinions will be extracted in a given domain. Besides, it contains a set of feature words for each feature. All these pairs (*feature,feature words*) are hierarchically organized: the object class itself is the root node of the taxonomy, with a set of features hanging on it. Each feature can be recursively decomposed into a set of subfeatures (see figure 1). The taxonomy hierarchy is useful to aggregate opinions to produce summaries.

The feature taxonomy is built in two steps. First, a list of feature words is generated from the corpus using an active-learning method. Then, an expert pro-

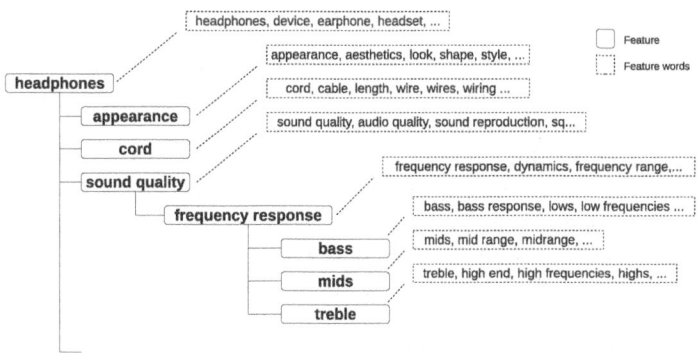

Fig. 1. An extract from the feature taxonomy for the *headphones* domain

duces the taxonomy, grouping feature words by feature and building a hierarchy. The whole process takes no more than a few minutes.

2.2 Domain-Specific Resources

A distinctive part of our approach is the definition of resources that capture knowledge about domains and the way people write reviews on them. To generate these resources, we start from a manual effort (although computer assisted) in order to describe a feature taxonomy and annotate opinions in a set of documents. Then we apply some algorithms in order to extract relevant information about key concepts of the annotated opinions. The resources include, between others, dependency patterns linking feature words and opinion words, opinion lexicons containing semantic orientation estimations for the opinion words more commonly used in the domain, and lists of lexical indicators to detect implicit features[1].

2.3 System Architecture

Our opinion extraction system is comprised of a set of independent abstract components, each one dealing with a different subtask. They can be combined in a wide variety of pipelines in order to complete the extraction task. This modular design together with the multiple implementations of each component make up an experimental setup that enables us to test different approaches.

Let us give a brief description of some of these components. The *feature word annotators* discover features explicitly mentioned in the input reviews. The *implicit feature annotators* discover implicitly mentioned features. Given some previously annotated feature words, the *opinion word linkers* intend to link them to related opinion words. The *opinion classifiers* decide if a previously annotated opinion is a positive or a negative one.

3 Experimentation

Some experiments were performed over a corpus of 587 reviews of headphones from *Epinions.com*. A feature taxonomy was built and the opinions appearing in the documents were annotated[2]. All the experiments were done using 10-fold cross-validation. We evaluated two subproblems: given a sentence, *opinion recognition* consists in identifying the existence of opinions, including determining the feature that opinion refers to; *opinion classification* consists in deciding the polarity of previously recognized opinions. We tested four different approaches (see table 1). In the first three experiments, domain-independent pipelines were

---

[1] A feature is implicit if it is not explicitly mentioned in the text.
[2] The annotated dataset is available in
http://www.lsi.us.es/~fermin/index.php/Datasets, including three different domains: headphones, hotels and cars.

Table 1. Experimental results for *headphones* domain

Experiment	Opinion Recognition			Opinion Classification
	p	r	$F_{\frac{1}{2}}$	accuracy
PMI-IR	0,6092	0,3039	0,5073	0,8706
WordNet	0,6756	0,3002	0,5405	0,8940
SentiWordnet	0,6744	0,3643	0,5763	0,8688
Resource-based	0,7869	0,5662	0,7300	0,9503

used. They all employ a window-based opinion word linker, and classify opinions using three different techniques from literature: the PMI-IR algorithm [7], an algorithm based on lexical distances in WordNet [4] and a state-of-art domain-independent opinion lexicon named SentiWordNet [1]. The fourth experiments were done using a domain-specific pipeline whose components make use of the domain-specific resources. The results obtained by the latter pipeline are far better than those obtained by the domain-independent pipelines. We also conducted some experiments to measure the impact of the number of annotated documents in the results; we found that just a few hours of annotation are enough to largely overcome the results obtained by the resource-free pipelines.

References

1. Baccianella, S., Esuli, A., Sebastiani, F.: Sentiwordnet 3.0: An enhanced lexical resource for sentiment analysis and opinion mining. In: Proceedings of the Seventh Conference on International Language Resources and Evaluation, LREC 2010. European Language Resources Association (ELRA), Valletta (2010)
2. Ding, X., Liu, B., Yu, P.S.: A holistic lexicon-based approach to opinion mining. In: WSDM 2008: Proceedings of the International Conference on Web search and Web Data Mining, pp. 231–240. ACM, New York (2008)
3. Hu, M., Liu, B.: Mining and summarizing customer reviews. In: Proceedings of the ACM SIGKDD Conference on Knowledge Discovery and Data Mining (KDD), pp. 168–177 (2004)
4. Kamps, J., Marx, M., Mokken, R.J., De Rijke, M.: Using wordnet to measure semantic orientation of adjectives. National Institute 26, 1115–1118 (2004)
5. Pang, B., Lee, L.: Opinion mining and sentiment analysis. Foundations and Trends in Information Retrieval 2(1-2), 1–135 (2008)
6. Popescu, A.-M., Etzioni, O.: Extracting product features and opinions from reviews. In: Proceedings of the Human Language Technology Conference and the Conference on Empirical Methods in Natural Language Processing, HLT/EMNLP (2005)
7. Turney, P.: Thumbs up or thumbs down? Semantic orientation applied to unsupervised classification of reviews. In: Proceedings of the Association for Computational Linguistics (ACL), pp. 417–424 (2002)

Combining Textual Content and Hyperlinks in Web Spam Detection

F. Javier Ortega[1], Craig Macdonald[2], José A. Troyano[1], Fermín L. Cruz[1], and Fernando Enríquez[1]

[1] Departamento de Lenguajes y Sistemas Informáticos
Universidad de Sevilla
Av. Reina Mercedes s/n 41012, Sevilla, Spain
{javierortega,troyano,fcruz,fenros}@us.es
[2] Department of Computing Science
University of Glasgow
Glasgow, G12 8QQ, UK
{craigm}@dcs.gla.ac.uk

Abstract. In this work[1], we tackle the problem of spam detection on the Web. Spam web pages have become a problem for Web search engines, due to the negative effects that this phenomenon can cause in their retrieval results. Our approach is based on a random-walk algorithm that obtains a ranking of pages according to their relevance and their spam likelihood. We introduce the novelty of taking into account the content of the web pages to characterize the web graph and to obtain an a priori estimation of the spam likelihood of the web pages. Our graph-based algorithm computes two scores for each node in the graph. Intuitively, these values represent how bad or good (spam-like or not) a web page is, according to its textual content and the relations in the graph. Our experiments show that our proposed technique outperforms other *link-based* techniques for spam detection.

Keywords: Information retrieval, Web spam detection, Graph algorithms, PageRank, web search.

1 Introduction

Web spam is a phenomenon where web pages are created for the purpose of making a search engine deliver undesirable results for a given query, ranking these web pages higher than they would otherwise [7]. Basically, there are two forms of spam intended to cause undesirable effects: Self promotion and Mutual promotion [4]. Self promotion tries to create a web page that gains high relevance for a search engine, mainly based on its content. This can be achieved through many techniques, such as word stuffing, in which visible or invisible keywords are inserted in the page, in order to improve the retrieved rank of the page for the

---

[1] Partially founded by Spanish Ministry of Education and Science (HUM2007-66607-C04-04).

R. Muñoz et al. (Eds.): NLDB 2011, LNCS 6716, pp. 266–269, 2011.
© Springer-Verlag Berlin Heidelberg 2011

most common queries. Mutual promotion is based on the cooperation of various sites, or the creation of a wide number of pages that form a *link-farm*, that is a large number of pages pointing one to another, in order to improve their scores by increasing the number of in-links to them.

There are different approaches to deal with the problem of web spam. They are usually classified into two groups, depending on the spam mechanism that they attempt to identify. *Content-based* techniques use the textual content of the web pages to classify them. These methods usually examine the distribution of statistics about the contents in spam and not-spam web pages, such as the number of words in a page, the HTML invisible content, the most common words in a page in relation with the ones in the entire corpus, etc. [4,5].

On the other hand, *link-based* techniques focus on the structure of the graph made up of the web pages and the hyperlinks among them. TrustRank [6] belongs to this group. It consists in a graph-based ranking algorithm that gives more weight in the computation to a set of hand-picked, trusted web pages. Another interesting work on this field is presented in [1].

Our contribution consists in a method that integrates concepts from both techniques for spam detection, combining the information from the content and the links of the web pages in order to build a ranking where spam web pages are demoted.

The organization of the rest of the paper is as follows. In the next section, we introduce the intuition behind our approach. The experimental design and results are shown in Section 3. Finally, we provide our conclusions concerning the present work, and discuss some ideas for future work.

2   Combining Links and Contents

Our approach consist of two parts: a graph-based ranking algorithm and a set of content-based heuristics. The aim of these metrics is to obtain some a piori information about the spam-likelihood of a web page, in accordance to its content. The values of the heuristics are included in the algorithm in order to introduce a bias in the computation of the ranking of web pages, in such way that the system reduces the final rank of every web page related to some (a priori) spam page, and vice versa. In our system we have implemented two content-based metrics: the average word length of each web page, and the number of repeated words. The ranking algorithm is an extension of PageRank [8], but it computes two scores for each node in the web graph: $PR^+(v_i)$, representing the relevance of node v_i in the graph, and $PR^-(v_i)$, that is the spam likelihood of web page v_i. The ranking of web pages is built regarding the difference between PR^+, and PR^+. These scores are computed following Equations (1) y (2).

$$PR^+(v_i) = (1 - d)e_i^+ + d \sum_{j \in In(v_i)} \frac{PR^+(v_j)}{|Out(v_j)|} \tag{1}$$

$$PR^-(v_i) = (1 - d)e_i^- + d \sum_{j \in In(v_i)} \frac{PR^-(v_j)}{|Out(v_j)|} \tag{2}$$

where v_i is a node in the web graph (a web page), $In(v_j)$ and $Out(v_j)$ are the set of inlinks and outlinks of v_j, respectively. The algorithm iterates over the graph applying the equations (1) y (2), until the highest difference between the scores of a node in two consecutive iterations is less than a certain threshold t. Vectors e_i^+ and e_i^- contain the information about the content-based metrics. In this way, we can give more relevance to some web pages over the rest, depending on the values of e_i^+ and e_i^-. These webs are the *seeds* of our algorithm. We have implemented three methods for the selection of these seeds:

- Most Positive and Negative web pages (MPN): the N web pages with highest and lowest content-based metrics are chosen as seeds. Each seed is initialised with a value of $e_i = 1/N$.
- Most Positive and Negative web pages with Metrics (MPN-M): similar to the previous method, but the values of the seeds are computed as follows: $e_i = Metrics_i/N$.
- All the web Pages as Seeds (APS): every web page in the collection are used as seeds, applying the same formula to obtain their initial weights.

3   Evaluation

Since our approach does not classify the web pages between spam or non-spam, it does not make sense to perform an evaluation in terms of classification accuracy. In our experiments, we use the PR-buckets evaluation method, also followed in other works on the application of graph-based algorithms to the spam detection task, such as [2,6]. Their intuition is that it is more important to correctly detect the spam in high PageRank valued sites, because they will often appear in the top positions in the search results for many queries. The aim of this evaluation method is to easily determine the number of spam web pages detected mainly in the highest positions of the ranking. We have used TrustRank as the baseline for our experiments. As mentioned above, this technique is based on a graph-based ranking algorithm that chooses by hand a set of a priori relevant web pages.

Table 1. Cumulated errors for each method: TrustRank (TR), Most Positive and Negative web pages (MPN), Most Positive and Negative web pages with Metrics (MPN-M) and All the web Pages as Seeds (APS). The best result for each case is highlighted.

#	Web pages	TR	MPN	MPN-M	APS
1	14	**0**	2	**0**	**0**
2	68	2	5	**1**	3
3	212	17	16	**4**	8
4	649	40	48	**21**	32
5	1719	73	104	**66**	101
6	3849	155	244	**124**	199
7	6513	254	392	**180**	297
8	9291	371	557	**255**	416
9	12102	448	742	**350**	537
10	14914	511	937	**440**	650

The experiments have been performed using the WEBSPAM-UK2006 Dataset [3], specifically compiled for the research on web spam detection. The corpus has about 98 million web pages and 120 million links. A subset of 11 thousand web hosts have been labelled as spam or not-spam, obtaining 10 million spam web pages in total. We have used the Terrier [2] system in order to efficiently index the corpus.

In Table 1 we show the results of TrustRank and our approach, with its three methods for the selection of seeds. The number of spam web pages in the 10 first buckets (the first positions of the ranking) are shown. MPN-M is the approach that obtains the best results for all the buckets.

4 Conclusions and Future Work

In this work we present a web spam detection system that takes into account both content and link-based information. The system has been evaluated with a standard corpus, achieving very good results and even outperforming a state-of-art technique, TrustRank.

We plan to further our work by implementing new content-based heuristics and studying the impact of these metrics in the overall performance of the system. It is also very interesting to prove the influence of the seed set in the final results of the system, and to experiment with alternative approaches for the selection of seeds.

References

1. Becchetti, L., Castillo, C., Donato, D., Baeza-Yates, R., Leonardi, S.: Link analysis for web spam detection. ACM Transactions on the Web 2(1), 1–42 (2008)
2. Benczur, A.A., Csalogany, K., Sarlos, T., Uher, M., Uher, M.: Spamrank - fully automatic link spam detection. In: Proceedings of the First International Workshop on Adversarial Information Retrieval on the Web (AIRWeb), pp. 25–38 (2005)
3. Castillo, C., Donato, D., Becchetti, L., Boldi, P., Leonardi, S., Santini, M., Vigna, S.: A reference collection for web spam. SIGIR Forum 40(2), 11–24 (2006)
4. Cormack, G.V., Smucker, M., Clarke, C.L.A.: Efficient and Effective Spam Filtering and Re-ranking for Large Web Datasets. Computing Research Repository, abs/1004.5 (2010)
5. Fetterly, D., Manasse, M., Najork, M.: Spam, damn spam, and statistics: using statistical analysis to locate spam web pages. In: Proceedings of the 7th International Workshop on the Web and Databases, pp. 1–6. ACM, New York (2004)
6. Gyöngyi, Z., Garcia-Molina, H., Pedersen, J.: Combating web spam with trustrank. In: Proceedings of the Thirtieth International Conference on Very Large Data Bases, vol. 30, pp. 576–587, VLDB Endowment, Toronto (2004)
7. Najork, M.: Web spam detection. In: Encyclopedia of Database Systems, pp. 3520–3523. Springer, US (2009)
8. Page, L., Brin, S., Motwani, R., Winograd, T.: The pagerank citation ranking: Bringing order to the web. World Wide Web Internet And Web Information Systems (66), 1–17 (1999)

---

[2] http://terrier.org/

Map-Based Filters for Fuzzy Entities in Geographical Information Retrieval

Fernando S. Peregrino, David Tomás, and Fernando Llopis

Department of Software and Computing Systems, University of Alicante
Carretera San Vicente del Raspeig s/n - 03690 Alicante Spain

Abstract. Many users employ vague geographical expressions to query Information Retrieval systems. These fuzzy entities do not appear neither in gazetteers nor in geographical databases. Searches such as "Ski resorts in north-central Spain" or "Restaurants near the Teatro Real of Madrid" often do not get the expected results, mainly due to the difficulty of disambiguating expressions like "north of" or "near". This paper presents a first approach to deal with this kind of fuzzy expressions, with the aim of improving the coverage and accuracy of traditional Information Retrieval systems. Our approach is based on the use of raster images as geographic filters to determine the relevance of the documents depending on the location of places referenced in them.

Keywords: Geographical Information Retrieval, Fuzzy entities, Vague entities, Imprecise regions, Vernacular geography.

1 Introduction

Geographical Information Retrieval (GIR) is the augmentation of Information Retrieval with geographic metadata. An example of query treated by these systems is "Politicians in exile in Germany".

GIR systems are an emerging research field in recent years due to the lack of good results in IR systems when dealing with spatial information. Several competitions have been organized around these systems. GeoCLEF[1], NTCIR GeoTime[2] task (combining GIR with time-based search), and GIR[3] are some of the most relevant.

IR is done in unstructured text, which makes it more difficult to obtain the place name referred. The typical structure of a query in a GIR system is <what_do_you_want> + <relationship> + <location>.

Many times, these geographical entities belong to a vague or diffuse definition. In most cases, these vague definitions are accompanied by expressions such as "close to", "in the north of", etc.

This paper presents a first approach to solve this kind of vague expressions using raster images of maps, creating filters to determine the relevance of spatial entities pertaining to fuzzy regions.

---

[1] http://ir.shef.ac.uk/geoclef/
[2] http://metadata.berkeley.edu/NTCIR-GeoTime/
[3] http://www.geo.uzh.ch/~rsp/gir10/

R. Muñoz et al. (Eds.): NLDB 2011, LNCS 6716, pp. 270–273, 2011.
© Springer-Verlag Berlin Heidelberg 2011

2 Description of the System

We have employed raster images with associated geographical information to create filters. These filters determine the relevance of a geo-located point in the map with respect to a fuzzy region.

To obtain geo-referenced images, such as those include in gazetteers, it is necessary to create a database of geographic entities.

The underlying algorithm starts with the image obtained from the IDEE[4], which is a geo-referenced raster image of a specific geographical region in a gazetteer. Then, we obtain another region, in this case a fuzzy one. Thus, if we want to get the north-central Spain, we will input to the algorithm a geo-referenced raster image of Spain, obtaining the filtered image as an output. In this way, we can identify the relevance of specific spatial entities with respect to a fuzzy region.

The filter consists of a grey scale image ranging between 0 and 255. It is calculated as follows:

- Assign 0 if the grid cell is not within the specific entity (raster input).
- Assign $\frac{\omega}{8}$ if it is within the specific entity but not in the fuzzy region (in the example of Figure 1, it would be the southern part of Spain).
- Assign the sum of the value $X + Y$ for the rest of grid cells, where X and Y are calculated according to the following formulas:

$$Y = V \cdot \left(\frac{\omega}{2} + \frac{\omega}{\alpha} \cdot \left(\frac{\alpha}{2} - y\right)\right) \tag{1}$$

$$X = (1 - V) \cdot 2x \cdot \frac{\omega}{\beta}, \text{if } x \le \frac{\beta}{2} \tag{2}$$

$$X = (1 - V) \cdot 2 \cdot \left(\omega - x \cdot \frac{\omega}{\beta}\right), \text{if } x > \frac{\beta}{2} \tag{3}$$

Where ω is the maximum value (255), α is the number of vertical grid cells which vary depending on the input image and its resolution analogously to β, which is the number of horizontal grid cells, and x and y are the coordinates on the grid cell. Finally, V is a variable between 0 and 1 to prioritize latitude ($V > 0.5$) or longitude ($V < 0.5$). Default value is 0.5.

If we focus in the example "Ski resorts in north-central Spain", we would get from a repository (the IDEE in our case) the specific raster image of the entities included in the query ("Spain" in the example above). It is also necessary to obtain from a database or a corpus the "what_do_you_want" part ("Ski resorts") with their geographical coordinates, obtaining the value of the raster cell that matches with the coordinates received in this step. Finally, a new gazetteer is created (in real time) with the entities found in the first part of the user query, ordered according to geographical relevance set by the filter.

---

[4] Infraestructura de Datos Espaciales de España: http://www.idee.es

3   Experiments and Evaluation

Continuing the example above, Figure 1 shows the resulting raster image with a gradient as we move away from the north-central of Spain. Different triangles represent several of the resorts in Spain and Andorra. Also, there are four ski resorts in France, two in the Pyrenees, on the border with Spain, and two in the Alps. These were included to test the accuracy of the algorithm when filtering. Due to these inclusions, Figure 1 shows an empty region, which reflect the absence of information as it is located outside the filter boundaries. Consequently, the ski resorts into this empty region will be assigned the lowest weight in the resulting gazetteer.

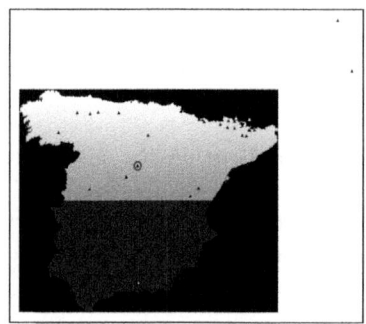

Fig. 1. Geo-referenced raster image of north-central Spain and ski resorts

In Table 1, we can see the results returned by the system. The results are grouped in three different sets: the first one shows ski resorts which are located further north-central of Spain; in the second set is where it can be seen a big drop in the score; in the last one, where there is a transition from the northern

Table 1. Geographical ranking by relevance. *Null* indicates that the point is outside the filter boundaries

Ski Resorts	Country	Score
San Isidro	Spain	232
Leitariegos	Spain	231
...	...	...
Port del compte	Spain	209
La Pinilla	Spain	179
...	...	...
Javalambre	Spain	148
Sierra Nevada	Spain	32
Vallnord	Andorra	0
Ascou Pailhères	France	0
Guzet	France	0
Grandvalira	Andorra	0
Les Houches	France	Null
Colmiane	France	Null

to the southern Spain, i.e., in the concise entity, from the fuzzy region to the region not requested. Between each of these groups, it has been introduced a line with ellipsis in each column representing the other existing ski resorts, from one group to the other, in order to simplify the view of the table.

Now we could skip all those ski resorts that have not obtained a high score. Thus, in Table 1, stations like "La Pinilla" (circled in Figure 1), which is in the northern Spain but they belongs to the central system, would be filtered out in the search.

4   Related Work

The previous work in the field of automatic definition of fuzzy geographical regions presents two main approaches. The first one focuses on obtaining information from real users, who are responsible for defining the region to study, [1] [2].

In the second approach, several works use the Web as a source of information, [3] [4].

5   Conclusions and Future Work

In this work, we have defined a map-based approach to identify fuzzy regions. This kind of of information can be an important part in a GIR system.

The main goal of this tool is to carry out a study to cover the situations in which information requested includes fuzzy entities such as "close to", "around", etc.

As future work, we plan to complement our system using the Web as a source for the definition of fuzzy entities [4] [5].

References

1. Evans, A.J., Waters, T.: Mapping vernacular geography: web-based gis tools for capturing fuzzy or vague entities. International Journal of Technology Policy and Management 7(2), 134–150 (2007)
2. Montello, D.R., Goodchild, M.F., Gottsegen, J., Fohl, P.: Where's downtown?: Behavioral methods for determining referents of vague spatial queries. Spatial Cognition Computation 3(2), 185–204 (2003)
3. Clough, P.: Extracting metadata for spatially-aware information retrieval on the internet, pp. 25–30. ACM Press, New York (2005)
4. Jones, C.B., Purves, R.S., Clough, P.D., Joho, H.: Modelling vague places with knowledge from the web. International Journal of Geographical Information Science 22(10), 1045–1065 (2008)
5. Pasley, R.C., Clough, P., Sanderson, M.: Geo-tagging for imprecise regions of different sizes. In: Proceedings of the 4th ACM Workshop on Geographical Information Retrieval GIR 2007, Lisbon, Portugal, pp. 77–82 (2007)

DDIExtractor: A Web-Based Java Tool for Extracting Drug-Drug Interactions from Biomedical Texts

Daniel Sánchez-Cisneros, Isabel Segura-Bedmar, and Paloma Martínez

University Carlos III of Madrid, Computer Science Department
Avd. Universidad 30, 28911 Leganés, Madrid, Spain
{dscisner,isegura,pmf}@inf.uc3m.es
http://labda.inf.uc3m.es/

Abstract. A drug-drug interaction (DDIs) occurs when one drug influences the level or activity of another drug. The detection of DDIs is an important research area in patient safety since these interactions can become very dangerous and increase health care costs. Although there are several databases and web tools providing information on DDIs to patients and health-care professionals, these resources are not comprehensive because many DDIs are only reported in the biomedical literature. This paper presents the first tool for detecting drug-drug interactions from biomedical texts called DDIExtractor. The tool allows users to search by keywords in the Medline 2010 baseline database and then detect drugs and DDIs in any retrieved document.

Keywords: Drug-Drug Interactions, Biomedical Information Extraction.

1 Background

Drug-drug interactions are common adverse drug reactions and unfortunately they are a frequent cause of death in hospitals. The management of DDIs is a critical issue due to the overwhelming amount of information available on them. We think that Information Extraction (IE) techniques can provide an interesting way of reducing the time spent by health care professionals on reviewing the literature. There are some examples of IE applications in different biomedical subdomains. iHOP (information Hyperlinked Over Proteins)[1] [1] is a web service that automatically extracts key sentences from MedLine documents where genes, proteins and chemical compounds terms are annotated and linked to MeSH terms by machine learning methods. Another example is the Reflect Tool [2], which is a free service tag of gene, protein, and small molecule names in any web page. The NCBO Resource Index system[2] allows to annotate and index texts with ontology concepts from more than twenty diverse biomedical resources.

---

[1] www.ihop-net.org/

[2] http://www.bioontology.org/resources-index

R. Muñoz et al. (Eds.): NLDB 2011, LNCS 6716, pp. 274–277, 2011.
© Springer-Verlag Berlin Heidelberg 2011

This paper describes a web-based tool for searching relevant documents from Medline 2010 database and analyzing them in order to highlight drugs and DDIs occurring in them. Drug name recognition and DDIs detection are performed by the DrugDDI system [3] which is based on a Shallow Linguistic Kernels to extract relationships between entities. The paper is organized as follows: Section 2 describes the functionality and the architecture of this tool; and the on-going and future work is drawn in Section 3.

2   DDIExtractor: A Web Tool to Detect Drugs and DDI

To make the reading of the biomedical literature easier and faster, we have developed a web-based java tool called DDIExtractor[3]. DDIExtractor allows users to find relevant documents entring specific keywords queries. Then, the tool can process any retrieved document highlighting its drug names and DDIs. For each drug, the tool can also show a popup including its description, drug categories and drug interactions among other useful information from Drugbank databases[4]. The tool works as is shown in Figure 1.

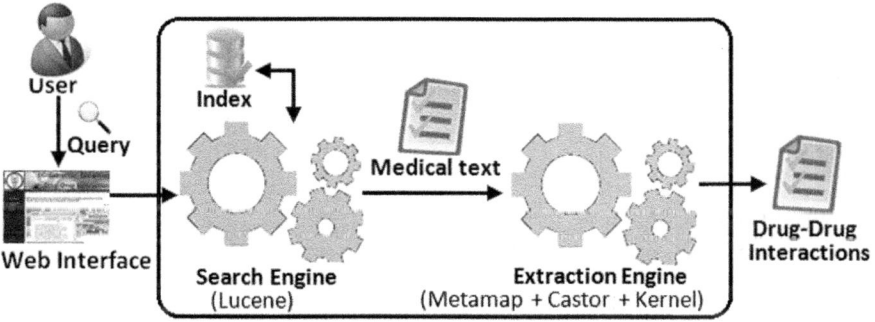

Fig. 1. Architecture of DDIExtractor

2.1   Search Engine

First, DDIExtractor allows searching relevant documents from the Medline 2010 database. This database contains a total of 18,502,916 records grouped by publication years. We used the Apache Lucene engine[5], which is a free open source information retrieval(IR) API originally implemented in Java.

This engine is a tradicional IR system that provide structured and unstructured information to the user, retriving a list of results related to the query searched (e.g.: Aspirin). Each result will specify its title, journal, authors, publication date, Medline identificator(PMID), etc. Then the user can click on a result to see the entire abstract or even to extract its drug-drug interactions.

---

[3] http://163.117.129.57:8080/ddiextractorweb/
[4] http://www.drugbank.ca/
[5] http://lucene.apache.org/

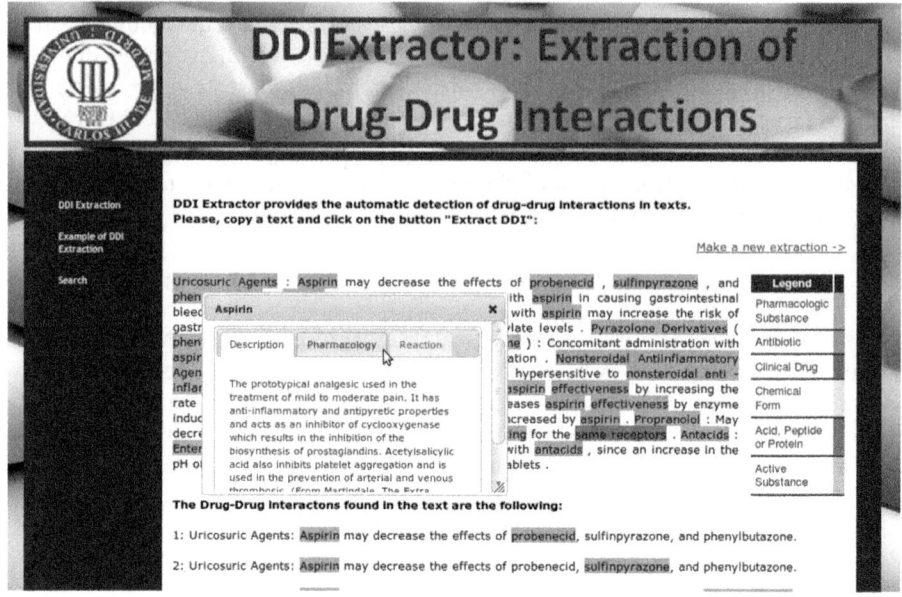

Fig. 2. The application highlights the drugs found in the text and shows a popup including related information for each drug

2.2 Extractor

To extract DDIs, the tool uses the DrugDDI system based on the tool java Simple Relation Extraction (JSRE)[6]. The following list shows the main steps to extract DDIs:

- Step 1: The tool allows users to select a text from the retrieved list of documents or introduce a new text.
- Step 2: The text is analyzed by the MMTx[7] tool which provides syntactic and semantic information. This information is converted into an input XML format compatible with our DrugDDI system. Then, the XML format is processed using the XML data binding framework Castor[8] in order to represent the information of the whole sentence as well as the local contexts around the drugs by their Part-of-Speech (PoS) tags, lemmas and drug names. A detailed description of this process can be found in [4]. Each pair of drugs in the same sentence is a possible DDI. Then, each candidate DDI is classified by the DrugDDI system in order to detect those pairs of drugs that are actually DDIs.
- Step 3: Drug names and DDIs are highlighted in texts using a different color according to their semantic types provided by the MMTx tool. For each

---

[6] http://hlt.fbk.eu/en/technology/jSRE
[7] http://metamap.nlm.nih.gov/
[8] http://www.castor.org/

drug identified, the tool can also show a popup including its definitions, drug categories and drug interactions among other useful information (see Figure 2).

3 On-Going and Future Work

DDIExtractor is an online tool that highlights drugs and DDIs from texts and represents a first step making the reading and interpretation of pharmacological texts easier and faster.

While several drug databases and web resources provide users structured information on DDIs, this tool retrieves structured and unstructured information in order to detect its DDIs.

Ideally, prescribed information about a drug should list its potential interactions, joined to the following information about each interaction: its mechanism, related doses of both drugs, time course, seriousness and severity, the factors which alter the individual's susceptibility to it, and the probability of its occurrence. However, in practice this information is rarely available. Our future work will focus on the improvement of our system for extracting DDIs, especially in the detection of these additional features. Also, we would like to design an evaluation methodology oriented toward user acceptance.

Acknowledgments. This work is supported by the projects MA2VICMR (S2009/TIC-1542) and MULTIMEDICA (TIN2010-20644-C03-01).

References

1. Hoffmann, R., Valencia, A.: Implementing the iHOP concept for navigation of biomedical literature. Bioinformatics 21(suppl. 2) (2005)
2. Pafilis, E., O'Donoghue, S., Jensen, L., Horn, H., Kuhn, M.: Reflect: augmented browsing for the life scientist. Nature Biotechnology 27(6), 508–510 (2009)
3. Segura-Bedmar, I., Martínez, P., de Pablo-Sánchez, C.: Using a shallow linguistic kernel for drug-drug interaction extraction. Journal of Biomedical Informatics (2011) (accepted)
4. Segura-Bedmar, I.: Application of information extraction techniques to pharmacological domain: extracting drug-drug interactions. PhD thesis, Universidad Carlos III de Madrid, Leganés, Madrid, Spain (2010)

Geo-Textual Relevance Ranking to Improve a Text-Based Retrieval for Geographic Queries

José M. Perea-Ortega, Miguel A. García-Cumbreras, L. Alfonso Ureña-López, and Manuel García-Vega

SINAI research group. Computer Science Department. University of Jaén. Spain
{jmperea,laurena,mgarcia,magc}@ujaen.es

Abstract. Geographic Information Retrieval is an active and growing research area that focuses on the retrieval of textual documents according to a geographic criteria of relevance. In this work, we propose a reranking function for these systems that combines the retrieval status value calculated by the information retrieval engine and the geographical similarity between the document and the query. The obtained results show that the proposed ranking function always outperforms text-based baseline approaches.

1 Introduction

In the field of Geographical Information Retrieval (GIR), a geographic query is structured as a triplet of $<theme><spatial\ relationship><location>$ and it is concerned with improving the quality of geographically-specific information retrieval with a focus on access to unstructured documents [2]. Thus, a search for *"castles in Spain"* should return not only documents that contain the word *"castle"*, also those documents which have some geographical entity within Spain.

One of the open research questions in GIR systems is how to best combine the textual and geographical similarities between the query and the relevant document [4]. For this reason, in this work we propose a reranking function based on these both similarities. Our experimental results show that the proposed ranking function can outperform text-based baseline approaches.

2 GIR System Overview

GIR systems are usually composed of three main stages: preprocessing of the document collection and queries, textual-geographical indexing and searching and, finally, reranking of the retrieved results using a particular relevance formula that combines textual and geographical similarity between the query and the retrieved document. The GIR system presented in this work follows the same approach, as can be seen in Figure 1.

The preprocessing carried out with the queries was mainly based on detecting their geographical scopes. This involves specifying the triplet $<theme><spatial\ relationship><location>$, which will be used later during the reranking process.

R. Muñoz et al. (Eds.): NLDB 2011, LNCS 6716, pp. 278–281, 2011.
© Springer-Verlag Berlin Heidelberg 2011

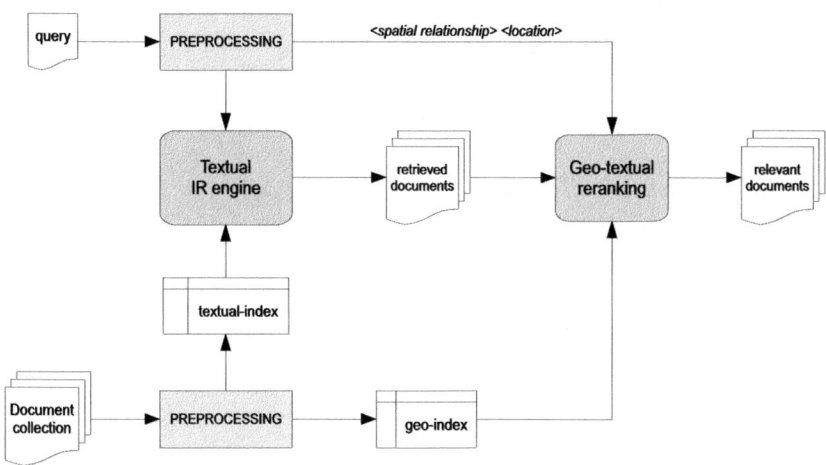

Fig. 1. Overview of the GIR system

To detect this triplet we have used a Part Of Speech tagger (POS tagger) like TreeTagger[1]. Moreover, the stop words were removed and the Snowball stemmer[2] was applied to each word of the query, except for the geographical entities. On the other hand, a similar preprocessing was carried out offline with the document collection. During this process, two indexes were generated:

- a **geographical index**, which contains the locations detected in each document. We have used Geo-NER[6] to recognize geographical entities in the collection and queries, a Named Entity Recognizer (NER) for geographical entities based on GeoNames[3] and Wikipedia.
- a **textual index**, which contains the preprocessed text (stemmer and stopper) of each document, including the geographical entities in their original form, i.e, without applying stemmer to them.

It is important to note that we consider several geographical scopes for a document, as many as geographical entities have been detected in it. Therefore, in the geographical index we can have different locations for the same document.

In the text retrieval process we obtain 3,000 documents for each query. We have applied three of the most used IR tools in these systems, according to a previous work [5]: Lemur[4], Terrier[5] and Lucene[6]. Then, the retrieved documents are used as a input in the geo-textual reranking process, which is reponsible for

---

[1] TreeTagger v.3.2 for Linux. Available in http://www.ims.uni-stuttgart.de/projekte/corplex/TreeTagger/DecisionTreeTagger.html

[2] Available in http://snowball.tartarus.org

[3] http://www.geonames.org

[4] Version 4.0, available in http://www.lemurproject.org

[5] Version 2.2.1, available in http://terrier.org

[6] Version 2.3.0, available in http://lucene.apache.org

modifying the Retrieval Status Value (RSV) of each document depending on the geographical similarity with the query. The geographical similarity between a document and a query is calculated using the following formula:

$$sim_{geo}(Q, D) = \frac{\sum_{i \in geoEnts(D)} match(i, GS, SR) \cdot frec(i, D)}{|geoEnts(D)|} \qquad (1)$$

where the function $match(i, GS, SR)$ returns 1 if the geographical entity i satisfies the geographical scope GS for the spatial relationship SR and 0 otherwise. Then, the geo-textual reranking process modifies the RSV of the retrieved document $(RSV(D))$ taking into account its geographical similarity $(sim_{geo}(Q, D))$ and its previous RSV using the following formula:

$$RSV'_D = RSV_D + log(RSV_D) + sim_{geo}(Q, D) \qquad (2)$$

Finally, the geo-textual reranking process returns the most relevant 1,000 documents (those with higher RSV' value) as a output of our GIR system.

3 Experiments and Results

In order to evaluate the proposed approach we have used the GeoCLEF framework [1,3], an evaluation forum for GIR systems held between 2005 and 2008 under the CLEF conferences. We have compared the results obtained applying the proposed geo-textual reranking process with those results obtained with the baseline system, i.e., the GIR system that only uses an IR engine to retrieve the relevant document for a geographical query. The Mean Average Precision (MAP) has been used as a evaluation measure. Table 1 shows the obtained results using different IR engines and weighting schemes.

As can be observed in Table 1, the results obtained applying the geo-textual reranking process always improve the results obtained using the baseline system, even with percentages of improvement over 40%.

Table 1. Summary of the experiments and results

Query set	IR engine	$MAP_{IR-engine}$	$MAP_{reranking}$	% of improv.
2005	Lemur (BM25+PRF)	0.3291	0.3402	3.37
	Lucene (BM25+QE)	0.1319	0.1857	40.79
	Terrier (inL2+Bo1)	0.3837	0.3874	0.96
2006	Lemur (BM25+PRF)	0.2040	0.2093	2.60
	Lucene (BM25+QE)	0.0597	0.0852	42.71
	Terrier (inL2+Bo1)	0.2602	0.2733	5.03
2007	Lemur (BM25+PRF)	0.2114	0.2175	2.89
	Lucene (BM25+QE)	0.0690	0.0777	12.61
	Terrier (inL2+Bo1)	0.2444	0.2600	6.38
2008	Lemur (BM25+PRF)	0.2609	0.2709	3.83
	Lucene (BM25+QE)	0.0773	0.1115	44.24
	Terrier (inL2+Bo1)	0.2690	0.2973	10.52

4   Conclusions and Further Work

In this work we propose a reranking formula for GIR systems based on the RSV calculated for the IR engine and the geographical similarity between the document and the geographic query. The comparison of the results obtained show that the proposed approach is always useful for GIR systems within the framework used. For future work, we will try to improve the reranking function analyzing the type of geographic query in each case, since for some queries the geographical scope is more bounded than for others.

Acknowledgments. This work has been partially supported by a grant from the Fondo Europeo de Desarrollo Regional (FEDER), project TEXT-COOL 2.0 (TIN2009-13391-C04-02) from the Spanish Government, a grant from the Andalusian Government, project GeOasis (P08-TIC-41999) and Geocaching Urbano research project (RFC/IEG2010).

References

1. Gey, F.C., Larson, R.R., Sanderson, M., Joho, H., Clough, P., Petras, V.: GeoCLEF: The CLEF 2005 Cross-Language Geographic Information Retrieval Track Overview. In: Peters, C., Gey, F.C., Gonzalo, J., Müller, H., Jones, G.J.F., Kluck, M., Magnini, B., de Rijke, M., Giampiccolo, D. (eds.) CLEF 2005. LNCS, vol. 4022, pp. 908–919. Springer, Heidelberg (2006)
2. Jones, C.B., Purves, R.S.: Geographical Information Retrieval. In: Encyclopedia of Database Systems, pp. 1227–1231. Springer, US (2009)
3. Mandl, T., Carvalho, P., Di Nunzio, G.M.D, Gey, F.C, Larson, R.R., Santos, D., Womser-Hacker, C.: GeoCLEF 2008: The CLEF 2008 Cross-Language Geographic Information Retrieval Track Overview. In: Peters, C., Deselaers, T., Ferro, N., Gonzalo, J., Jones, G.J.F., Kurimo, M., Mandl, T., Peñas, A., Petras, V. (eds.) CLEF 2008. LNCS, vol. 5706, pp. 808–821. Springer, Heidelberg (2009)
4. Martins, B., Calado, P.: Learning to rank for geographic information retrieval. In: Proceedings of the 6th Workshop on Geographic Information Retrieval, GIR 2010, pp. 21:1–21:8. ACM, New York (2010)
5. Perea-Ortega, J.M., García-Cumbreras, M.Á., García-Vega, M., Ureña-López, L.A.: Comparing several textual information retrieval systems for the geographical information retrieval task. In: Kapetanios, E., Sugumaran, V., Spiliopoulou, M. (eds.) NLDB 2008. LNCS, vol. 5039, pp. 142–147. Springer, Heidelberg (2008)
6. Perea-Ortega, J.M., Martínez-Santiago, F., Montejo-Ráez, A., Ureña-López, L.A.: Geo-NER: un reconocedor de entidades geográficas para inglés basado en GeoNames y Wikipedia. Sociedad Española para el Procesamiento del Lenguaje Natural (SEPLN) 43, 33–40 (2009)

BioOntoVerb Framework: Integrating Top Level Ontologies and Semantic Roles to Populate Biomedical Ontologies

Juana María Ruiz-Martínez, Rafael Valencia-García, and Rodrigo Martínez-Béjar

Facultad de Informática. Universidad de Murcia.
Campus de Espinardo. 30100 Espinardo (Murcia). España
Tel.: +34 968398522,
Fax: +34 968364151
{jmruymar,valencia,rodrigo}@um.es

Abstract. Ontology population is a knowledge acquisition activity that relies on (semi-) automatic methods to transform un-structured, semi-structured and structured data sources into instance data. A semantic role is a relationship between a syntactic constituent and a predicate that defines the role of a verbal argument in the event expressed by the verb. In this work, we describe a framework where top level ontologies that define the basic semantic relations in biomedical domains are mapped onto semantic role labeling resources in order to develop a tool for ontology population from biomedical natural language text. This framework has been validated by using an ontology extracted from the GENIA corpus.

1 Introduction

Within the scope of the Semantic Web several resources have been developed with the aim of discovering, accessing, and sharing biomedical knowledge. Bio-ontologies have been used for organizing and sharing biological knowledge as well as integrating different sources of knowledge in order to provide interoperability among different research communities [1-3]. However, since most of medical advances are disseminated into scientific literature, specific resources for text mining have been created such as annotated corpora [4].

Ontology Population (OP) is a knowledge acquisition activity that relies on (semi-) automatic methods to transform un-structured, semi-structured and structured data sources into instance data. The population of ontologies with new knowledge is a relevant step towards the provision of valuable ontology-based knowledge services, allowing researchers to access massive amounts of online information and to keep abreast of biomedical knowledge.

In this work, we describe the BioOntoVerb framework, where top level ontologies that define the basic semantic relations in biomedical domains are mapped onto semantic role labeling resources such as FrameNet[1]. This has been ideated in order to develop a framework for ontology population from biomedical natural language text.

---

[1] http://framenet.icsi.berkeley.edu/

R. Muñoz et al. (Eds.): NLDB 2011, LNCS 6716, pp. 282–285, 2011.
© Springer-Verlag Berlin Heidelberg 2011

A semantic role is the relationship between a syntactic constituent and a predicate. It defines the role of a verbal argument in the event expressed by the verb [5]. In FrameNet roles are defined for each semantic frame, that is, a schematic representation of situations involving various participants, props, and other conceptual roles.

2   BioOntoVerb Framework

The framework described in this work is based on the integration of ontological and linguistic resources into three different layers of application, as it is shown in Figure 1. The framework is intended to be used as an architecture for populating biomedical ontologies. Next, the layers are explained in detail.

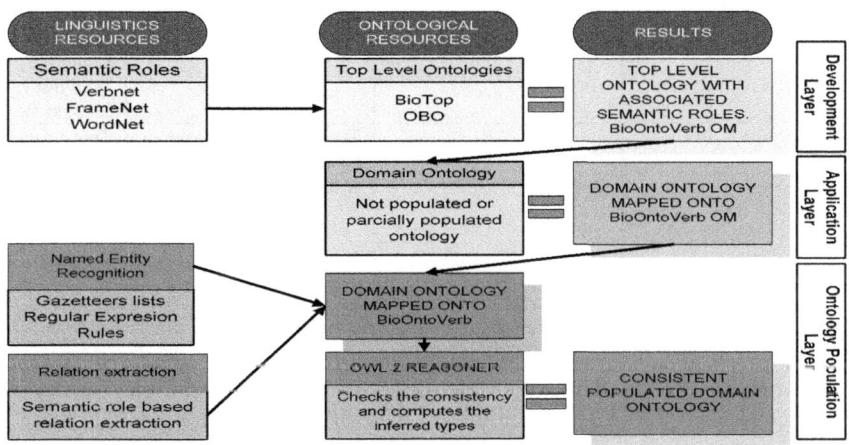

Fig. 1. Architecture of the BioOntoVerb framework

Development Layer. In this layer, a top level ontological model based on the different types of relations described in OBO Relations Ontology [1] and the relationships defined in the BioTop ontology [2] is defined. This ontological model, based in a previous a work [6] is called BioOntoVerb OM and it allows to define domain ontologies based on the semantic relationships that are widely used in biomedical domains. Here relationships are expressed by means of Object Properties in OWL 2 and some property axioms have been assigned for each relation including transitivity, asymmetry or symmetry. For each of these Object Properties, a set of semantic roles extracted from FrameNet has been assigned in order to detect ontological semantic relations between instances from natural language texts. Each semantic role is associated with a set of verbal expressions or lexical units representing the relationships in a textual level. These verbal expressions have been obtained from Wordnet[2] and VerbNet[3]. The result is an ontological model integrated

---

[2] http://wordnet.princeton.edu/
[3] http://verbs.colorado.edu/~mpalmer/projects/verbnet.html

into the populating framework. In Table 1 an example of the mappings between the ontological relationships, their property axioms, the lexical units and the corresponding semantic frames are shown. It can be seen that the relationship *"contained_in"* is defined as a transitive, irreflexive and asymmetric relationship and the semantic frame containing has been associated to this relationship. Furthermore, some verbs such as "contain", "include" and "comprise" are assigned to this semantic frame.

Table 1. Mappings between ontological relationships and semantic frames

Ontological Relationship	Property Axioms	Semantic Frame	Lexical Units
PhysicallyContainedIn (BioTop) Contained_in (OBO)	Transitive Irreflexive Asymmetric	Containing: **Container** holds within its physical boundaries the **Contents**.	Contain Include Have as a component Comprise Hold inside Incorporate
Transformation_of (OBO)	Transitive	Transform: An **Agent** or **Cause** causes an **Entity** to change	Transform Modify in Mutate in/into Change into Convert into Turn into

Application Layer. In this layer, the domain ontology, i.e the ontology to be populated, is defined using BioOntoVerb OM. The relations of the domain ontology have to be mapped onto the relations defined in the top level ontological model of the development layer [6]. The semantic roles defined in the development layer for each relation are used to extract and obtain the relations between instances of the domain ontology from a corpus.

Ontology Population Layer. This layer implements the ontology population process through two stages. Firstly the Named Entities (NEs) candidates are identified using the GATE Framework[4]. A combination of JAPE rules and lists of Gazetters are also used to perform the processes associated with this phase. Jape is a rich and flexible regular expression- based rule mechanism offered by the GATE framework. On the other hand, the gazetteer consists of a list of entities that are relevant in the domain under question. Several lists containing biological terms extracted from GeneOntology[5] and UMLS[6] have been created.

Each type of NE is associated with a Class of the Ontology, so that those NE mentions which have been classified into a type of NE become candidates for Ontology Instances. These candidates are provisionally inserted in the corresponding class or classes.

---

[4] http://gate.ac.uk/
[5] http://www.geneontology.org/
[6] http://www.nlm.nih.gov/research/umls/

Once the NEs have been extracted, the semantic role labeling frames are detected in the text in order to extract the possible relationships between these NEs. Regarding the relationships between the individuals, they are represented by means of object properties between classes in the domain ontology. For each of these relationships, the participating individuals are obtained from the ontology and the system checks if they are already related by an object property of the same type (i.e. part_of, located_in, derives_from, etc). The relationship is only inserted if the relationship does not exist in the ontology yet.

After that, a reasoner such as Hermit is executed in order to (1) check for the consistency of the ontology and (2) compute inferred types. If the ontology is inconsistent, the last relationship inserted into the ontology is removed. In case the ontology is consistent, and the reasoner has inferred that one individual belonging to the relationship can be classified into a new class, this new classification is done.

3 Evaluation

The framework has been validated by using an ontology extracted from the xGENIA ontology [3], which is an OWL-DL ontology based on the GENIA taxonomy that was developed as a result of manual annotation of the GENIA corpus. Some changes in its properties have been done in the application layer, in order to represent them as a subset of the relations proposed in the ontological model [6]. With all, it achieved a precision value of 79.02% and a recall value of 65.4%. These values are significant because (1) the domain is a quite specific one, and (2) the semantic roles used have been designed for the biomolecular domain.

Acknowledgments. This work has been possible thanks to the Spanish Ministry for Science and Innovation under project TIN2010-21388-C02-02. Juana María Ruiz-Martínez is supported by the Fundación Séneca through grant 06857/FPI/07.

References

1. Smith, B., Ceusters, W., Klagges, B., Kohler, J., Kumar, A., Lomax, J., Mungall, C.J., Neuhaus, F., Rector, A., Rosse, C.: Relations in Biomedical Ontologies. Genome Biol. 6, R46 (2005)
2. Beißwanger, E., Schulz, S., Stenzhorn, H., Hahn, U.: BioTop: An Upper Domain Ontology for the Life Sciences - A Description of its Current Structure, Contents, and Interfaces to OBO Ontologies. Applied Ontology 3(4), 205–212 (2008)
3. Rak, R., Kurgan, L., Reformat, M.: xGENIA: A comprehensive OWL ontology based on the GENIA corpus. Bioinformation 1(9), 360–362 (2007)
4. Kim, J., Ohta, T., Tateisi, Y., Tsujii, J.: GENIA corpus - a semantically annotated corpus for bio-textmining. Bioinformatics (19), 180–182 (2003)
5. Moreda, P., Llorens, H., Saquete, E., Palomar, M.: Combining semantic information in question answering. Information Processing and Management (2010) (article in press, corrected proof)
6. Ruiz-Martinez, J.M., Valencia-Garcia, R., Martinez-Bejar, R., Hoffmann, A.: Populating biomedical ontologies from natural language texts. In: Proceedings of KEOD 2010, Valencia, Spain (2010)

Treo: Best-Effort Natural Language Queries over Linked Data

André Freitas[1], João Gabriel Oliveira[1,2], Seán O'Riain[1],
Edward Curry[1], and João Carlos Pereira da Silva[2]

[1] Digital Enterprise Research Institute (DERI)
National University of Ireland, Galway
[2] Computer Science Department
Universidade Federal do Rio de Janeiro

Abstract. Linked Data promises an unprecedented availability of data on the Web. However, this vision comes together with the associated challenges of querying highly heterogeneous and distributed data. In order to query Linked Data on the Web today, end-users need to be aware of which datasets potentially contain the data and the data model behind these datasets. This query paradigm, deeply attached to the traditional perspective of structured queries over databases, does not suit the heterogeneity and scale of the Web, where it is impractical for data consumers to have an a priori understanding of the structure and location of available datasets. This work describes *Treo*, a best-effort natural language query mechanism for Linked Data, which focuses on the problem of bridging the semantic gap between end-user natural language queries and Linked Datasets.

Keywords: Natural Language Queries, Linked Data.

1   Introduction

End-users querying Linked Data on the Web should be able to query data spread over potentially a large number of heterogeneous, complex and distributed datasets. However, Linked Data consumers today still need to have a previous understanding of the available datasets and vocabularies in order to execute expressive queries over Linked Datasets. In addition, in order to query Linked Data, end-users need to cope with the syntax of a structured query language. These constraints represent a concrete barrier between data consumers and datasets, strongly limiting the visibility and value of existing Linked Data.

In this scenario, natural language queries emerge as a simple and intuitive way for users to query Linked Data [1]. However, unrealistic expectations of achieving the precision of structured query approaches in a natural language query scenario and in the Web scale, brings the risk of overshadowing short-term opportunities of addressing fundamental challenges for Linked Data queries. With the objective of reaching the balance between precision, flexibility and usability, this work focuses on the construction of a *best-effort natural language query*

R. Muñoz et al. (Eds.): NLDB 2011, LNCS 6716, pp. 286–289, 2011.
© Springer-Verlag Berlin Heidelberg 2011

approach for Linked Data. Search engines for unstructured text are an example of the success of best-effort approaches on the Web. One assumption behind best-effort approaches is the fact that part of the search/query process can be delegated to the cognitive analysis of end-users, reducing the set of challenges that need to be addressed in the design of the solution. This assumption, applied to natural language queries over Linked Data, has the potential to enable a flexible approach for end-users to consume Linked Data.

This work presents *Treo*, a query mechanism which focuses on addressing the challenge of bridging the semantic gap between user and Linked datasets, providing a precise and flexible best-effort semantic matching approach between natural language queries and distributed heterogeneous Linked Datasets.

2   Description of the Approach

In order to address the problem of building a best-effort natural language query mechanism for Linked Data, an approach based on the combination of *entity search*, a *Wikipedia-based semantic relatedness measure* and *spreading activation* is proposed. The center of the approach relies on the use of a Wikipedia-based semantic relatedness measure as a key element for matching query terms to dataset terms. Wikipedia-based relatedness measures address limitations of existing works which are based on similarity measures/term expansion based on WordNet [2].

The query processing starts with the determination of key entities present in the user natural language query, using named entity recognition. Key entities are entities which can be potentially mapped to instances or classes in the Linked Data Web. After detected, key entities are sent to the entity search engine which determines a list of pivot entities in the Linked Data Web. A pivot entity is an URI which represents an entry point in the Linked Data Web for the spreading activation search. After the entities present in the user natural language query are determined, the query is analyzed in the query parsing module. The output of this module is a *partial ordered dependency structure* (PODS), which is a reduced representation of the query, targeted towards maximizing the matching probability between the structure of the terms present in the query and the subject, predicate, object structure of RDF.

The PODS and the list of pivot entities are used as the input for the spreading activation search algorithm. Starting from the URI representing the first pivot entity in the list, Treo dereferences the pivot entity URI, fetching its associated RDF description. The RDF description contains a set of properties and objects associated with the entity URI. For each associated pair (*property, object*), *Treo* calculates the semantic relatedness measure between the pair and the next query term. The Wikipedia-based relatedness measure allows the computation of the semantic proximity between the query terms and the terms present in the datasets (in both terminological and instance level). This step represents the core of the flexibility introduced in the semantic matching process. Objects associated with pairs with high relatedness discrimination are further explored

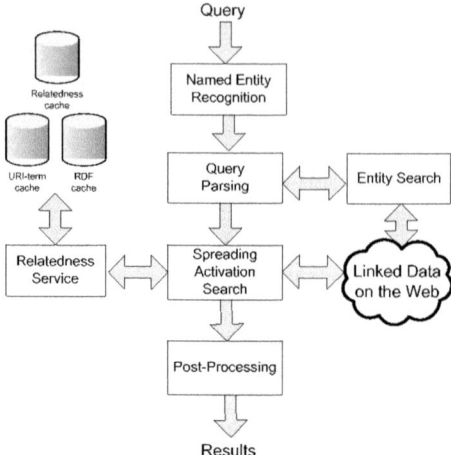

Fig. 1. Components of Treo's architecture

(i.e. their URIs are dereferenced) and the semantic relatedness of the properties and objects in relation to the next query term is computed. The process continues until the end of the query is reached. The process of node search using semantic relatedness defines a spreading activation search where the activation function is defined by the semantic relatedness measure. The spreading activation search returns a set of ranked triple paths, i.e. connected triples starting from the pivot entities which are the result of the search process. The set of triple paths are further post-processed and merged into a graph which is the output displayed to users. The characteristic of the approach of using semantic relatedness to determine the direction of navigation on the Linked Data Web and the set of returned triple paths, defined the name Treo for the prototype, the Irish word for path and direction.

The components of the prototype's architecture are depicted in figure 1. In order to minimize the query execution time, three caches are implemented. The relatedness cache stores the values of previously computed semantic relatedness measures between pairs of terms. The RDF cache stores local copies of the RDF descriptions for the dereferenced URIs, in order to minimize the number of HTTP requests. The URI-term cache stores labels for each URI which are used in the semantic relatedness computation process. An important characteristic of the proposed approach, which was not implemented in the current version of Treo, is the natural level of parallelization that can be achieved in the relatedness-based spreading activation process.

3   Query Example

Figure 2 shows the output for the query 'From which university did the wife of Barack Obama graduate?'. The output shows an example of merged triple paths containing the correct answer for the user query (the URIs for *Princeton*

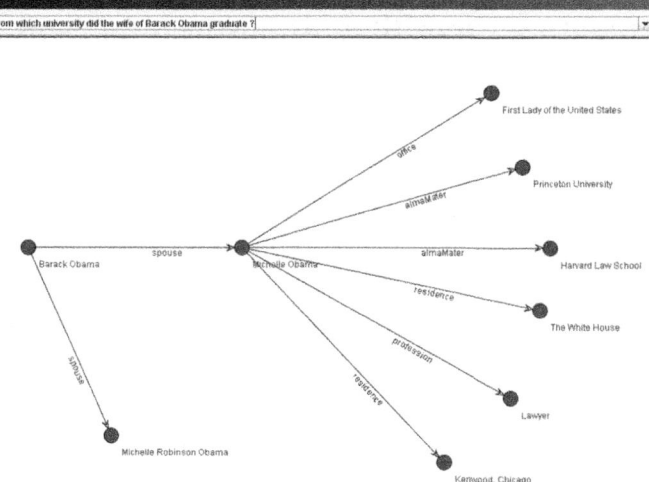

Fig. 2. Treo result set for the query *'From which university did the wife of Barack Obama graduate?'*

University and *Harvard Law School*) together with triples which are not part of the desired answer set. The best-effort nature of the approach delegates to the user the final cognitive validation of the answer set. In order to support users in the validation process, the relationships from the pivot entity (in this query, *Barack Obama*) to the final answer resources/values are displayed. This example emphasizes the flexibility introduced by the semantic relatedness measure in the query-dataset matching, where the query term *'wife'* is matched with the dataset term *'spouse'* and the query terms *'graduate'* is matched with the *'university'* type in the dataset. Additional results for different queries can be found in [3].

Acknowledgments. The work presented in this paper has been funded by Science Foundation Ireland under Grant No. SFI/08/CE/I1380 (Lion-2).

References

1. Kaufmann, E., Bernstein, A.: Evaluating the usability of natural language query languages and interfaces to Semantic Web knowledge bases. In: Web Semantics: Science, Services and Agents on the World Wide Web, vol. 8, pp. 377–393 (2010)
2. Freitas, A., Oliveira, J.G., O'Riain, S., Curry, E., da Silva, P.: Querying Linked Data using Semantic Relatedness: A Vocabulary Independent Approach. In: Proceedings of the 16th International Conference on Applications of Natural Language to Information Systems, NLDB (2011)
3. Treo Query Examples (2011), http://treo.deri.ie/gallery/nldb2011.htm

Evaluating EmotiBlog Robustness for Sentiment Analysis Tasks[*]

Javi Fernández, Ester Boldrini, José Manuel Gómez, and Patricio Martínez-Barco

University of Alicante –Spain-, GPLSI, Department of Language and Computying Systems
{Javifm,eboldrini,jmgomez,patricio}@dlsi.ua.es

Abstract. *EmotiBlog* is a corpus labelled with the homonymous annotation schema designed for detecting subjectivity in the new textual genres. Preliminary research demonstrated its relevance as a Machine Learning resource to detect opinionated data. In this paper we compare *EmotiBlog* with the *JRC* corpus in order to check the *EmotiBlog* robustness of annotation. For this research we concentrate on its coarse-grained labels. We carry out a deep ML experimentation also with the inclusion of lexical resources. The results obtained show a similarity with the ones obtained with the *JRC* demonstrating the *EmotiBlog* validity as a resource for the SA task.

Keywords: Sentiment Analysis, EmotiBlog, Machine Learning Experiments.

1 Introduction

The exponential growth of the subjective information on the Web and the employment of new textual genres originated an explosion of interest in Sentiment Analysis (SA). This is a task of Natural Language Processing (NLP) in charge of identifying the opinions related to a specific target (Liu, 2006). Subjective data has a great potential. It can be exploited by business organizations or individuals, for ads placements, but also for the Opinion Retrieval/Search, etc (Liu, 2007). Thus, our research is motivated by the lack of resources, methods and tools to properly treat subjective data. In this paper is we demonstrate that the EmotiBlog annotation schema can be successfully employed to overcome the challenges of SA because of its reliability in terms of annotation. EmotiBlog allows a double level of annotation, coarse and fine-grained and in this paper we test the reliability of the coarse-grained labels. In order to achieve this, we train a Machine Learning (ML) system with two domain-specific corpora annotated with EmotiBlog (EmotiBlog Kyoto[1] and EmotiBlog Phones[2]). We carry out the same experiments with the well-known JRC corpus in order to compare their performances and understand if they are comparable under similar conditions. We carry out a deep study using basic NLP techniques

---

[*] This work has been partially founded by the TEXTMESS 2.0 (TIN2009-13391-C04-01) and Prometeo (PROMETEO/2009/199) projects and also by the complimentary action from the Generalitat valenciana (ACOMP/2011/001).

[1] The *EmotiBlog* corpus is composed by blog posts on the Kyoto Protocol, Elections in Zimbabwe and USA election, but for this research we only use the ones about the Kyoto Protocol (*EmotiBlog Kyoto*). Available on request from authors.

[2] The *EmotiBlog Phones* corpus is composed by users' comments about mobile phones. extracted from Amazon UK (http://www.amazon.co.uk). Available on request from authors.

R. Muñoz et al. (Eds.): NLDB 2011, LNCS 6716, pp. 290–294, 2011.
© Springer-Verlag Berlin Heidelberg 2011

(stemmer, lemmatiser, term selection, etc.) also integrating SentiWordNet (Esuli and Sebastiani, 2006) and WordNet (Miller, 1995) as lexical resources. In previous works EmotiBlog has been applied successfully to Opinionated Question Answering (OQA) (Balahur et al. 2009 c and 2010a,b), to Automatic Summarization of subjective content (Balahur et al. 2009a), but also to preliminary ML experiments (Boldrini et al. 2010). Thus, the first objective of our research is to demonstrate that our resource is reliable and valuable to train ML systems for NLP applications. As a consequence, our second aim is to show that the next step of our research will consist in a deeper text classification for the SA task –and its applications-. In fact, alter having demonstrated the validity of the coarse-grained annotation, we will test the EmotiBlog fine-grained annotation. We believe there is a need for positive/negative text categories, but also emotion intensity (high/medium/low), emotion type (Boldrini et al, 2009a) and the annotation of the elements that give the subjectivity to the discourse, contemplated by EmotiBlog, not just at sentence level.

2   Related Work

The first step of SA consists in building lexical resources of affect, such as WordNet Affect (Strapparava and Valitutti, 2004), SentiWordNet (Esuli and Sebastiani, 2006), Micro-WNOP (Cerini et. Al, 2007) or "emotion triggers" (Balahur and Montoyo, 2009). All these lexicons contain single words, whose polarity and emotions are not necessarily the ones annotated within the resource in a larger context. The starting point of research in emotion is represented by (Wiebe 2004), who focused the idea of subjectivity around that of private states setting the benchmark for subjectivity analysis. Furthermore, authors show that the discrimination between objective/ subjective discourses is crucial for the sentiment task, as part of Opinion Information Retrieval (last three editions of the TREC Blog tracks[3] competitions, the TAC 2008 competition[4]), Information Extraction (Riloff and Wiebe, 2003) and QA (Stoyanov et al., 2005) systems. Related work also includes customer review classification at a document level, sentiment classification using unsupervised methods (Turney, 2002), ML techniques (Pang and Lee, 2002), scoring of features (Dave, Lawrence and Pennock, 2003), using PMI, or syntactic relations and other attributes with SVM (Mullen and Collier, 2004). Research in classification at a document level included sentiment classification of reviews (Ng, Dasgupta and Arifin, 2006), on customer feedback data (Gamon, Aue, Corston-Oliver, Ringger, 2005). Other research has been conducted in analysing sentiment at a sentence level using bootstrapping techniques (Riloff, Wiebe, 2003), considering gradable adjectives (Hatzivassiloglou, Wiebe, 2000), (Kim and Hovy, 2004), or determining the semantic orientation of words and phrases (Turney and Littman, 2003). Other work includes (Mcdonald et al. 2007) who investigated a structured model for jointly classifying the sentiment of a text at varying levels of granularity. Neviarouskaya (2010) classified texts using finegrained attitude labels basing its work on the compositionality principle and an approach based on the rules elaborated for semantically distinct verb classes, while Tokuhisa (2008) proposed a data-oriented method for inferring the emotion of a speaker conversing with a dialogue system from the semantic content of an utterance. They divide the emotion classification into two steps: sentiment polarity and emotion

---

[3] http://trec.nist.gov/data/blog.html

[4] http://www.nist.gov/tac/

classification. Our work starts from the conclusions drawn by (Boldrini et al 2010) in which authors performed several experiments on three different corpora, aimed at finding and classifying both the opinion, as well as the expressions of emotion they contained. They showed that the fine and coarse-grained levels of annotation that EmotiBlog contains offers important information on the structure of affective texts, leading to an improvement of the performance of systems trained on it. Thanks to EmotiBlog- annotated at sentence, as well as element level- we have the possibility of carrying out ML experiments of different nature proposing a mixed approach based on the ML training but also the lexical resources integration.

3 Training Corpora

The corpora (in English) we employed for our experiments are EmotiBlog Kyoto extended with the collection of mobile phones reviews extracted from Amazon (EmotiBlog Phones). It allows the annotation at document, sentence and element level (Boldrini et al. 2010), distinguishing between objective and subjective discourses. The list of tags for the subjective elements is presented in (Boldrini et al, 2009a). We also use the JRC quotes[5], a set of 1590 English language quotations extracted automatically from the news and manually annotated for the sentiment expressed towards entities mentioned inside the quotation. For all of these elements, the common attributes are annotated: polarity, degree and emotion. As we want to compare the two corpora, we will consider entire sentences and evaluate the polarity, to adapt to the JRC annotation schema. Table 1 presents the size of all the corpora in sentences divided by its classification.

Table 1. Corpora size in sentences

	EB Kyoto	EB Phones	EB Full	JRC
Objective	347	172	519	863
Subjective	210	246	456	427
Positive	62	198	260	193
Negative	141	47	188	234
Total	557	418	975	1290

4 Machine Learning Experiments and Discussion

In order to demonstrate that *EmotiBlog* is valuable resource for ML, we perform a large number of experiments with different approaches, corpus elements and resources. As features for the ML system we use the classic *bag of words* initially. To reduce the dimensionality we also employ techniques such as *stemming*, *lemmatization* and *dimensionality reduction by term selection* (TSR) methods. For TSR, we compare two approaches, *Information Gain* (IG) and *Chi Square* (X2), because they reduce the dimensionality substantially with no loss of effectiveness (Yang and Pedersen, 1997). For weighting these features we evaluate the most common methods: *binary weighting*, *tf/idf* and *tf/idf normalized* (Salton and Buckley, 1988). As supervised learning method we use *Support Vector Machines* (SVM) due to its good results in text categorization (Sebastiani, 2002) and the promising results

---

[5] http://langtech.jrc.ec.europa.eu/JRC_Resources.html

obtained in previous studies (Boldrini et al. 2009b). We also evaluate if grouping features by their semantic relations increases the coverage in the test corpus and reduces the samples dimensionality. The challenge at this point is Word Sense Disambiguation (WSD) due the poor results that these systems traditionally obtain in international competitions (Agirre et al. 2010). Choosing the wrong sense of a term would introduce noise in the evaluation and thus a low performance. But we believe that if we include all senses of a term in the set of features the TSR will choose only the correct ones. For example, using all *WordNet* (WN) senses of each term as learning features, the TSR methods could remove the non-useful senses to classify a sample in the correct class. In this case this disambiguation methods would be adequate. As lexical resources for these experiments we employ WN, but also *SentiWordNet* (SWN), since the use of this specific OM resource demonstrated to improve the results of OM systems. It assigns to some of the synsets of WN three sentiment scores: *positivity*, *negativity* and *objectivity*. As the synsets in SWN are only the opinionated ones, we want to test if expanding only with those ones can improve the results. In addition, we want to introduce the sentiment scores into the ML system by adding them as new attributes. For example, if we get a synset S with a positivity score of 0.25 and a negativity score of 0.75, we add a feature called S (with the score given by the weighting technique) but also two more features: S-*negative* and S*positive* with their negative and positive scores respectively. The experiments with lexical resources have been carried out with five different configurations using: **i)** only SWN synsets (experiment s), **ii)** only WN synsets (experiment w), **iii)** both SWN and WN synsets (experiment sw), **iv)** only SWN synsets including sentiment scores (experiment ss) and **v)** both SWN and WN synsets including sentiment scores (experiment sws). In case a term is not found in any of the lexical resources, then its lemma is left. Moreover, to solve the ambiguity, two techniques have been adopted: including all its senses and let the TSR methods perform the disambiguation (experiments s, w, sw, ss and sws), or including only the most frequent sense for each term (experiments $s1$, $w1$, $sw1$, $ss1$ and $sws1$). We made an exhaustive combination of all the possible parameters (tokenization, weighting, feature selection and use of lexical resources), with the different classifications and corpus, which is summarized in Table 2, where we only show the best results for each pair classification-corpus because of space reasons.

The lower results belong to the *JRC* and *EmotiBlog Full* corpora, although they are the bigger ones. They are not domain-specific so is more difficult for the ML system to create a specialized model. But their best results using similar techniques are ver similar. This fact shows us that the annotation schema and process is valid. On the other hand, experiments with *EmotiBlog Kyoto* and *Phones* obtain the best results because they are domain-specific. This makes *EmotiBlog* more usable in real-word applications, which demand higher performance and usually belong to a specific domain. Regarding the evaluation of the different techniques, we can see that the best results include the lexical resources. Having a look to the totality of experiments, they are always in the top positions. Moreover, in Table 2 we can see that SWN is present in 6 of the 8 best results shown, and the sentiment scores in 5 of them. This encourages us to continue using SWN in our following experiments and find new

Table 2. Best configuration and result for each pair classification-corpus

| | EB Kyoto | | EB Phones | | EB Full | | JRC | |
	Conf	F1	Conf	F1	Conf	F1	Conf	F1
Objectivity	**sws**	**0.6647**	sws1	0.6405	sw	0.6274	w1	0.6088
Polarity	ss1	0.7602	**ss**	**0.8093**	ss1	0.6374	w1	0.5340

ways to take advantage of the sentiment information it provides. The other mentioned techniques (tokenization, weighting and feature selection) affect the results but not in significant way. Their improvements are very small and do not seem to follow any pattern. As future work we propose to experiment with other well-known corpora and combinations of different ones, in order to evaluate if the improvements depend on the type and the size of the corpus. We will also continue evaluating the EmotiBlog robustness. Specifically we will test the reliability of its fine-grained annotation.

References

1. Agirre, E., Lopez de Lacalle, O., Fellbaum, C., Hsieh, S., Tesconi, M., Monachini, M., Vossen, P., Segers, R.: SemEval-2010 Task 17: All-words Word Sense Disambiguation on a Specific Domain (2010)
2. Balahur, A., Boldrini, E., Montoyo, A., MartÃnez-Barco, P.: Opinion and Generic Question Answering systems: a performance analysis. In: Proceedings of ACL, Singapore (2009c)
3. Balahur, A., Boldrini, E., Montoyo, A., MartÃnez-Barco, P.: A Unified Proposal for Factoid and Opinionated Question Answering. In: Proceedings of the COLING Conference (2010a)
4. Balahur, A., Boldrini, E., Montoyo, A., MartÃnez-Barco, P.: Opinion Question Answering: Towards a Unified Approach. In: Proceedings of the ECAI Conference (2010b)
5. Balahur A., Lloret E., Boldrini E., Montoyo A., Palomar M., MartÃnez-Barco P.: Summarizing Threads in Blogs Using Opinion Polarity. In: Proceedings of ETTS Workshop. RANLP (2009a)
6. Balahur A., Montoyo A.: Applying a Culture Dependent Emotion Triggers Database for Text Valence and Emotion Classification. In: Proceedings of the AISB 2008 Symposium on Affective Language in Human and Machine, Aberdeen, Scotland (2008)
7. Boldrini, E., Balahur, A., MartÃnez-Barco, P., Montoyo, A.: EmotiBlog: a finer-grained and more precise learning of subjectivity expression models. In: Proceedings of LAW IV, ACL (2010)
8. Boldrini, E., Balahur, A., MartÃnez-Barco, P., Montoyo, A.: EmotiBlog: an Annotation Scheme for Emotion Detection and Analysis in Non-traditional Textual Genres. In: Proceedings of DMIN, Las Vegas (2009a)
9. Cerini, S., Compagnoni, V., Demontis, A., Formentelli, M., Gandini, G.: Micro-WNOp: A gold standard for the evaluation of automatically compiled lexical resources for opinion mining. In: Angeli, F. (ed.) Language Resources and Linguistic Theory: Typology, Second Language Acquisition, English Linguistics, Milano, IT (2007) (forthcoming)
10. Dave, K., Lawrence, S., Pennock, D.: Mining the Peanut Gallery: Opinion Extraction and Semantic Classification of Product Reviews. In: Proceedings of WWW 2003 (2003)
11. Esuli, A., Sebastiani, F.: SentiWordNet: A Publicly Available Resource for Opinion Mining. In: Proceedings of the 6th International Conference on Language Resources and Evaluation, LREC 2006, Genoa, Italy (2006)
12. Gamon, M., Aue, S., Corston-Oliver, S., Ringger, E.: Mining Customer Opinions from Free Text. In: Famili, A.F., et al. (eds) IDA 2005. LNCS, vol. 3646, pp. 121–132. Springer, Heidelberg (2005)
13. Hatzivassiloglou, V., Wiebe, J.: Effects of adjective orientation and gradability on sentence subjectivity. In: Proceedings of COLING (2000)
14. Kim, S.M., Hovy, E.: Determining the Sentiment of Opinions. In: Proceedings of COLING 2004 (2004)
15. Liu, B.: Web Data Mining book. Ch. 11 (2006)
16. Liu, B.: Web Data Mining. In: Exploring Hyperlinks, Contents and Usage Data, 1st edn. Springer, Heidelberg (2007)
17. McDonald, R., Hannan, K., Neylon, T., Wells, M., Reynar, J.: Structured Models for Fine-to-Coarse Sentiment Analysis. In: Proceedings of the 45th Annual Meeting of the Association of Computational Linguistics (2007)
18. Miller, G.A.: WordNet: A Lexical Database for English (1995)

Syntax-Motivated Context Windows of Morpho-Lexical Features for Recognizing Time and Event Expressions in Natural Language

Hector Llorens, Estela Saquete, and Borja Navarro

University of Alicante, Alicante, Spain
{hllorens,stela,borja}@dlsi.ua.es

Abstract. We present an analysis of morpho-lexical features to learn SVM models for recognizing TimeML time and event expressions. We evaluate over the TempEval-2 data, the features: word, lemma, and PoS in isolation, in different size static-context windows, and in a *syntax-motivated dynamic-context windows* defined in this paper. The results show that word, lemma, and PoS introduce complementary advantages and their combination achieves the best performance; this performance is improved using context, and, with dynamic-context, timex recognition is improved to reach state-of-art performance. Although more complex approaches improve the efficacy, the morpho-lexical features can be obtained more efficiently and show a reasonable efficacy.

1 Introduction

Within temporal information processing, this paper focuses on the recognition of temporal expressions (timexes) and events, as illustrated in (1).

(1) He <u>arrived</u> to Alicante <u>last Friday</u>.

Recently, in TempEval-2 [5], different timex and event recognition approaches were evaluated following TimeML standard [3]. All the approaches used morpho-lexical features in addition to more complex features, but in different ways: some used part-of-speech (PoS), others used lemmas, some used context windows, etc; The contribution of these morpho-lexical features were not analyzed.

The objective of this paper is to analyze the contribution of the following uses of morpho-lexical features to timex and event recognition:

- Basic features: the use of word, lemma, PoS, and their combinations.
- Static context: 1-window (no context), 3-window (-1,1), 5-window (-2,2), 7-window (-3,3), and 9 window (-4,4).
- *Dynamic context*: Our syntax-motivated dynamic-context.

This analysis will show the efficacy of different morpho-lexical features, that can be obtained more efficiently than more complex ones.

R. Muñoz et al. (Eds.): NLDB 2011, LNCS 6716, pp. 295–299, 2011.
© Springer-Verlag Berlin Heidelberg 2011

Table 1. TempEval-2 ML approaches for English

System	Morpho-lexical features	timex F1	event F1
TRIOS [4]	*Word* (1-window)	0.85	0.77
TIPSem [2]	*Lemma, PoS* (5-window)	0.85	0.83
KUL [1]	*Word, Lemma, PoS* (3-window)	0.84	-

2   Related Work

Different data-driven approaches addressing the automatic recognition of timexes and events for English were presented in the TempEval-2 evaluation exercise. Table 1 summarizes the best data-driven approaches detailing their morpho-lexical features and the scores they obtained.

Each system proposes a different use of morpho-lexical features and morpho-lexical context. However, these approaches also include more complex and computationally more costly features (e.g., semantics).

3   Proposal: Analyzing Morpho-Lexical Features

We present a SVM-based[1] approach, which uses morpho-lexical features, to determine whether a token is part of one of these temporal entities:

– **Timex.** A timex is a linguistic representation of a time point or period. Timexes are normally denoted by nouns, adjectives, adverbs, or complete phrases (e.g., Monday, two weeks, 8 o'clock, every year).
– **Event.** Events are things that that *happen or hold* and are generally expressed by verbs, nominalizations, adjectives, predicative clauses or prepositional phrases (e.g, see, war, attack, alive).

The following **morpho-lexical features** are included in different models:

– **The basic features**: These are *Word, Lemma* and *PoS*, and their combinations via addition (e.g., *Lemma+PoS*).
– **The static context**: For example, tokens like "last" (last, last, JJ) can be part of a timex like in Fig. 1 or not like in "last song". Therefore, context is required to disambiguate. We analyze symmetric 1-window (no context), 3-window (-1,1), 5-window (-2,2), 7-window (-3,3), and 9 window (-4,4).
– **The *dynamic context***: We observed that the static-size context-windows sometimes introduce ***irrelevant information***. In Fig. 1, the static 3-window for "last" includes [***won***, last, year]. However, the fact that "last" is preceded by "won" is irrelevant and may introduce ambiguity. To tackle this problem, we define a *syntax-motivated dynamic n-window* as: an n-window in which the elements that are outside the syntactic argument of the token are replaced by "-". In the case of verbs, the window includes the verbal form

---

[1] YamCha SVM was used (http://chasen.org/˜taku/software/YamCha/).

Word	Lemma	PoS	Syntax	
He	he	PRP	(NP*)	
won	win	VBD	(VP*	
last	last	JJ	(NP*	1-window
year	year	NN	*)	3-window

dynamic
3-window

Fig. 1. Features and Context Windows Illustration

(simple or complex). In the case of nested phrases, the window includes the complete nesting. Then, the dynamic 3-window for "last" excludes "won": [-,last,year]. Our hypothesis is that this will lead to a better performance.

4 Evaluation: TempEval-2 Test

The results for the morpho-lexical features are shown in the table below.

Table 2. No-context results for morpho-lexical features

(a) TIMEX

Feature	P	R	$F_{\beta=1}$
Word	0.14	0.74	0.23
Lemma	0.21	0.65	0.31
PoS	0.00	0.00	0.00
Word+PoS	0.83	0.57	0.68
Lemma+PoS	0.83	0.54	0.66
Word+Lemma+PoS	0.83	0.64	0.72

(b) EVENT

Feature	P	R	F1
Word	0.42	0.88	0.57
Lemma	0.51	0.88	0.65
PoS	0.77	0.75	0.76
Word+PoS	0.80	0.80	0.80
Lemma+PoS	0.79	0.80	0.80
Word+Lemma+PoS	0.80	0.82	0.81

For timex recognition, first, word and lemma alone achieve reasonably high results given the model simplicity (one feature). Secondly, the PoS alone is completely useless to determine whether a token is a timex. The reason is that there are not parts-of-speech denoting timexes more often than non-timexes. Finally, the combination of these features leads to a performance improvement indicating that they are complementary.

For event recognition, first, word and lemma alone again achieve good results. Secondly, unlike for timexes, the PoS isolated is very useful because most verbs denote events. Finally, like for timex, the combination of features offers the best performance, verifying that these are complementary.

Adding Context
Given the previous analysis, the context is analyzed for the feature combination Word+Lemma+PoS. The results of static and dynamic context are shown in Fig. 2 incrementally from no context (1-window) to a context-window size of 9. For timexes, it is observed that using a 3-window introduces a great improvement over the no-context approach (1-window) – from 0.72 to 0.84. Further size

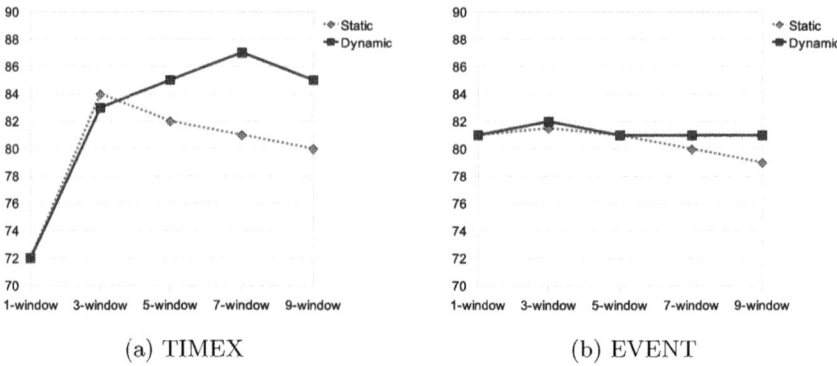

(a) TIMEX   (b) EVENT

Fig. 2. Static and Dynamic context F1 results for different sizes

increasings in the static-context decrease the results. However, in the dynamic window, the results improve until the 7-window (0.87). This improves the maximum F1 obtained by static windows (0.84), and is comparable to the state of the art results obtained in TempEval-2 (0.86). Therefore, the *dynamic-context* is valuable for timex recognition.

For events, using a 3-window context improves the performance of no-context approach but less than for timexes – from 0.81 to 0.82. The *dynamic-context* does not improve the results and only smooths the performance drop that static-context suffers in larger windows. This is because events, mainly consist of one token, and the relevant morpho-lexical context seems to match with the average phrase-size of the corpus – about three tokens. Regardless the type of context used (dynamic or static), the data-driven approach automatically learns that the features outside the 3-window context do not give discriminative information. The highest F1 obtained by a 3-window context (0.82) is comparable to the state of the art performance (0.83).

5   Conclusions and Future Work

This paper presented an analysis of the different uses of morpho-lexical features in TimeML timex and event recognition. Different SVM-based models have been presented and evaluated for these tasks over the TempEval-2 data.

The four main insights of this paper are: (i) word, lemma, and PoS introduce complementary advantages and their combination is useful; (ii) the use of context (3-window) improves the performance; (iii) the dynamic-context avoids considering irrelevant information and outperforms static-context in timex recognition; and (iv) a good use of the morpho-lexical features offers a performance comparable to the state-of-art with a lower computational complexity.

Given this conclusions, as further work, we propose to empirically measure the efficiency of the presented approaches vs. more complex state of the art approaches for studying if the approaches based only on morphosyntax can be relevant in applications where the efficiency takes priority over the efficacy.

References

1. Kolomiyets, O., Moens, M.-F.: Kul: Recognition and normalization of temporal expressions. In: Proceedings of SemEval-5, pp. 325–328. ACL (2010)
2. Llorens, H., Saquete, E., Navarro-Colorado, B.: TIPSem (English and Spanish): Evaluating CRFs and Semantic Roles in TempEval-2. In: Proceedings of SemEval-5, pp. 284–291. ACL (2010)
3. Pustejovsky, J., Castaño, J.M., Ingria, R., Saurí, R., Gaizauskas, R., Setzer, A., Katz, G.: TimeML: Robust Specification of Event and Temporal Expressions in Text. In: IWCS-5 (2003)
4. UzZaman, N., Allen, J.F.: Trips and trios system for tempeval-2: Extracting temporal information from text. In: Proceedings of SemEval-5, pp. 276–283. ACL (2010)
5. Verhagen, M., Saurí, R., Caselli, T., Pustejovsky, J.: Semeval-2010 task 13: Tempeval-2. In: Proceedings of the 5th International Workshop on Semantic Evaluation, pp. 57–62. ACL, Uppsala (2010)

Tourist Face: A Contents System Based on Concepts of Freebase for Access to the Cultural-Tourist Information

Rafael Muñoz Gil, Fernando Aparicio, Manuel de Buenaga, Diego Gachet,
Enrique Puertas, Ignacio Giráldez, and Mª Cruz Gaya

Grupo de Sistemas Inteligentes, Univ. Europea de Madrid.
Villaviciosa de Odón, Madrid, Spain, 28670
{rafael.munoz,fernando.aparicio,
buenaga,gachet,enrique.puertas,ignacio.giraldez,mcruz}@uem.es

Abstract. In more and more application areas large collections of digitized multimedia information are gathered and have to be maintained (e.g. in tourism, medicine, etc). Therefore, there is an increasing demand for tools and techniques supporting the management and usage of digital multimedia data. Furthermore, new large collections of data are available through it every day. In this paper we are presenting Tourist Face, a system aimed at integrating text analyzing techniques into the paradigm of multimedia information, specifically tourist multimedia information.

Particularly relevant components to its the development are *Freebase*, a large collaborative base of knowledge, and *General Architecture for Text Engineering* (GATE), a system for text processing. The platform architecture has been built thinking in terms of scalability, with the following objectives: to allow the integration of different natural language processing techniques, to expand the sources from which information extraction can be performed and to ease integration of new user interfaces.

Keywords. Information Extraction, Multimedia Data, Text Processing, Scalable Architecture, Freebase, GATE.

1 Introduction

In recent years, the information published online has not only increased but diversified, appearing in various formats such as text, audio or video, which is known as multimedia information. Nowadays, there are tools for searching and managing these collections..Nevertheless, the need for tools able to extract hidden useful knowledge embedded in multimedia data is becoming pressing for many decision-making applications. The tools needed today are those aimed at discovering relationships between data items or segments within images, classifying images based on their content, extracting patterns from sound, categorizing speech and music, recognizing and tracking objects in video streams, etc.

R. Muñoz et al. (Eds.): NLDB 2011, LNCS 6716, pp. 300–304, 2011.
© Springer-Verlag Berlin Heidelberg 2011

In this context, automatic text analyzing may undoubtedly help to optimize the treatment of electronic documentation and to tailor it to the user's needs. In order to understand the depth of the problem we may consider the following questions: What are the problems encountered when extracting data in Multimedia? What are the specific issues raised in pattern extraction from text, sound, video...?

The system proposed in this paper covers several aspects regarding the challenges mentioned above. It is a system whose architecture enables to integrate the knowledge about different travel concepts contained in Freebase (expandable to other sources), which are processed with computational linguistics techniques through the GATE system. It provides a Web user interface, potentially expandable to other clients such as mobile devices, whose utility to physicians has already been mentioned (Buenaga M. et al, 2008), or even other servers in order to increase their knowledge base. The system allows the users to search for three kinds of concepts, i.e. tourist attraction, accommodation and travel destination, in a tourist text extracted from a video source, thus helping to optimize the time devoted to comprehend the text. As a result, the different concepts appearing in the different sections of the video source are automatically detected and linked to related information in FreeBase.

2   System Architecture and Information Extraction from Freebase

As regards the system architecture, we have adopted a client/server scheme to process data on a server offering a simple interface for HTTP communication. This scheme is shown in Figure 1.

The modularization of the components has the following specific objectives:

- The access module lets the system be connected through other clients, such as mobile devices or other applications that require the use of another server.
- The natural language processing module enables the system to use other tools or libraries, keeping the same structure and making possible the interaction between different components possible.
- The information retrieval module extracts the information from the runtime system, optimizing the online response time.

Freebase (Bollacker K. et al, 2008) is a large collaborative database of knowledge, published in 2007 by Metaweb and recently acquired by Google. It structures the content in "topics", which is typically a Wikipedia article. A topic can be related to different aspects, introducing the concept of "types", which in turn can have sets of "properties". The types are merged into "domains" (like travel, medicine, biology, location, astronomy, music, sports, etc.) to which an identifier directly related to the Web link to these contents (i.e. /travel, /medicine, /biology, /astronomy, /music, /sports, etc.) is assigned. Types and properties also have a unique identifier, consisting of concatenating the type or the topic name, as it is shown in the following example. The *tourist attraction* type belongs to the *travel* domain, thus its identifier is /travel/tourist attraction. As far as we know, Freebase has already been used in other research areas, like pieces of work related to software in the context of the Web 2.0 and 3.0 (Sören A. et al, 2007; Bizer C. et al, 2009), tools for names disambiguation by their profession, query classifications (Brenes D. J. et al, 2009), geo-tagging extensions (Weichselbraun A., 2009) or as an example for faceted navigation (Hearst M. A., 2008).

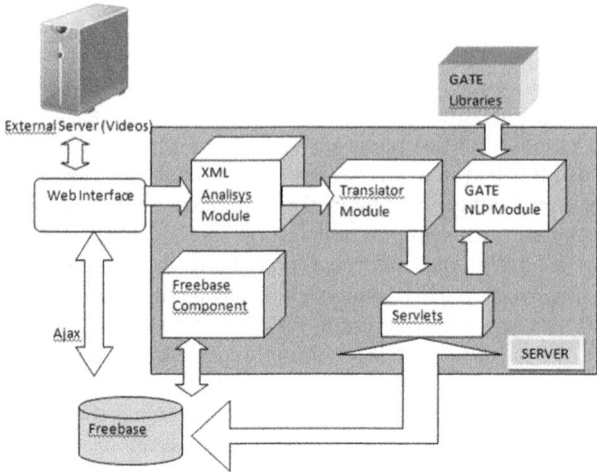

Fig. 1. System Architecture Diagram

In our proposal, Freebase is used to retrieve tourist concepts lists for the recognition of previously mentioned entities (Nadeau D. and Sekine S., 2007) in tourist texts and to link them with semantically-related content. The system uses some lists of *tourist attraction, accommodation, travel destination* extracted from FreeBase for the subsequent recognition using GATE tool (Aparicio F. 2011). Although Freebase is not a formal ontology of knowledge, the information quality is high precisely because of the manual input.

3   Tourist and Cultural Information Processing and User Interface

The type of domain we are facing has certain characteristics which make it especially interesting as a case study for automatic information extraction system.

First of all, tourism lexicon uses a broad terminology taken from various areas (geography, economics, art history, etc.). We can find words relating to hotels and restaurants, transport, leisure and entertainment, history and art, etc. The specific type of information we hold (videos on monuments or tourist destinations) is mainly characterized by the wide variety of information presented. In general, when describing a monument or a city, this description includes information on the type of place we are referring to (eg. a church or a fortress), location, date of construction or foundation, historical and artistic information about it, details of other monuments and places of interest around, etc.

The context of tourism is enshrined by the goal of recovery and the extraction of information, with particular emphasis on personalized information retrieval in context, embraced in the Mavir consortium, where the plan is named: "Improving access, analysis and information visibility and multilingual content and multimedia network for Madrid Community". Mavir http://www.mavir.net

The videos have been developed by the Spanish Tourism Institute (TURESPAÑA) which is the organization of the General Administration for the external promotion of Spain as a tourist destination (http://www.tourspain.es).

The present application includes a text extraction module in the video content analysis, this text is the started point of Tourist Face, being analyzed by our system. GATE (Cunningham H. et al, 2002) is open source software developed by Sheffield University, which provides a framework for text processing. Among the different programmatic options to integrate this software, we have selected one that admits the development and testing with the GUI Developer and reuses it from the NLP module logic. GATE is distributed including an information extraction system called ANNIE, A Nearly-New Information Extraction System, Gazetteer being one of its components. In the design of our system we have given special importance to this component, which, based on predefined lists, allows the recognition of previously mentioned entities. The lists used include the definition of relevant tourist attractions, accommodations and travel destinations. These lists allow the inclusion of features for each entity, which in our proposal are primarily used to store the Freebase identifier.

To illustrate the use, we can select videos from the Web or give them a format designed to our better comprehension. The system results after processing the text are:

- Tourist Attraction: A tourist attraction is a place or feature that you might visit as a tourist. Examples include monuments, parks, museums, eg: Albuquerque Aquarium, Alcatraz Island, Basilica di San Lorenzo…
- Acommodation: This type is intended to be used for hotels, bed and breakfasts, backpacker hostels, or any other such place where you might stay while travelling, eg: Palace Hotel San Francisco, Hotel Bintumani, Gateway Motel…
- Travel Destination: A travel destination is a location that you might go to for a vacation/holiday, eg. Paris or Bali.

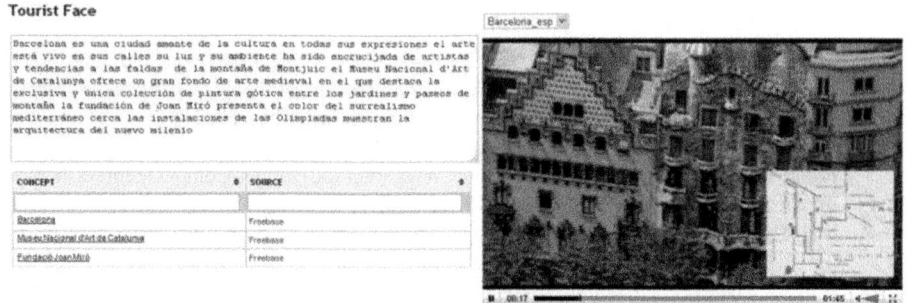

Fig. 2. Tourist Face Interface

We can go further in the meaning of any of the listed *items*. First of all, a tourist attraction's description is shown along with two other ways to get more information: (a) linking directly to the Freebase website or (b) retrieving some information from the system itself. All these results are linked to the associated Web page in Freebase.

7 Conclusion

In this paper, we have presented a system characterized by the incorporation of previously mentioned entities from Freebase so as to recognize tourist concepts in input system tourist videos, thus helping the users to increase the extracted information. It has been also shown the way to link semantic information from Freebase. In addition, we have described both the architecture of the platform and a systematic development method to optimize the integration of new functionalities based on GATE. Finally, we have exemplified the usefulness of the system.

In future works we will try to include a systematical evaluation by several groups of users as well as other computational linguistic strategies and the integration with other system in a common scalable architecture.

Acknowledgements. This research has been supported by the Spanish Ministry of Science and Technology (TIN-2009-14057-C03-01) and the Regional Government of Madrid under the Research Network MA2VICMR (S2009/TIC-1542).

References

1. Aparicio, F.: TMT: A Scalable Platform to Enrich Translational Medicine Environments (2011)
2. Bizer, C., et al.: DBpedia - A Crystallization Point for the Web of data. In: Web Semantics: Science, Services and Agents on the World Wide Web, vol. 7(3), pp. 154–165 (2009)
3. Bollacker, K., et al.: Freebase: A Collaboratively Created Graph Database For Structuring Human Knowledge. In: Proceedings of the ACM SIGMOD International Conference on Management of Data, Vancouver, Canada, pp. 1247–1250 (2008)
4. Brenes, D.J., et al.: Survey and evaluation of query intent detection methods. In: Proceedings of the Workshop on Web Search Click Data, Barcelona, Spain, pp. 1–7 (2009)
5. Buenaga, M., et al.: Clustering and Summarizing Medical Documents to Improve Mobile Retrieval. In: SIGIR Workshop on MobIR, Singapore, pp. 54–57 (2008)
6. Cunningham, H., et al.: GATE: A Framework and Graphical Development Environment for Robust NLP Tools and Applications. In: Proceedings of the 40th Anniversary Meeting of the Association for Computational Linguistics, Philadelphia (2002)
7. Hearst, M.A.: UIs for Faceted Navigation: Recent Advances and Remaining Open Problems. In: Proceedings of the Second Workshop on Human-Computer Interaction and Information Retrieval, Redmond, pp. 13–17 (2008)
8. Nadeau, D., Sekine, S.: A survey of named entity recognition and classification. Linguisticae Investigationes 30, 3–26 (2007)
9. Auer, S., Bizer, C., Kobilarov, G., Lehmann, J., Cyganiak, R., Ives, Z.G.: DBpedia: A Nucleus for a Web of Open Data. In: Aberer, K., Choi, K.-S., Noy, N., Allemang, D., Lee, K.-I., Nixon, L.J.B., Golbeck, J., Mika, P., Maynard, D., Mizoguchi, R., Schreiber, G., Cudré-Mauroux, P. (eds.) ASWC 2007 and ISWC 2007. LNCS, vol. 4825, pp. 722–735. Springer, Heidelberg (2007)
10. Weichselbraun, A.: A Utility Centered Approach for Evaluating and Optimizing Geo-Tagging. In: First International Conference on Knowledge Discovery and Information Retrieval, Madeira, Portugal, pp. 134–139 (2009)

Person Name Discrimination in the Dossier–GPLSI at the University of Alicante

Isabel Moreno, Rubén Izquierdo, and Paloma Moreda

Natural Language Processing Research Group,
University of Alicante. Alicante, Spain
{imoreno,ruben,moreda}@dlsi.ua.es

Abstract. We present the Dossier–GPLSI, a system for the automatic generation of press dossiers for organizations. News are downloaded from online newspapers and are automatically classified. We describe specifically a module for the discrimination of person names. Three different approaches are analyzed and evaluated, each one using different kind of information, as semantic information, domain information and statistical evidence. We demonstrate that this module reaches a very good performance, and can be integrated in the Dossier–GPLSI system.

Keywords: Person Name Discrimination, LSA, Semantic Information, WordNet Domains, Statistical Evidence.

1 Introduction

A dossier is a set of press clippings that have appeared in the media and are related to a particular topic. Companies, specifically, look for news about themselves in the media in order to determine what is being said about them.

Newspapers usually use predefined templates to develop their web pages. This let the easy development of wrappers to extract the information contained on newspapers websites. This automation combined with an easy human supervision implies a considerable gain of time for this task. We have develop a system taking into consideration this idea. Our system is called Dossier–GPLSI. Some of its advantages are a considerable gain of time and more exhaustive and accurate searches.

Dossier–GPLSI is an application that automatically collects news about the University of Alicante. One of the modules of the classification engine is based on a list of name entities (NE) that are relevant for the University of Alicante. This classification process needs a discrimination module in order to distinguish different NE with the same name. The kind of NE considered in the this paper are person names.

The rest of the paper is organized as follows. In section 2 we describe our Dossier–GPLSI system, its components and how it works. Different approaches for NE discrimination module in Dossier–GPLSI are shown in section 3 and evaluated in section 4. Finally, we show how we are planning to improve Dossier–GPLSI in section 5.

R. Muñoz et al. (Eds.): NLDB 2011, LNCS 6716, pp. 305–308, 2011.
© Springer-Verlag Berlin Heidelberg 2011

2   Dossier–GPLSI Architecture

As we mentioned previously, Dossier–GPLSI is an application that automatically generates press dossiers and allows you to retrieve published news in different newspapers online edition. The system has been developed using InTime [2], a platform for the integration of NLP resources.

The main module of the Dossier–GPLSI system is the **Newsdownloader**. It consists of a crawler and a classifier. The crawler downloads all news from online digital newspapers websites. The classifier filter these news in order to select those news relevant for the organization. The classification module is an hybrid system that joins a machine learning SVM–based module, and a statistic module based on a list of Named Entities.

Another relevant module is the **Web Dossier**. This module is the web front–end of Dossier–GPLSI that allows you to manage users, review which news will be published and produce the dossier in PDF or e-mail formats.

Finally, the **Dossearch** [1] module is a search engine implemented as a web application that lets you search across the entire newspaper library including filtering between dates or newspapers.

3   NE Discrimination Approaches in Dossier–GPLSI

We have realized that some ambiguity appears for some Named Entities relevant for the organization. Two different people can share the same name. For example, in the University of Alicante we have a vice-rector called Cecilia Gómez and in the newspapers gossip section there is another Cecilia Gómez known for dating a Spanish bullfighter and being a dancer. Dossier–GPLSI needs a NE Discrimination module in order to distinguish different entities with equal name.

In this section, we are going to describe different approaches to accomplish this task. We can consider that we have a retrieved article related with a relevant entity for the University of Alicante, but we must know if the entity is the correct one (i.e. Cecilia Gómez as the vice-rector). Our experiments have been done with a gold standard of 16 articles for the vice-rector Cecilia Gómez, and 16 for the other Cecilia Gómez (the dancer).

NE Discrimination using LSA [4]. We use Latent Semantic Analysis (LSA[1]) in order to calculate the semantic similarity score for the downloaded pages. First, we train a model using the 32 articles. Then for a new Cecilia Gómez related article (the *type* of Cecilia Gómez is still unknown), LSA obtains the semantic similarity of the article with each article of the gold standard. The final entity assigned to the article is the one of the top ten high–scoring documents.

NE Discrimination using Wordnet Domains (WND) [3]. Basically, in this approach we obtain a domain vector (vice–rector DV) for the vice-rector and one domain vector (dancer DV) for the dancer. Each DV is obtained from

---

[1] http://infomap-nlp.sourceforge.net

the monosemic nouns[2] in the corresponding articles of our gold standard (16 for each entity type). Freeling[3] has been employed to obtain the lemmas of the words.

For classifying a new Cecilia Gómez article, first we obtain the DV from the monosemic nouns of the article. Then we compute the similarity of the article DV with each of the pre–computed DVs (vice-rector DV and dancer DV), using the Cosine measure between vectors. The category selected is the one associated with that pre–computed DV that is more similar to the article DV according to the Cosine value.

NE Discrimination using a Bag of Words (BoW). In this approach, we have two BoW for each type of Cecilia Gómez entity (vice-rector and dancer). Each BoW contains the most frequent words associated with the corresponding entity. Each word has a normalized weight, which takes into consideration the frequency distribution of the word in both gold standard articles set (vice-rector and dancer).

For classifying a new article, all lemmas are extracted, and a value of association with each BoW is obtained. It is calculated considering the weight of the overlapping words between the article and each BoW. The entity belonging to that BoW that reaches the higher result is selected for the article.

4   Result Analysis

In order to evaluate LSA approach, we have tested this module with 6 articles. Out of these 6, 4 articles referred to the vice-rector Cecilia Gómez and the remainder ones were from the dancer Cecilia Gómez. We have created different models with different size windows context and different size content bearing words. Any combination of size windows context and size content bearing words, gets the right NE (100% accuracy). Except in one article, when using 2000 content bearing words size and 10 words as windows context size.

For the evaluation of WND and BoW approaches, we have previously calculated two DVs and two BoW for NE Cecilia Gómez using 38 articles (20 for the vice-rector and 18 for the dancer). For testing these approaches, we use 12 articles about the previous vice-rector, Maria Jose Frau, and other 12 for the Spanish dancer, Sara Baras, despite of they don't share names.

When using WND, the discrimination module fail to establish the right entity type in 6 documents out of 24 (71,43% accuracy). 3 Spanish dancer documents were classified as vice-rector and 3 vice-rector documents where classified as Spanish dancer. These failures happened for several reasons, mainly due to there are not many monosemic words in those articles; moreover there are domains that are not discriminative, like person or factotum; there are domains that overlap each other, like sport domain that contains coach or court, referred to the vice-rector, and bullfighter or matador, related to the Spanish dancer.

---

[2] In this case a word is monosemic if all its WordNet[6] senses belong to the same WordNet Domain.

[3] http://nlp.lsi.upc.edu/freeling

In the case of BoW approach, the discrimination module fail to establish the right NE in only 2 documents out of 24 (93,75% accuracy). Two Spanish dancer documents were classified as vice-rector. This module fails because the documents contains words that are usually associated with vice-rector documents, like president or teacher.

5   Conclusions and Future Work

In this paper we have presented the Dossier–GPLSI, a system for generating automatic press dossiers, which also provides exhaustive and accurate searches. In order to provide accurate searches, we have evaluated three approaches for discriminate Named Entities and found that LSA has an excellent precision but needs a manual process in order to obtain the classification. So the best and fastest alternative for Discrimination NE is the BoW approach.

At the moment, we are working on the article summarization in order to provide each search result with its summary, so the user will decide easier which news correspond with his query. In the future, we want to modify the *NewsDownloader* crawler behavior in order to allow the Internet users opinion extraction in forums and social networks. In this way, companies will have the customer opinion.

Acknowledgements. This paper has been partially supported by the Valencian Region Government under projects PROMETEO/2009/119 and ACOMP/2011/ 001 and the Spanish Government under the project TEXT MESS 2.0 (TIN2009-13391-C04-01).

References

1. Fernádez, J., Gómez, J.M., Martinez-Barco, P.: Evaluación de sistemas de recuperación de información web sobre dominios restringidos. Journal Sociedad Española para el Procesamiento del Lenguaje Natural (SEPLN) 45, 273–276 (2010)
2. Gómez, J.M.: InTiMe: Plataforma de Integración de Recursos de PLN. Journal Sociedad Española para el Procesamiento del Lenguaje Natural (SEPLN) 40, 83–90 (2010)
3. Magnini, B., Cavaglia, G.: Integrating subject fields codes into WordNet. In: Proceedings of the Second International Conference on Language Resources and Evaluation (LREC 2000), Atenes (2000)
4. Kozareva, Z., Vázquez, S., Montoyo, A.: UA-ZSA: Web Page Clustering on the basis of Name Disambiguation. In: Proceedings of the 4th International Workshop on Semantic Evaluations (SemEval), Prague, Czech Republic, June 23-24, pp. 338–341 (2007)
5. Kozareva, Z., Vázquez, S., Montoyo, A.: The Influence of Context during the Categorization and Discrimination of Spanish and Portuguese Person Names. Journal Sociedad Española para el Procesamiento del Lenguaje Natural (SEPLN) 39, 81–88 (2007)
6. Fellbaum, C.: WordNet. An Electronic Lexical Database. MIT Press, Cambridge (1998)

MarUja: Virtual Assistant Prototype for the Computing Service Catalogue of the University of Jaén

Eugenio Martínez-Cámara, L. Alfonso Ureña-López, and José M. Perea-Ortega

SINAI research group
Computer Science Department
University of Jaén
{emcamara,laurena,jmperea}@ujaen.es

Abstract. The information and web services that many organizations offer through its web pages are increasing every day. This makes that navigation and access to information becomes increasingly complex for visitors of these web pages, so, it is necessary to facilitate these tasks for users. In this paper we present a prototype of a Virtual Assistant, which is the result of applying a methodology to develop and set this kind of systems with a minimum cost.

Keywords: Virtual Assistant, search engine, text to speech.

1 Introduction

A Virtual Assistant (VA) is a virtual entity, usually represented by an avatar, which is be able to respond queries using natural language.

Virtual Assistants use speech recognition techniques, which allow them to understand the meaning of the questions, choose the most suitable answer, suggest other responses and lead users to the web page where they can find the information needed.

Virtual Assistants can be characterized by the following features:

- They are designed to understand natural language.
- They could be programmed to speak in different languages.
- They are very useful to obtain information about users.
- They could be designed to understand the conversation context.
- They know how to build responses. A VA analyzes the user query and builds the best answer for it.
- The answer could be formed by web links or other kinds of electronics resources, and not only by plain text.
- The response format depends on the type of user, the kind and the scope of the user query.

VA can be considered a new kind of search engines, which may be more or less complex according to the requirements of the problem. There are two groups of VAs:

R. Muñoz et al. (Eds.): NLDB 2011, LNCS 6716, pp. 309–312, 2011.
© Springer-Verlag Berlin Heidelberg 2011

1. VA with a high NLP level: Those which are usually composed by semantic search engines, dialog managers, user profiles, or other NLP components.

2. VA with a low NLP level: Those that use a small number of NLP components. The most outstanding are the so-called FAQ Smarts. These ones are an information retrieval system over a question-answers database about the relevant information.

Although these two groups of VA have several differences, both have a humanoid animation, and provide a more natural interaction than the traditional search engines.

2   Project Reasons

The Computing Service of the University of Jaén provides several services to the university community. The web page of the Computing Service has a FAQ[1] to solve the users information needs about some of the services that it offer. There are several problems: very few people of the university community know the existence of this FAQ, its access is complicated and finally it is very difficult to find the answer of the questions. Thus, the motivation of this project is to provide users a easy way to resolve their doubts about the technological services of the University of Jaén. Therefore, we have developed a VA with a low NLP level for the Computing Service of the University of Jaén.

3   Components

One of the objectives of the project was the fact of minimize the development of new components and attempt to reuse NLP components.

The system is mainly composed of two software components:

1. Syntactic search engine: We have used the search server Apache Solr[2]. Solr [1] is an open source enterprise search server which is used in some of the most important web pages, such us Netflix[3]. It is written in Java and its kernel is the text search library Lucene [2].

2. Text to speech synthesis (TTS): We have used the TTS of the Festival project[4]. This project does not provide a Spanish voice, so we had to find a Spanish voice. We founded the female voice of the free software project Hispavoces[5].

---

[1] http://faq.ujaen.es/
[2] http://lucene.apache.org/solr/
[3] https://www.netflix.com/
[4] http://www.cstr.ed.ac.uk/projects/festival/
[5] http://forja.guadalinex.org/projects/hispavoces/

4   Architecture

Other objectives of the project were the development of a product with a simple architecture which could be easily and quickly adapted to any domain, reducing the development costs. The Figure 1 shows the system architecture of the MarUja prototype.

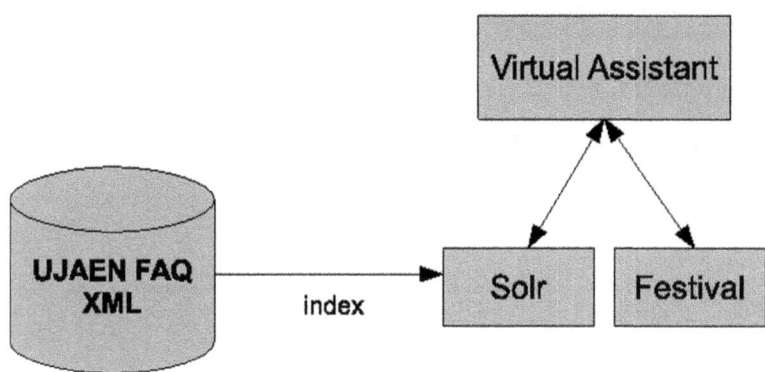

Fig. 1. System architecture

The system development started with the automatic extraction of questions and answers from the Computing Service's FAQ of the University of Jaén. A XML file was generated for each question of the FAQ, containing several fields such as the question, the answer and the question categories. These XML files were used to generate the Solr index.

The VA module is written in Java, which is the responsible for coordinating communication between Solr and Festival, so its main functions are:

- Take the user questions.
- Build the Solr query.
- Take the text of the best Solr result for that query, and give it to Festival to generate the sound file of the VA response.
- Build the system response with sound generated by Festival, and three questions related to the user query as a suggestion.

5   Conclusion

In this work we have described briefly the MarUja prototype, a Virtual Assistant developed in the University of Jaén for the Computing Services. Its main goal is to provide a simple interface in natural language to respond to any information need related to the services offered by such department. Moreover, we have shown the methodology followed to generate the prototype as well as the main components of MarUja. Figure 2 and Figure 3 show some screenshots of the proposed prototype.

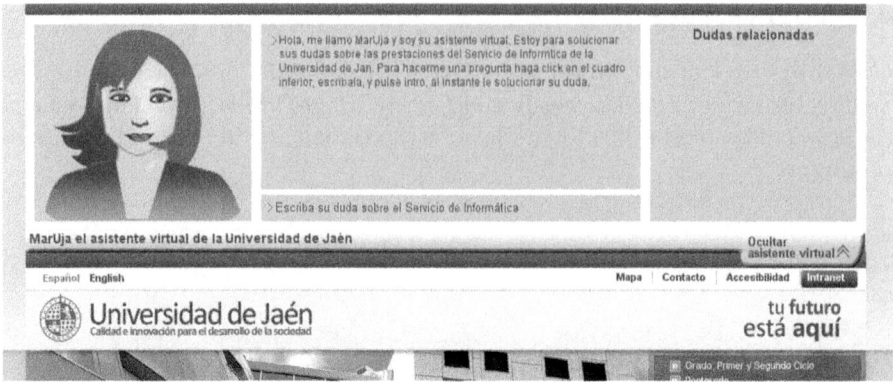

Fig. 2. MarUja is waiting an user query

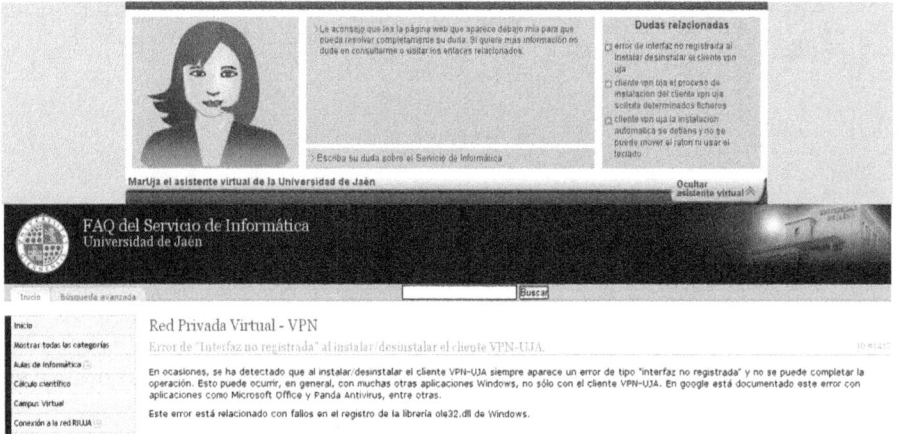

Fig. 3. MarUja has aswered an user query

References

1. Smiley, Pugh, D.: Solr 1.4 Enterprise Search Server. Packt Publishing Ltd. (2009)
2. Hatcher, E., Gospodneti, O., McCandless, M.: Lucene in Action. In: MEAP Edition, 2nd edn.

Processing Amazighe Language

Mohamed Outahajala[1, 2]

[1] Royal Institut for Amazighe Culture, Rabat, Morocco
[2] Ecole Mohammadia d'Ingénieurs, Université Med V, Rabat, Morocco
outahajala@ircam.ma

Abstract. Amazighe is a language spoken by millions of people in north Africa in majority, however, it is suffering from scarcity resources. The aim of this PhD thesis is to contribute to provide elementary resources and tools to process this language. In order to achieve this goal, we have achieved an annotated corpus of ~20k tokens and trained two sequence classification models using Support Vector Machines (SVMs) and Conditional Random Fields (CRFs). We have used the 10-fold technique to evaluate our approach. Results show that the performance of SVMs and CRFs are very comparable, however, CRFs outperformed SVMs on the 10 folds average level (88.66% vs. 88.27%). For future steps, we are planning to use semi-supervised techniques to accelerate part-of-speech (POS) annotation in order to increase accuracy, afterwards to approach base phrase chunking, for future work.

Keywords: POS tagging, Amazighe language, supervised learning, semi-supervised learning.

1 Introduction

Amazighe language belongs to the Hamito-Semitic/"Afro-Asiatic" languages[1], with rich templatic morphology [4]. In linguistic terms, the language is characterized by the proliferation of dialects due to historical, geographical and sociolinguistic factors. It is spoken in Morocco, Algeria, Tunisia, Libya, and Siwa (an Egyptian Oasis); it is also spoken by many other communities in parts of Niger and Mali. It is used by tens of millions of people in North Africa mainly for oral communication and has been introduced in mass media and in the educational system in collaboration with several ministries in Morocco.

In Morocco, for instance, one may distinguish three major dialects: Tarifit in the North, Tamazight in the center and Tashlhit in the southern parts of the country; it is a composite of dialects which none have been considered the national standard. According to the last governmental population census of 2004, the Amazighe language is spoken by some 28% of the Moroccan population (millions).

In Natural Language Processing (NLP) terms, Amazighe, like most non-European languages, still suffers from the scarcity of language processing tools and resources.

---

[1] http://en.wikipedia.org/wiki/Berber_languages

R. Muñoz et al. (Eds.): NLDB 2011, LNCS 6716, pp. 313–317, 2011.
© Springer-Verlag Berlin Heidelberg 2011

In line with this, and since corpora constitute the basis for human language technology research, we have constructed an annotated corpus by POS information; the corpus consists of over than 20k Moroccan Amazighe tokens. It is the first one to deal with Amazighe. This resource even though small, was very useful for training two sequence classification models using SVMs and CRFs.

The rest of the paper is organized as follows: in Section 2 I present my motivation behind the proposed research. Then, in Section 3 I describe related work done until now in the perspective of achieving that goal. Finally, in Section 4 I present current work and I discuss the proposed work to be done in the future.

2 Motivation

Due to the use of the different dialects in its standardization (Tashlhit, Tarifit and Tamazight the three more used ones) as well as its complex morphology, the Amazighe language presents interesting challenges for NLP researchers which need to be taken into account. Some of these characteristics are:

1- It does not support capitalization in its script.

2- It is written with its own alphabet called Tifinaghe [11] [13].

2- It is a complex morphology language.

3- Nouns, quality names (adjectives), verbs, pronouns, adverbs, prepositions, focalizers, interjections, conjunctions, pronouns, particles and determinants consist of a single word occurring between blank spaces or punctuation marks. However, if a preposition or a parental noun is followed by a pronoun, both the preposition/parental noun and the following pronoun make a single whitespace-delimited string [2]. For example: ⴴⲟ (ɣr) "to, at" + ⵣ (i) "me (personal pronoun)" results into ⴴⲟⵣ/ⴴⵊⲟⵣ (ɣari/ɣuri) "to me, at me, with me".

4 – Same surface form might be tagged with a different POS tag depending on how it has been used in the sentence. For instance, ⵣⵅⵓⲟⴰ (ig°ra^2) may have many meanings: as a verb, it means 'lag behind' while as a noun it refers to the plural noun of ⴰⵅⲟⵊ (agru) meaning 'a frog'. Some stop words such as "ⴷ" (d) might function as a preposition, a coordination conjunction, a predicate particle or an orientation particle.

5 - Like most of the languages which have only recently started being investigated for NLP tasks, Amazighe lacks annotated corpora and tools and still suffers from the scarcity of language processing tools and resources.

For all this reasons and given that I am a native speaker of this language, together with my advisors we have planned to work in this research project in order to contribute to provide elementary resources and tools to process this language.

3 Related Works in Amazighe

Very few linguistic resources have been developed so far for Amazighe, such as a web dictionary[3] [6], a concordance tool [3] and some tools and resources achieved by

---

[2] The amazighe Latin transliteration used in this paper is the one defined in [8].
[3] http://81.192.4.228/

LDC/ELDA under a relationship of partnership with IRCAM [1] as an encoding converter, a word and sentence segmenter, and a named-entity tagger and tagged text. The development of a POS tagger tool is the first step needed for automatic text processing. In the following subsections, we present an annotated corpus and some preliminary results in POS tagging in 3.1 and 3.2 respectively.

3.1 Corpus

The developed corpus consists of a list of texts extracted from a variety of sources such as: Amazighe version of IRCAM's web site[4], the periodical "Inghmisn n usinag" (IRCAM newsletter) and primary school textbooks, annotated using AnCoraPipe tool [9]. Annotation speed of this corpus was between 80 and 120 tokens/hour. Randomly chosen texts were revised by different annotators. On the basis of the revised texts the Kappa coefficient is 0.88. Common remarks and corrections were generalized to the whole corpora in the second validation by a different annotator [10].

Table 1. A synopsis of the features of the Amazighe POS tag set with their attributes

POS	Attributes and sub attributes with number of values
Noun	gender(3), number(3), state(2), derivative(2), POSsubclassification(4), person(3), possessornum(3), possessorgen(3)
name of quality(Adjective)	gender(3), number(3), state(2), derivative(2), POS subclassification(3)
Verb	gender(3), number(3), aspect(3), negative(2), form(2), derivative(2), voice(2)
Pronoun	gender(3), number(3), POS subclassification(7), deictic(3), person(3)
Determiner	gender(3), number(3), POS subclassification(11), deictic(3)
Adverb	POS subclassification(6)
Preposition	gender(3), number(3), person(3), possessornum(3), possessorgen(3)
Conjunction	POS subclassification(2)
Interjection	
Particle	POS subclassification(7)
Focus	
Residual	POS subclassification(5), gender(3), number(3)
Punctuation	punctuation mark type(16)

3.2 POS Tagging Amazighe

Amazighe tag set might contains 13 elements, each element has attributes and sub attributes (see Table 1), with two common attributes to each one: "wd" for "word" and "lem" for "lemma", whose values depend on the lexical item they accompany.

In order to build an automatic POS-tagger, we have used sequence classification techniques based on two state-of-art machine learning approaches, SVMs [7] [13] and CRFs [8]. The used feature set that gave the best results contains in a window of -/+2:

---

[4] http://www.ircam.ma/

1- the surrounding words and their correspondent POS-tags;
2- the n-gram feature which consists of the last and first i character n-gram, with i spanning from 1 to 4

As illustrated in Table 2, SVMs slightly outperformed CRFs on the fold level (91.66% vs. 91.35%) and CRFs outperformed SVMs on the 10 folds average level (88.66% vs. 88.27%). These results are very promising considering that we have used a corpus of only ~20k tokens with a derived tag set of 15 tags.

Table 2. 10-fold cross validation results

Fold#	SVMs (with lexical features)	CRFs (with lexical features)
0	86,86	86,95
1	83,86	84,98
2	**91,66**	90,86
3	88,34	88,58
4	88,24	88,87
5	89,99	90,48
6	85,38	85,38
7	86,6	87,96
8	91,38	91,14
9	90,41	**91,35**
AVG	**88,27**	**88,66**

4 Current Research and Future Work

I am currently trying to improve the performance of the POS tagger by using additional features and more annotated data based on hand annotation. Regarding future work:

- We are planning to use semi-supervised and active learning techniques to obtain more annotated data [5];
- to improve POS-tagger accuracy by looking for Moroccan dialect corpus and in general the use of lexicons containing POS information and lexicons of name entities, since many Arabic name entities are used in the Amazighe language;
- to develop a tokenizer together with a tool for the correction of some usual errors such as the misplacing of the letter "ⵓ" (e[5]). For instance, to correct the word "ⵓⵛⵓⵎⴰⵣⵓ" (amaziGe) to "ⵓⵛⵓⵎⴰⵣ" (amaziG). Overall, the character"ⵓ" have to be used in two cases only: when we have the character repeated twice or three times, for instance: tettr (she asked);

---

[5] The amazighe Latin transliteration used in this paper is the one defined in [9].

- to approach base phrase chunking by hand labeling the already annotated corpus with morphology information;
- to develop a base phrase chunker by using supervised learning techniques.

Acknowledgements. I would like to thank my advisors Prof. Lahbib Zenkouar (Ecole Mohammadia d'Ingénieurs) and Dr. Paolo Rosso (Universidad Politécnica de Valencia) for their assistance and guidance. Also, I thank Dr. Yassine Benajiba for his feedbacks on my research and the many discussions about my PhD thesis and my colleagues in IRCAM for helping me in Amazighe linguistic issues.

References

1. Aït Ouguengay, Y., Bouhjar, A.: For standardised Amazigh linguistic resources. In: Proceedings of the 7th International Conference on Language Resources and Evaluation, LREC-2010, Malta, May 17-23, pp. 2699–2701 (2010)
2. Ameur, M., Bouhjar, A., Boukhris, F., Boukouss, A., Boumalk, A., Elmedlaoui, M., Iazzi, E.: Graphie et orthographe de l'Amazighee. Publications de l'IRCAM (2006)
3. Bertran, M., Borrega, O., Recasens, M., Soriano, B.: AnCoraPipe: A tool for multilevel annotation. Procesamiento del lenguaje Natural (41) (2008)
4. Boulaknadel, S.: Amazigh ConCorde: an appropriate concordance for Amazigh. In: Proc. Of 1st Symposium International Sur le Traitement Automatique de la Culture Amazighe, Agadir, Morocco, pp. 176–182 (2009)
5. Chafiq, M.: درسـا وأربعـون وأربعـة الأمازيغيــة فــى (Forty four lessons in Amazighe). Arabo-africaines (éd.) (1991)
6. Guzmán, R., Montes, M., Rosso, P., Villaseñor, L.: Using the Web as Corpus for Self-training Text Categorization. Information Retrieval 12(3), 400–415 (2009), Special Issue on Non-English Web Retrieval, doi:10.1007/s10791-008-9083-7
7. Iazzi, E., Outahajala, M.: Amazigh Data Base. In proceedings of HLT & NLP within the Arabic world: Arabic language and local languages processing status updates and prospects. In: 6th International Conference on Language Resources and Evaluation, LREC-2008, Morocco, pp. 36–39 (2008)
8. Kudo, T., Matsumoto, Y.: Use of Support Vector Learning for Chunk Identification (2000)
9. Lafferty, J., McCallum, A., Pereira, F.: Conditional Random Fields: Probabilistic Models for Segmenting and Labeling Sequence Data. In: Proceedings of ICML 2001, 282–289 (2001)
10. Outahajala, M., Zenkouar, L., Rosso, P., Martí, A.: Tagging Amazighe with AncoraPipe. In: Proc. Workshop on LR & HLT for Semitic Languages, 7th International Conference on Language Resources and Evaluation, LREC-2010, Malta, May 17-23, pp. 52–56 (2010)
11. Outahajala, M., Zenkouar, L., Rosso, P.: Building an annotated corpus for Amazighe. In: Will appear in Proceedings of 4th International Conference on Amazighe and ICT, Rabat, Morocco (2011) (to be published in the conference post proceedings)
12. Outahajala, M., Zenkouar, L.: La norme du tri, du clavier et Unicode. In: Proceedings of la typographie entre les domaines de l'art et l'informatique, Rabat, Morocco, pp. 223–238 (2008)
13. Vapnik, V.N.: The Nature of Statistical Learning Theory. Springer, New York (1995)
14. Zenkouar, L.: Normes des technologies de l'information pour l'ancrage de l'écriture Amazighe. Revue Etudes et Documents Berbères (27), 159–172 (2008)

How to Extract Arabic Definitions from the Web?
Arabic Definition Question Answering System

Omar Trigui

ANLP Research Group- MIRACL Laboratory, University of Sfax, Tunisia
omar.trigui@fsegs.rnu.tn

Abstract. The Web is the most interest information resource available for users, but the issue is how to obtain the precise and the exact information easily and quickly. The classic information retrieval system such as Web search engines can just return snippets and links to Web pages according to a user query. And it is the role of the user to fetch these results and identify the appropriate information. In this paper, we propose dealing with the results returned by Web search engines to return the appropriate information to a user question. The solution proposed is integrated in an Arabic definition question answering system called 'DefArabicQA'. The experiment was carried out using 140 Arabic definition questions and 2360 snippets returned by divers Web search engines. The result obtained so far is very encouraging and it can be outperformed more in the future.

Keywords: Natural language processing, information extraction, information retrieval, Web mining, Arabic language.

1 Introduction

Today the Web is the largest resource of knowledge, therefore, sometimes this makes it difficult to find precise information. Current search engines can only return ranked snippets containing the effective answers to a user query. However, they can not return an exact information. The Question Answering (QA) field has appeared to resolve the issue of the identification of exact information in a collection of documents. Many researches are interested in this task in diverse competitions (e.g., TREC[1] and CLEF[2]). Typically, QA systems are designed to retrieve the exact answers from a set of knowledge resources to the user question. The main evaluation conferences classify the questions to different types: factual, definition, list, etc. The definition question type is typically a question about information related to an organization, a person or a concept. The different forms of a definition question are mentioned in table 1. Our research takes place in this context, we have set the development of an approach to facilitate the returning of an exact information to a definition user from Web knowledge resource as a goal. In this paper, we are presenting our current and future works to deal with this goal. This paper is structured

---

[1] Text Retrieval Conference http://trec.nist.gov/
[2] Cross-Language Evaluation Forum http://clef-campaign.org/

R. Muñoz et al. (Eds.): NLDB 2011, LNCS 6716, pp. 318–323, 2011.
© Springer-Verlag Berlin Heidelberg 2011

in 6 sections. Section 2 presents our motivation to deal with the definition QA for Arabic language. Section 3 presents the related work. Section 4 presents our current work where we explain the approach proposed and the experiment carried out. While section 5 presents the future works. And finally, section 6 presents the conclusion.

Table 1. Different forms of a definition question

Definition question forms		Expected answers
Who+be+<topic> ?	من هو/ من هي <الموضوع>؟	Information about a person
What+be+<topic> ?	ما هو/ ما هي< الموضوع>؟	Information about an organization or a concept

2 Motivations

In spite of the interesting role given by the QA systems to different fields, the QA research for Arabic language is largely unexplored. The problem becomes even more difficult with the absence of standard corpus and tools which are available for other languages. Unfortunately, most evaluation platforms of QA task in the main evaluation conferences do not include the Arabic language. For these reasons, we have set developing the definition QA research for Arabic language as a goal. This is not an easy matter, but we have decided to deal with the lack of corpus and tools by exploiting the Web as a resource and using superficial language processing techniques. Knowing that more than 65 million users use the Arabic language in the Web[3]. Our research will give these users the ability to obtain important information from the Web about a particular person, an organization or a concept according to a question in natural language such as ما هو الاتحاد العام التونسي للشغل؟ - 'What is the Tunisian General Union of Labor?' -.

3 Related Works

Researches on the identification of definition information have appeared in many fields such as the Question Answering (QA) and the automatic building of glossaries. In the area of definition Question Answering (QA), most researches start by identifying the question topic and then search the sentences or snippets containing the definition information related to the question topic. A big part of definition QA researches are based on the evaluation corpus made by the international evaluation conference such as TREC and CLEF where systems respond to definition questions by returning definition answers extracted from these corpora. Otherwise, there are other researches which use the wealth of information on the web to return answers. Jijkoun and Rijke, have used the frequently asked questions (FQA) pages on the web to retrieve answers [1]. Agichtein et al., have introduced a method to learn retrieving documents containing answers on the web [2].

---

[3] http://internetworldstats.com/stats7.htm

The methods used to identify the definition answers often rely on pattern approach [3]. The main methods used to build patterns are the method of Ravichandran and Hovy [4] based on the Web and the method of Cui et al. [5] based on the unsupervised learning. The rank of definition answers deduced are done by using statistical methods and machines learning techniques. Various methods have been used to rank definition answers. Yang et al., have used the centroid based ranking method using external knowledge resource [6]. Other researchers have used mining external definitions method to rank the definitions [3][5][7]. This method consists to calculate the similarity between candidate answers and definitions from external Web knowledge resources. There are other researches which have not used pattern approach to identify definition answers such as [7][8]. Prager et al. have proposed other approach to answer definition questions based on the decomposition of a definition question into a series of factoid questions. And then from the answers to the factual questions, they deduce definition answers [8]. Zhang et al. have identified and selected definition sentences from document collections by using Web knowledge bases [7].

To our knowledge, no research has been done on Arabic definitional QA systems except our work. However, there are some attempts to build factual QA systems (e.g. Hammo et al, [9], Benajiba, [10] and Brini et al., [11]).

4 Current Work

In this section, we are detailing our current work in three parts. We start by explaining the approach proposed in the first part. Then, we are detailing the integration of our approach in a definition QA system. And finally, we are showing the experiment carried out.

4.1 Approach Proposed

In this part, we propose an approach to identify definition answers to a definition question. Our approach is divided mainly in three steps: (1) fetching snippets from Web-based search engines, (2) extracting definition answers, (3) and ranking these answers to identify the top appropriate ones.

At the first step, we extract the question topic and then we collect snippets containing information related to the topic from Web search engines. In the second step, we identify the definition answers. We have combined two methods for this identification task where the first method is the hard matching and the second method is the tolerant matching. The hard matching method consists of making the alignment between a snippet and a lexical pattern slot-by-slot without any tolerance. However, tolerant matching method consists of exploiting the snippets missed by the hard matching process. It handles tokens that are not expected in patterns [12]. The final step of this approach is the ranking of candidate definitions. We are using four criteria relied to statistical approach in this ranking process. All these criteria present divers features of each candidate definition. These information permit to make the ranking of the different definition answers collected. The first criterion is the "lexical coverage criterion" which its score presents the ratio of the number of words of the frequency substrings in a definition answer to the total number of words of this definition answer. The second criterion is the "pattern weight criterion", where the score of this criterion represents the precision of the pattern which has been used to identify a definition answer. The third criterion is the "snippets position criterion", where its score is the position of the snippet containing a definition answer in the collection of snippets. The last criterion is the "word frequency criterion" where its score represents the sum of the frequencies of the words occurring in a definition answer. We aggregate these four criteria by proceeding the normalization of their scores.

4.2   The DefArabicQA System

In this second part, we are showing the integration of this approach in a definitional QA system for Arabic language called DefArabicQA [13]. Figure 1 shows the general architecture of this system.

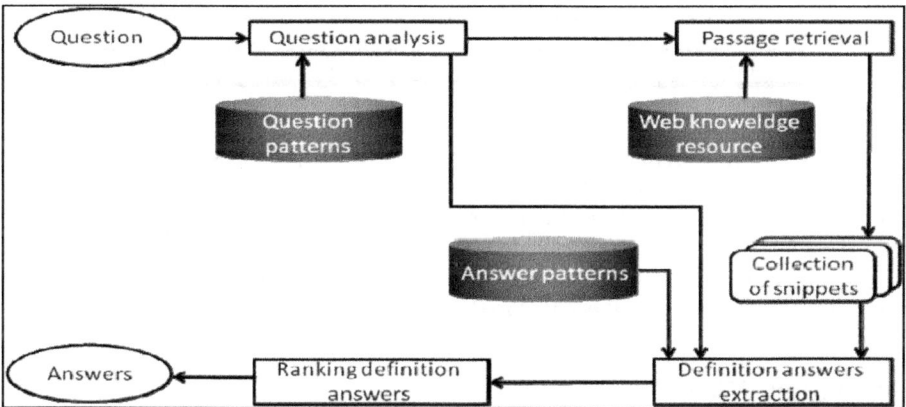

Fig. 1. Architecture of DefArabicQA

This architecture is made up of four modules: question analysis; passage retrieval; definition answers extraction; and ranking definition answers. DefArabicQA system starts by analyzing a user definition question in the module question analysis. The result of this module is the identification of the question topic and the question type: person, organization or concept definition. In the module passage retrieval, the question topic is reformulated in a query and submitted to Web search engines. The top snippets returned by each Web search engines are collected. In the module definition answers extraction, a set of lexical answer patterns are used to identify the definition answers from the set of snippets collected. Finally the definition answers are filtered and ranked based on an average of the score of four criteria (detailed in the section 4.1).

4.3 Experiment

This part describes the results of the experiment carried out using the DefArabicQA system. The experiment has been carried out using 140 definition questions and 2360 snippets collected from the Arabic version of the Web search engine Google[4] and the Web Search engine of Wikipedia[5], with an average of 17 snippets by question[6].

Table 2. Repartition of the questions

	Questions having an appropriate answer	Questions not having an appropriate answer
Questions	78 (**55.71%**)	62 (44.29%)

Table 3. Division of the questions having an appropriate answer in the top five rank

Top five Ranking	Questions
Top 1	**71 (91%)**
Top 2, 3, 4 and 5	7 (9 %)
Total	78 (100%)

Out of the 140 questions in the test collection, 78 questions (55.71%) have been answered correctly by at least one appropriate definition answer in the top 5 answers (Table 2). 71 questions of them (91%) have obtained an appropriate definition answer in the top one ranking answers, and only 7 questions (9%) have not obtained an appropriate definition in the top one ranking answers (Table 3). The MRR metric is equal to 0.53 and the recall is equal to 0.71.

5 Future Work

In the future works, we are planning to improve the result obtained by improving the criteria used in the ranking process of definition answers. We are planning also to determine the suitable weights attributed to these criteria. We are studying the efficiency of making a combination with these criteria and the machine learning approach. To deal the questions not having an appropriate answer noticed in the experiment carried out (table 2), we have planned to integrate the normalization and the extension of question. We are also going to create patterns to generate definition answers and to study the combination of the use of Wikipedia info-box and the method of Prager et al. [8] to decompose a definition question into a series of factual questions.

---

[4] http://www.google.com/intl/ar/
[5] http://ar.wikipedia.org/w/index.php?title=خاص:بحث&search=&go=اذهب
[6] Resources available for research purpose at:
   http://sites.google.com/site/omartrigui/downloads

6   Conclusion

In this paper, we have detailed our current work and our future works to identify definition answers from the Web to a user question. We have presented the approach proposed and its integration in a definition QA system. The result of the experiment carried out with this system has obtained a MRR equal to 0.53 where the best definition answers are ranked in the top one for 91% of the total questions having an appropriate answer.

Acknowledgements. I would like to thank my advisors Dr. Lamia Hadrich Belguith (University of Sfax-Tunisia) and Paolo Rosso (Universidad Politécnica de Valencia-Spain) for their assistance and their guidance.

References

1. Jijkoun, V., de Rijke, M.: Retrieving Answers from Frequently Asked Questions Pages on the Web. In: Proc. 14th ACM International Conference on Information and Knowledge Management, Bremen, Germany, pp. 76–83 (2005)
2. Agichtein, E., Lawrence, S., Gravano, L.: Learning to find answers to questions on the web. ACM Trans. Inter. Tech. 4(2), 129–162 (2004)
3. Xu, J., Licuanan, A., Weischedel, R.M.: TREC 2003 QA at BBN: Answering definitional questions. In: 12th TREC, Washington, DC, pp. 98–106 (2003)
4. Ravichandran, D., Hovy, E.H.: Learning surface text patterns for a Question Answering system. In: ACL 2002, Philadelphia, PA, pp. 41–47 (2002)
5. Cui, H., Kan, M.Y., Chua, T.S.: Soft Pattern Matching Models for Definitional Question. Answering. ACM Transactions on Information Systems (TOIS) 25(2) (2007)
6. Yang, H., Cui, H., Maslennikov, M., Qiu, L., Kan, M.Y., Chua, T.S.: QUALIFIER in TREC-12 QA main task. In: TREC 2003, Gaithersburg, MD, pp. 480–488 (2003)
7. Zhang, Z., Zhou, Y., Huang, X., Wu, L.: Answering definition questions using web knowledge bases. In: Dale, R., Wong, K.-F., Su, J., Kwong, O.Y. (eds.) IJCNLP 2005. LNCS (LNAI), vol. 3651, pp. 498–497. Springer, Heidelberg (2005)
8. Prager, J.M., Chu-Carroll, J., Czuba, K., Welty, C., Ittycheiach, A., Mahindru, R.: IBM's PIQUANT in TREC 2003. In: 12th Text REtreival Conference, Gathersburg, MD, pp. 283–292 (2003)
9. Hammou, B., Abu-salem, H., Lytinen, S., Evens, M.: QARAB: A question answering system to support the Arabic language. In: Proceedings of the Workshop on Computational Approaches to Semitic Languages, ACL, pp. 55–65 (2002)
10. Benajiba, Y.: Arabic Question Answering. Diploma of advanced studies. Technical University of Valencia, Spain (2007)
11. Brini, W., Ellouze, M., Trigui, O., Mesfar, S., Belguith H.L., Rosso, P.: Factoid and definitional Arabic Question Answering system. In: Post-Proc. NOOJ-2009, Tozeur, Tunisia, June 8-10, pp.243-255 (2009)
12. Trigui, O., Belguith, H.L., Rosso, P.: An Automatic Definition Extraction in Arabic Language. In: Hopfe, C.J., Rezgui, Y., Métais, E., Preece, A., Li, H. (eds.) NLDB 2010. LNCS, vol. 6177, pp. 240–247. Springer, Heidelberg (2010)
13. Trigui, O., Belguith, H.L., Rosso, P.: DefArabicQA: Arabic Definition Question Answering System. In: Workshop on Language Resources and Human Language Technologies for Semitic Languages, 7th LREC, Valletta, Malta, pp. 40–45 (2010)

Answer Validation through Textual Entailment

Partha Pakray

Computer Science and Engineering Department,
Jadavpur University, Kolkata, India
parthapakray@gmail.com

Abstract. Ongoing research work on an Answer Validation System (AV) based on Textual Entailment and Question Answering has been presented. A number of answer validation modules have been developed based on Textual Entailment, Named Entity Recognition, Question-Answer type analysis, Chunk boundary module and Syntactic similarity module. These answer validation modules have been integrated using a voting technique. We combine the question and the answer into the Hypothesis (H) and the Supporting Text as Text (T) to identify the entailment relation as either "VALIDATED" or "REJECTED". The important features in the lexical Textual Entailment module are: WordNet based unigram match, bi-gram match and skip-gram. In the syntactic similarity module, the important features used are: subject-subject comparison, subject-verb comparison, object-verb comparison and cross subject-verb comparison. The precision, recall and f-score of the integrated AV system on the AVE 2008 English annotated test set have been observed as 0.66, 0.65 and 0.65 respectively that outperforms the best performing system at AVE 2008 in terms of f-score.

Keywords: Answer Validation Exercise (AVE); Textual Entailment (TE); Named Entity (NE); Chunk Boundary; Syntactic Similarity; Question Type.

1 Introduction

Answer Validation Exercise (AVE) task is aimed at developing systems that decide the correctness of the answer to a question given a supporting text. More specifically, AVE systems receive a set of triplets (Question, Answer and Supporting Text) and return a judgment of "SELECTED", "VALIDATED" or "REJECTED" for each triplet. One of the "VALIDATED" answers is identified as "SELECTED". There were three AVE competitions: AVE 2006 [1], AVE 2007 [2] and AVE 2008 [3].

Textual Entailment [4] is defined as a directional relationship between pairs of text expressions, denoted by T - the entailing "Text", and H- the entailed "Hypothesis". If the meaning of H can be inferred from the meaning of T then T entails H. There were six Recognizing Textual Entailment (RTE) competitions RTE-1 in 2005, RTE-2 in 2006, RTE-3 in 2007, RTE-4 in 2008, RTE-5 in 2009 and RTE-6 in 2010. We participated in the TAC RTE-5 and TAC RTE-6 Challenge.

Related works are described in Section 2. The answer validation system is described in Section 3. The experimental results have been presented in Section 4. The conclusions are drawn in Section 5.

R. Muñoz et al. (Eds.): NLDB 2011, LNCS 6716, pp. 324–329, 2011.
© Springer-Verlag Berlin Heidelberg 2011

2   Related Work

Most of the AVE systems in various AVE Challenges use some sort of lexical match-ing, e.g., simple word overlap, n-gram match and longest Common subsequence. A number of systems represent the text as parse trees (e.g., syntactic, dependency) before the actual task. Some of the systems use semantic relations (e.g., logical infe-rence, Semantic Parsing) for solving the AVE problem.

Use of Textual Entailment recognition techniques for answer validation has shown a great success [5]. The system [6] utilizes a Textual Entailment (TE) system as a component to validate answers. The rules followed in building the patterns for ques-tion transformation, the generation of the corresponding hypothesis and final answer ranking are described in [7]. The AVE task was cast into a Textual Entailment recog-nition problem in [8] and an existing RTE system was used to validate answers. Addi-tional information from named-entity (NE) recognizer, question analysis component and other sources are considered to make the final decision. The scoring technique is based on applying voting principle on the outputs generated from answer validation modules based on the TE system, the NER system, Question Answer type analysis, Chunk Boundary system and the Syntactic Similarity system. Their approach is clos-est to the method used in the present work. But, a different scoring mechanism and a different set of features have been used in the present work.

3   Answer Validation System

The architecture of the proposed Answer Validation (AV) system is described in Figure 1.

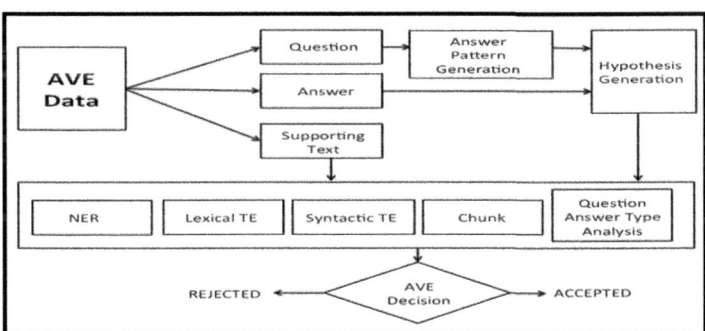

Fig. 1. Answer Validation System

A number of answer validation modules have been developed based on Textual Entailment, Named Entity Recognition, Question Answer type analysis, Chunk Boundary system and Syntactic Similarity system. These answer validation modules have been integrated using voting technique. We combine the question and the answer into Hypothesis (H) and the Supporting Text as Text (T) to check the entailment rela-tion as either "VALIDATED" or "REJECTED". The details of the answer validation system are presented in [9].

3.1 Answer Pattern Generation Module

At first we convert each question into an affirmative answer pattern with the answer slot. For example, the answer pattern generated for the question 'What is X?' is 'X is </answer>'.

3.2 Hypothesis Generation Module

The hypothesis is generated for each alternative answer to a given question by considering the affirmative answer pattern sentence as produced by the pattern generation module.

3.3 NER Module

If the number of Named Entities (NE) in the generated hypothesis matches the number of NEs in the supporting text, then the text-hypothesis pair as generated from the AVE pair is considered as validated.

3.4 Lexical Textual Entailment (TE) Module

The lexical TE module is based on three types of matching, i.e., WordNet based Unigram Match, bigram matching and Skip-bigram Matching. The details of the system are presented in [10]. The textual entailment decision is mapped to the decision of the answer validation module.

3.5 Syntactic Textual Entailment (TE) Module

The syntactic TE module compares the dependency relations identified in the text and the hypothesis. Some important comparisons are: Subject-Verb Comparison, WordNet Based Subject-Verb Comparison, Subject-Subject Comparison, Object-Verb Comparison, WordNet Based Object-Verb Comparison, Cross Subject-Object Comparison, Noun Comparison. The details of the system are presented in [11].

3.6 Chunk Module

The question sentences are pre-processed using Stanford dependency parser [12]. The words along with their part of speech (POS) information are passed through a Conditional Random Field (CRF) based chunker[1] to extract phrase level chunks of the questions. A rule-based module is developed to identify the chunk boundaries. Key chunks are identified for each question. The answer validation decision depends on the presence of these key chunks in the supporting text. The details of the chunk module have been presented in [9].

3.7 Question Answer Type Analysis Module

The question keyword generally identifies the question type and the expected answer type. For example, if the question type is "When", the expected answer type is a "DATE/TIME". The RASP Parser [13] parses the answer string "<a_str>". If the

---

[1] http://crfchunker.sourceforge.net/

RASP parser generates the tag "<timex type=date>" then the answer string is "VA-LIDATED", otherwise it is "REJECTED". The details of the module have been presented in [14].

3.8 Answer Validation – Voting Technique

Voting technique has been used to integrate the decisions from each of the answer validation modules. The voting technique has been detailed in [14]. At first, the generated Hypothesis (H) is checked for the presence of any named entities (NE). If any NE is present and the NER module generates "VALIDATED" tag to the answer, then the results of the different answer validation modules are checked. If all these modules generate the "VALIDATED" result, the answer is tagged as "VALIDATED". Otherwise, the answer is tagged as "REJECTED". If no NE is present in the generated hypothesis (H), then the results of the different answer validation modules are checked. If all these modules generate the "VALIDATED" result, the answer is tagged as "VALIDATED". Otherwise, the answer is tagged as "REJECTED".

4 Experiment Results

The Answer Validation system has been tested on the AVE 2008 Development Set for English. The AVE 2008 development set consists of 195 pairs of which only 21 are positives (10.77% of the total number of pairs). The AVE 2008 English annotated test set consists of 1055 pairs and the number of correct "VALIDATED" answer is 79 (7.5% of the total). The recall, precision and f-measure values of AVE development set and test set obtained over correct answers are shown in Table 1.

Table 1. AVE 2008 Development and Test Set Result

	AVE Development Set Result	AVE Test Set Result
"VALIDATED" in the Set	21	79
"VALIDATED" in the proposed AV system	34	78
"VALIDATED" match	15	52
Precision	0.44	0.66
Recall	0.71	0.65
F-score	0.54	0.65

5 Discussion

We compare our results with other systems that participated in the CLEF 2008@AVE track. The results of the participating systems are shown in Table 2.

The results obtained by our AV system on the CLEF 2008@AVE track are shown in Table 2 in bold. It is observed that the proposed AV system has outperformed the participating systems in the AVE-2008 on the basis of F-Score value.

The AV system performance may be improved in future works by incorporating new models of textual entailment. A two-way textual entailment (TE) recognition

Table 2. Comparison of the AV System Result with AVE 2008 Participating Systems

Group Name	F-Score	Precision	Recall
DFKI	0.64	0.54	0.78
UA	0.49	0.35	0.86
UNC	0.21	0.13	0.56
Our AV Result	**0.65**	**0.66**	**0.65**

system that uses semantic features has been developed. The semantic features are based on the Universal Networking Language (UNL). The proposed TE system compares the UNL relations in both the text and the hypothesis to arrive at the two-way entailment decision. The details of the system are presented in [15]. Another TE system that is based on the Support Vector Machine and uses features for Lexical method, Lexical distance and the output tag from a rule based syntactic two-way TE system as another feature have been presented in [16].

Acknowledgements. The author acknowledges his Doctoral supervisors Professor Sivaji Bandyopadhyay, India and Prof. Alexander Gelbukh, Mexico and the support of the DST India-CONACYT Mexico project "Answer Validation through Textual Entailment".

References

1. Peñas, A., Rodrigo, Á., Sama, V., Verdejo, F.: Overview of the answer validation exercise 2006. In: Peters, C., Clough, P., Gey, F.C., Karlgren, J., Magnini, B., Oard, D.W., de Rijke, M., Stempfhuber, M. (eds.) CLEF 2006. LNCS, vol. 4730, pp. 257–264. Springer, Heidelberg (2007)
2. Peñas, A., Rodrigo, Á., Verdejo, F.: Overview of the answer validation exercise 2007. In: Peters, C., Jijkoun, V., Mandl, T., Müller, H., Oard, D.W., Peñas, A., Petras, V., Santos, D. (eds.) CLEF 2007. LNCS, vol. 5152, pp. 237–248. Springer, Heidelberg (2008)
3. Rodrigo, Á., Peñas, A., Verdejo, F.: Overview of the answer validation exercise 2008. In: Peters, C., Deselaers, T., Ferro, N., Gonzalo, J., Jones, G.J.F., Kurimo, M., Mandl, T., Peñas, A., Petras, V. (eds.) CLEF 2008. LNCS, vol. 5706, pp. 296–313. Springer, Heidelberg (2009)
4. Dagan, I., Glickman, O., Magnini, B.: The PASCAL Recognising Textual Entailment Challenge. In: Proceedings of the PASCAL Challenges Workshop on RTE (2005)
5. Rodrigo, Á., Peñas, A., Verdejo, F.: UNED at Answer Validation Exercise 2007. In: Peters, C., Jijkoun, V., Mandl, T., Müller, H., Oard, D.W., Peñas, A., Petras, V., Santos, D. (eds.) CLEF 2007. LNCS, vol. 5152, pp. 404–409. Springer, Heidelberg (2008)
6. Wang, R., Neumann, G.: DFKI–LT at AVE 2007: Using Recognizing Textual Entailment as a Core Engine for Answer Validation. In: Peters, C., Jijkoun, V., Mandl, T., Müller, H., Oard, D.W., Peñas, A., Petras, V., Santos, D. (eds.) CLEF 2007. LNCS, vol. 5152, pp. 387–390. Springer, Heidelberg (2008)
7. Iftene, A., Balahur-Dobrescu, A.: Answer Validation on English and Romanian Languages. In: Peters, C., Deselaers, T., Ferro, N., Gonzalo, J., Jones, G.J.F., Kurimo, M., Mandl, T., Peñas, A., Petras, V. (eds.) CLEF 2008. LNCS, vol. 5706, pp. 448–451. Springer, Heidelberg (2009)

8. Wang, R., Neumann, G.: Information Synthesis for Answer Validation. In: Peters, C., Deselaers, T., Ferro, N., Gonzalo, J., Jones, G.J.F., Kurimo, M., Mandl, T., Peñas, A., Petras, V. (eds.) CLEF 2008. LNCS, vol. 5706, pp. 472–475. Springer, Heidelberg (2009)
9. Pakray, P., Pal, S., Bandyopadhyay, S., Gelbukh, A.: Automatic Answer Validation System on English Language. In: IEEE ICACTE 2010, vol. 6, pp. 329–333 (2010)
10. Pakray, P., Bandyopadhyay, S., Gelbukh, A.: Lexical based two-way RTE System at RTE-5. System Report. In: TAC Recognizing Textual Entailment Track Notebook (2009)
11. Pakray, P., Gelbukh, A., Bandyopadhyay, S.: A Syntactic Textual Entailment System Based on Dependency Parser. In: Gelbukh, A. (ed.) CICLing 2010. LNCS, vol. 6008, pp. 269–278. Springer, Heidelberg (2010)
12. Marneffe, M., MacCartney, B., Manning, C.D.: Generating Typed Dependency Parses from Phrase Structure Parses. In: 5th International Conference on Language Resources and Evaluation, LREC (2006)
13. Briscoe, E., Carroll, J., Watson, R.: The Second Release of the RASP System. In: Proceedings of the COLING/ACL 2006 Interactive Presentation Session (2006)
14. Pakray, P., Gelbukh, A., Bandyopadhyay, S.: Answer Validation Using Textual Entailment. In: Gelbukh, A. (ed.) CICLing 2011, Part II. LNCS, vol. 6609, pp. 353–364. Springer, Heidelberg (2011)
15. Pakray, P., Poria, S., Gelbukh, A., Bandyopadhyay, S.: Semantic Textual Entailment using UNL. In: 12th CICLing-2011, Japan (2011) (will be published in IJCLA Journal)
16. Pakray, P., Gelbukh, A., Bandyopadhyay, S.: A Hybrid Textual Entailment System using Lexical and Syntactic Features. In: 9th IEEE ICCI 2010, pp. 291–295 (2010)

Analyzing Text Data for Opinion Mining[*]

Wei Wei

Department of Computer and Information Science
Norwegian University of Science and Technology
wwei@idi.ntnu.no

Abstract. Opinion mining has become a hot topic at the crossroads of information retrieval and computational linguistics. In this paper, we propose to study two key research problems of designing an opinion mining system, i.e., entity-related opinion detection problem and sentiment analysis problem. For the entity-related opinion detection problem, we want to use sophisticated statistical models, e.g., probabilistic topic models and statistical rule generation methods, to achieve better performance than existing baselines. For the sentiment analysis problem, we have proposed a novel HL-SOT approach and reported its feasibility in an academic publication. Since the kernel classifier utilized in the HL-SOT approach is a linear function, we are working on developing a multi-layer neural network kernel algorithm which results in a non-linear classifier and is expected to improve the performance of the original HL-SOT approach to sentiment analysis.

Keywords: Opinion Mining, Entity-related Opinion Detection, Sentiment Analysis, Sentiment Ontology Tree.

1 Introduction

As the internet reaches almost every corner in this world, more and more people are used to accessing information on the World Wide Web (WWW). At the meantime, with rapid expansion of Web 2.0 technologies that facilitate people to write reviews and share opinions on products, a large amount of review texts are generated online. As the number of product reviews grows, it becomes difficult for a user to manually learn the panorama of an interesting topic from existing online information. Faced with this problem, research issues, e.g., [10,13,16], on opinion mining were proposed and have become a popular research topic at the crossroads of information retrieval and computational linguistics.

The research on opinion mining problems was proposed to deal with not only what topics are talking about but also what opinions and sentiments are expressed in texts [7]. A system that has features of opinion mining can benefit many application areas, for examples, 1) an opinion mining system can help potential consumers to learn experiences on products from other customers; 2) manufacturers can use an opinion mining system to follow customers' sentiments and opinions on their products; 3) politicians can

---

[*] Ph.D. advisor: Prof. Jon Atle Gulla, Department of Computer and Information Science, Norwegian University of Science and Technology, jag@idi.ntnu.no

R. Muñoz et al. (Eds.): NLDB 2011, LNCS 6716, pp. 330–335, 2011.
© Springer-Verlag Berlin Heidelberg 2011

utilize an opinion mining system to know their images in peoples' opinions; 4) organizations can make use of an opinion mining system to summarize opinions on some social issues.

In order to introduce key problems of designing opinion mining systems in an intuitive way, let's imagine an example of using an opinion mining system. When considering to buy a NIKON digital camera, a prospective user might issue a query like "how is Nikon camera" to an opinion mining system. Upon receiving the query, the system at least needs to do the following two steps: 1) it needs to gather all texts that mainly contain opinion information related to Nikon cameras; 2) it needs to analyze sentiment information within the texts gathered in step one. The research problem involved in the first step is called **entity-related opinion detection problem**. The research problem involved in the second step is called **sentiment analysis problem**.

In this Ph.D. research proposal, the above two key research problems are planed to be studied. The rest of this proposal is organized as follows. Section 2 discusses the related work. In section 3, we present our proposed research work. In section 4, we conclude this research proposal.

2   Related Work

2.1   Entity-Related Opinion Detection

Entity-related opinion detection concerns both entity-related search and opinion detection simultaneously. Compared with the feature-based opinion mining and summarization framework proposed in [10], the entity-related opinion detection can be deemed as pre-processed inputs which collects opinion-rich texts related to a given entity for further sentiment analysis. Although there already exist a lot of research work related to general opinion detection, without focusing on a given entity most of them mainly aim at extracting texts that contain any opinion information. In [23], Wiebe et al. present a procedure for automatically formulating a single best tag when there are multiple judges who disagree. Subsequently, Hatzivassiloglou and Wiebe report in [9] that gradable adjectives are a useful feature for subjectivity classification. With understanding that presence of one or more adjectives is useful for predicting that a sentence is subjective, using the performance of the simple adjective feature as a baseline, Wiebe propose in [22] to involve adjective features based on similarity clusters which is to identify higher quality adjective features using the results of a method for clustering words according to distributional similarity [12], and further refined with the addition lexical semantic features of adjectives, specifically *polarity* and *gradability* [8]. In [24], a Bayesian classifier to separate documents containing primarily opinions from documents that report mainly facts was presented and three different approaches (similarity approach, Naive Bayes classifier, and multiple Naive Bayes classifiers) to classify opinions from facts at the sentence level are developed. In [10], Hu and Liu use adjectives as opinion words and limit the opinion words extraction to those sentences that contain one or more product features. In [11], a method for obtaining opinion-bearing and non-opinion-bearing words is introduced. The obtained words are then used to recognize opinion-bearing sentences. In [2], Bethard et al. propose to find propositional opinions as many verbs contain the actual opinion rather than full opinion sentences. In [4], the authors presented a global inference approach to identify two types of entities: entities that express opinions and entities that denote sources of opinions.

2.2 Sentiment Analysis

The task of sentiment analysis was originally performed to extract overall sentiment from the target texts. Document overall sentiment analysis is to summarize the overall sentiment in documents, e.g., *word-level sentiment annotation* [9,1,6,8,5,24] and *phrase-level sentiment annotation* [21]. However, in [19], as the difficulty shown in the experiments, the whole sentiment of a document is not necessarily the sum of its parts. Then there came up with research works shifting focus from overall document sentiment to attributes-based sentiment analysis, which is to analyze sentiment based on each attribute of a product. In [10], mining product features was proposed together with sentiment polarity annotation for each opinion sentence. In that work, sentiment analysis was performed on product attributes level. In [13], a system with framework for analyzing and comparing consumer opinions of competing products was proposed. The system made users be able to clearly see the strengths and weaknesses of each product in the minds of consumers in terms of various product features. In [17], Popescu and Etzioni not only analyzed polarity of opinions regarding product features but also ranked opinions based on their strength. In [14], Liu et al. proposed Sentiment-PLSA that analyzed blog entries and viewed them as a document generated by a number of hidden sentiment factors. These sentiment factors may also be factors based on product attributes. In [15], Lu et al. proposed a semi-supervised topic models to solve the problem of opinion integration based on the topic of a product's attributes. The work in [18] presented a multi-grain topic model for extracting the ratable attributes from product reviews. In [16], the problem of rated attributes summary was studied with a goal of generating ratings for major aspects so that a user could gain different perspectives towards a target entity. All these research works concentrated on attribute-based sentiment analysis. However, the main difference with our work is that they did not sufficiently utilize the hierarchical relationships among a product's attributes. Although a method of ontology-supported polarity mining, which also involved ontology to tackle the sentiment analysis problem, was proposed in [25], that work studied polarity mining by machine learning techniques that still suffered from a problem of ignoring dependencies among attributes within an ontology's hierarchy. In the contrast, our proposed work solves the sentiment analysis problem as a hierarchical classification problem that fully utilizes the hierarchical relationships among attributes of an entity in training and classification process.

3   Proposed Research Work

The research issues in this research proposal respectively concern the entity-related opinion detection problem and the sentiment analysis problem. Specifically, we plan the following research work.

Entity Related Opinion Detection. This work will tackle the problem of searching texts that not only relate to the entity described by a given query but also contain opinion information. The results of this work can be used as input for further sentiment analysis. This problem can be solved by a baseline method in two steps. First, traditional searching techniques can be used to obtain a set of entity-related texts. Then, utilizing rule-based methods using NLP techniques, those texts that contain any opinion information can be extracted from the text collection produced by the first step. Since the traditional keyword-based searching strategy failed to focus on semantic aspects of an entity, we propose to utilize a more sophisticated probabilistic topic model that treats each entity as a semantic topic to improve the performance of entity-related search in step one. Since the rule-based methods developed for entities in one domain is not easy to be generalized and applied to entities in other domains, we propose a statistical rule generation method that generates rules based on training data so that our proposed approach can be easily generalized across different entity domains.

Sentiment Analysis with Sentiment Ontology Tree. This work aims at achieving two sub-tasks of the sentiment analysis problem, i.e., 1) to analyze which aspects of an entity are mentioned in texts; 2) to analyze the corresponding sentiments expressed in texts. When we look into example opinion-rich texts that relate to an entity, e.g., product reviews on an camera, we find that there are some intrinsic properties [20] that suggest us to utilize the knowledge of ontology to analyze review-like texts of an entity. Inspired by a hierarchical classification algorithm proposed in [3], we propose to use a tree-like ontology structure SOT, i.e., Sentiment Ontology Tree, to formally describe the hierarchial relationships among aspects/attributes of an entity. With the defined SOT, we develop a hierarchical classification algorithm to achieve the two sub-tasks of the sentiment analysis problem in one hierarchical classification process.

Multi-layer Neural Network Kernel for Sentiment Analysis with Sentiment Ontology Tree. The kernel hierarchical classification algorithm developed in [20] is a linear classifier which is deemed as not competent enough for classifying complicated instances in high dimension space. In this work, we plan to utilize a multi-layer neural network to develop the kernel hierarchical classification algorithm. The multi-layer neural network is a non-linear classifier which is expected to improve the performance of sentiment analysis approach proposed in [20].

Current Progress. Research on the problem of "sentiment analysis with sentiment ontology tree" has been conducted. In the paper [20], we studied the problem of sentiment analysis on product reviews through a novel method, called the HL-SOT approach, namely Hierarchical Learning (HL) with Sentiment Ontology Tree (SOT). In the proposed HL-SOT approach [20], each target text is to be indexed by a vector $x \in X, X = \mathbb{R}^d$. Weight vectors $w_i (1 \leq i \leq N)$ define linear-threshold classifiers of each node i in SOT so that the target text x is labeled true by node i if x is labeled true by i's parent node and $w_i \cdot x \geq \theta_i$. The parameters w_i and θ_i are learned from the training data set: $D = \{(r, l)|r \in X, l \in \mathcal{Y}\}$, where $\mathcal{Y}$ denotes the set of label vectors. In the training process, when a new instance r_t is observed, each row vector $w_{i,t}$ is updated by a regularized least squares estimator given by:

$$w_{i,t} = (I + S_{i,Q(i,t-1)} S_{i,Q(i,t-1)}^{\top} + r_t r_t^{\top})^{-1} \times S_{i,Q(i,t-1)} (l_{i,i_1}, l_{i,i_2}, ..., l_{i,i_{Q(i,t-1)}})^{\top}, \qquad (1)$$

where I is a $d \times d$ identity matrix, $Q(i, t-1)$ denotes the number of times the parent of node i observes a positive label before observing the instance r_t, $S_{i,Q(i,t-1)} = [r_{i_1}, ..., r_{i_{Q(i,t-1)}}]$ is a $d \times Q(i, t-1)$ matrix whose columns are the instances $r_{i_1}, ..., r_{i_{Q(i,t-1)}}$, and $(l_{i,i_1}, l_{i,i_2}, ..., l_{i,i_{Q(i,t-1)}})^{\top}$ is a $Q(i, t-1)$-dimensional vector of the corresponding labels observed by node i. The Formula 1 restricts that the weight vector $w_{i,t}$ of the classifier i is only updated on the examples that are positive for its parent node. Then the label vector $\hat{y}_{r_t}$ is computed for the instance r_t, before the real label vector l_{r_t} is observed. Then the current threshold vector θ_t is updated by:

$$\theta_{t+1} = \theta_t + \epsilon(\hat{y}_{r_t} - l_{r_t}), \qquad (2)$$

where ϵ is a small positive real number that denotes a corrective step for the current threshold vector θ_t. After the training process for each node of SOT, each target text is to be labeled by each node i parameterized by the weight vector w_i and the threshold θ_i in the hierarchical classification process. The performance evaluation shows that the HL-SOT approach outperforms two baselines: the HL-flat and the H-RLS approach, which respectively confirms two intuitive motivations based on which the approach was developed.

Currently, with success of the HL-SOT approach, our work on developing the multi-layer neural network kernel hierarchical classification algorithm is entering the experimental stage. For the problem of "entity related opinion detection", since there is a lack of standard data sets we are working on collecting data and label them manually to create a standard bench data set. After this stage, we will develop and test our approaches and report the results in future publications.

4   Conclusion

This Ph.D. research proposal focuses on the research problems of opinion mining. The proposed research problems will be respectively addressed and presented in academic publications. In each publication, we will present our proposed solutions and comparatively study and discuss them in the context of the existing state-of-the-art approaches. Through this systematic research process, we aim at deeply understanding the studied problems and develop significant solutions respectively and therefore make good contributions to the research community of opinion mining.

References

1. Andreevskaia, A., Bergler, S.: Mining wordnet for a fuzzy sentiment: Sentiment tag extraction from wordnet glosses. In: Proceedings of 11th Conference of the European Chapter of the Association for Computational Linguistics, EACL 2006 (2006)
2. Bethard, S., Yu, H., Thornton, A., Hatzivassiloglou, V., Jurafsky, D.: Automatic extraction of opinion propositions and their holders. In: Proceedings of the AAAI Spring Symposium on Exploring Attitude and Affect in Text: Theories and Applications (2004)

3. Cesa-Bianchi, N., Gentile, C., Zaniboni, L.: Incremental algorithms for hierarchical classification. Journal of Machine Learning Research (JMLR) 7, 31–54 (2006)
4. Choi, Y., Breck, E., Cardie, C.: Joint extraction of entities and relations for opinion recognition. In: Proceedings of EMNLP 2006, pp. 431–439 (2006)
5. Devitt, A., Ahmad, K.: Sentiment polarity identification in financial news: A cohesion-based approach. In: Proceedings of ACL 2007, Prague, Czech Republic (2007)
6. Esuli, A., Sebastiani, F.: Determining the semantic orientation of terms through gloss classification. In: Proceedings of CIKM 2005, Bremen, Germany (2005)
7. Esuli, A., Sebastiani, F.: Determining term subjectivity and term orientation for opinion mining. In: Proceedings of EACL 2006, Trento, Italy (2006)
8. Hatzivassiloglou, V., McKeown, K.R.: Predicting the semantic orientation of adjectives. In: Proceedings of ACL 1997, Madrid, Spain (1997)
9. Hatzivassiloglou, V., Wiebe, J.M.: Effects of adjective orientation and gradability on sentence subjectivity. In: Proceedings of COLING 2000, Saarbrüken, Germany (2000)
10. Hu, M., Liu, B.: Mining and summarizing customer reviews. In: Proceedings of 10th ACM SIGKDD Conference on Knowledge Discovery and Data Mining, KDD 2004 (2004)
11. Kim, S.-M., Hovy, E.: Automatic detection of opinion bearing words and sentences. In: Proceedings of IJCNLP 2005, pp. 61–66 (2005)
12. Lin, D.: Automatic retrieval and clustering of similar words. In: Proceedings of COLING-ACL, pp. 768–774 (1998)
13. Liu, B., Hu, M., Cheng, J.: Opinion observer: analyzing and comparing opinions on the web. In: Proceedings of WWW 2005 (2005)
14. Liu, Y., Huang, X., An, A., Yu, X.: ARSA: a sentiment-aware model for predicting sales performance using blogs. In: Proceedings of SIGIR 2007 (2007)
15. Lu, Y., Zhai, C.: Opinion integration through semi-supervised topic modeling. In: Proceedings of 17th International World Wide Web Conference, WWW 2008 (2008)
16. Lu, Y., Zhai, C., Sundaresan, N.: Rated aspect summarization of short comments. In: Proceedings of 18th International World Wide Web Conference, WWW 2009 (2009)
17. Popescu, A.-M., Etzioni, O.: Extracting product features and opinions from reviews. In: Proceedings of Human Language Technology Conference and Empirical Methods in Natural Language Processing Conference (HLT/EMNLP 2005), Vancouver, Canada (2005)
18. Titov, I., McDonald, R.T.: Modeling online reviews with multi-grain topic models. In: Proceedings of 17th International World Wide Web Conference, WWW 2008 (2008)
19. Turney, P.D.: Thumbs up or thumbs down? semantic orientation applied to unsupervised classification of reviews. In: Proceedings of ACL 2002, Philadelphia, USA (2002)
20. Wei, W., Gulla, J.A.: Sentiment learning on product reviews via sentiment ontology tree. In: Proceedings of 48th Annual Meeting of the Association for Computational Linguistics, ACL 2010, Uppsala, Sweden (2010)
21. Whitelaw, C., Garg, N., Argamon, S.: Using appraisal taxonomies for sentiment analysis. In: Proceedings of 14th ACM Conference on Information and Knowledge Management CIKM 2005, Bremen, Germany (2005)
22. Wiebe, J.M.: Learning subjective adjectives from corpora. In: Proceedings of the 2000 National Conference on Artificial Intelligence. AAAI, Menlo Park (2000)
23. Wiebe, J.M., Bruce, R.F., O'Hara, T.P.: Development and use of a gold-standard data set for subjectivity classifications. In: Proceedings of ACL 1999, pp. 246–253 (1999)
24. Yu, H., Hatzivassiloglou, V.: Towards answering opinion questions: Separating facts from opinions and identifying the polarity of opinion sentences. In: Proceedings of 8th Conference on Empirical Methods in Natural Language Processing, EMNLP 2003 (2003)
25. Zhou, L., Chaovalit, P.: Ontology-supported polarity mining. Journal of the American Society for Information Science and Technology (JASIST) 59(1), 98–110 (2008)

On the Improvement of Passage Retrieval in Arabic Question/Answering (Q/A) Systems

Lahsen Abouenour

Ecole Mohammadia d'Ingénieurs, Med Vth University-Agdal, Rabat, Morocco
abouenour@yahoo.fr

Abstract. The development of advanced Information Retrieval (IR) applications is of a particular priority in the context of the Arabic language. In this PhD thesis, our aim is improving the performances of Arabic Question/Answering (Q/A) systems. Indeed, we propose an approach which is composed of three levels. We have showed through experiments conducted on a set of 2,264 translated CLEF and TREC questions that the accuracy, the Mean Reciprocal Rank (MRR) and the number of answered questions are enhanced using a Query Expansion (QE) module based on Arabic Wordnet (AWN) in the first level and a structure-based Passage Retrieval (PR) module in the second level. In order to evaluate the impact of the AWN coverage on the performances, we have automatically extended its content in terms of Named Entities (NE), Nouns and Verbs. The next step consists in developing a semantic reasoning process based on Conceptual Graphs (CGs) as a third level.

Keywords: Question/Answering, Arabic Wordnet, Query Expansion, Conceptual Graphs.

1 Introduction

Arabic is among the most spoken semitic languages since more than 280 million people use it as a first language. This explains the rapid growth of the Arabic content on the Web. The amount of information and knowledge contained in this content is of great interest. Unfortunately, in comparison with other languages such as English, there are few Arabic Natural Language Processing (ANLP) applications which can help users in order to reach and access relevant information. The classic Search Engines (SEs) as well as the basic Information Retrieval (IR) systems are of limited usefulness when they are faced to the particularities of the Arabic language (short vowels, absence of capital letters, complex morphology, etc.). These systems are also inadequate when they come to more advanced requirements. This is the case when a user looks for a precise answer to a question which is formulated in natural language rather than a list of snippets to be checked one by one.

The research described in this paper contributes a three-level approach which aims improving the performances reached in the context of the Arabic Question/Answering (Q/A) task. The rest of the paper is organized as follows: Section 2 presents the motivation behind being interesting in Arabic Q/A systems and describes the research problem to

R. Muñoz et al. (Eds.): NLDB 2011, LNCS 6716, pp. 336–341, 2011.
© Springer-Verlag Berlin Heidelberg 2011

be solved. After that, Section 3 describes related works by addressing some attempts in the context of the Arabic Q/A. Section 4 describes the methodology followed in this research and presents the current state of the research as well as future works.

2 Motivation and Research Problem

The motivation behind being interested in the Arabic Q/A field is to contribute in the development of more convenient systems for users of the Arabic language which is less concerned by NLP researches than English and other European languages. Moreover, the modules and linguistic resources that will be built in the context of the current research can be used in other ANLP domains, as components of more advanced or cross-language processes.

Generally, a Q/A system is composed of three main modules:

(i) *Question analysis and classification module.* In this module a question is analyzed in order to extract its keywords, identify the class of the question and the structure of the expected answer, form the query to be passed to the PR module, etc.

(ii) Passage Retrieval *(PR) module.* This module is a core component of a Q/A system. The quality of the results returned by such system depends mainly on the quality of the PR module it uses [17]. Indeed, this module uses the query formed by the previous module and extracts a list of passages from an IR process (generally a SE such as Google[1] or Yahoo[2]). Thereafter, this module has to perform a ranking process in order to improve the relevance of the candidate passages according to the user question.

(iii) *Answer Extraction (AE) module.* This module tries to extract the answer from the candidate passages provided by the previous module. In advanced Q/A systems, this module can be designed to formulate the answer from one or many passages.

In this research, we focus on enhancing the relevance of the candidate passages provided by the PR module since the performance of the whole Q/A system depends on it. Our research consists in proposing a new approach which improves PR from the Web for questions formulated in Arabic language and belonging to different domains. The intended work concerns the development of processes as well as resources that are required for the implementation of the proposed approach.

3 Related Works

To our knowledge, there are few attempts in terms of building Arabic Q/A systems. We have found five systems, namely: QARAB [10], AQAS [11], ArabiQA [9], QAS-AL [7] and AJAS [12]. These systems are limited in terms of the covered domain, the nature of the processed data (structured or unstructured), the lack of

---

[1] http://www.google.com
[2] http://www.yahoo.com

complete experiments with a significant number of questions and/or the number of integrated modules. In our work, our aim is overcoming these limitations.

4 Proposed Methodology

Our approach is based on three levels that can be described as follows [6]:

- Keyword-based level: this level retrieves the passages where the question keywords as well as their Arabic Wordnet (AWN) [2] synonyms, hyponyms and hypernyms appear;
- Structure-based level: this level ranks the passages that have been retrieved in the previous level according to the N-gram density of keywords as well as the related AWN words. Actually, the first passages are those having a structure which is similar to the question.
- Semantic reasoning level: this level applies a new ranking to the passages retrieved and ranked using the two first levels. The new ranking is made by comparing the Conceptual Graphs (CGs) of the expected structure of the answer with that of the passages.

To summarize, since a semantic reasoning process is a time consuming task, the main objective of the first and second levels is to develop the quality of the passages to be processed at the third level.

In order to show the usefulness of our approach, we have adopted an experimental process which is composed of three steps. The following sub sections describe the work done in the context of each step.

4.1 Preliminary Experiments

The first step consists in showing preliminarily that the use of a QE in the first level combined with the use of JIRS[3] (which is the N-gram density based PR module) in the second level improves the considered measures, namely: the accuracy, the Mean Reciprocal Rank[4] (MRR) and the number of answered questions. Therefore, we have manually translated a collection of 2,264 TREC[5] and CLEF[6] question into the Arabic language. For each question of the collection, we apply the AWN-based QE module. After that, using the Yahoo API, we extract from the snippets that match the queries formed from the questions keywords as well as the words generated by QE. The answer is checked in the best five snippets according firstly to the SE ranking and secondly to the one of JIRS combined with QE.

The preliminary experiments showed that the performances are improved after applying the keyword-based and structure-based levels in comparison with using only the SE API. Indeed, the accuracy is passed from 9.66% to 20.20%, the MRR from

---

[3] http://sourceforge.net/projects/jirs
[4] Mean Reciprocal Rank (MRR) is defined as the average of the reciprocal ranks of the results for a sample of queries (the reciprocal rank of a query response is the multiplicative inverse of the rank of the correct answer).
[5] Text REtrieval Conference, http://trec.nist.gov/data/qa.html
[6] Cross Language Evaluation Forum, http://www.clef-campaign.org

3.41 to 9.22 and the number of answered questions from 20.27% to 26.74%. The statistical significance of these results has been proofed through a t-test [9, 4].

4.2 Linguistic Resource Enrichment

The preliminary experiments also show that the coverage of AWN has to be extended. Indeed, for a high number of non answered questions, the corresponding keywords cannot be found in AWN and therefore the QE module could not be applied. In this step, we focus on the enrichment of AWN which is a key resource in the three levels of our approach.

The extension of AWN has been done with respect to three lines: (i) adding new NEs content, (ii) adding new verbs and (iii) adding new nouns and enriching the hyponymy relation between concepts.

The questions containing a NE as a keyword or whose answer is a NE represent a high percentage in the considered TREC and CLEF questions. Hence, we relied on the YAGO[7] ontology [8] to add around 300,000 NEs in AWN. In order to achieve this task, the YAGO content has been automatically translated and two kinds of mappings have been set [3]: Wordnet-based and heuristic-based. This helped us extending more questions and improving the performances [5] (Acc.=23.53%, MRR=9.59 and answered questions=31.37%).

After that, we have been interested in increasing the coverage of AWN in terms of verbs. This is due to the fact that verbs are key elements in our semantic reasoning process to be developed in the context of the third level. Therefore, we have translated the English Verbnet content into the Arabic language. Three heuristics out of eight existing have been applied on the graphs that have been constructed from the Arabic verbs and AWN synsets (concepts). The other heuristics have not been considered due to their low accuracy (even tough their recall is higher). Thanks to this technique, 2,569 new verbs have been added and connected with the corresponding AWN synsets.

The last task in the context of this step consists in improving the coverage and the hierarchy of nouns in AWN. The process that we have developed for automatically performing this task is based on the two following stages:

- Stage 1: Identifying hyponymy patterns over snippets that are retrieved from the Web. The Maximal Frequent Sequences of words has been adopted for pattern discovery over text ;
- Stage 2: Instantiating the identified patterns. The instantiation is performed by searching for hypernym/hyponym pairs which match the given pattern;

This technique which is inspired from a previous work done for the Spanish language [1] has been adapted to the Arabic language by using the Broken Plural (BP) form as well as the regular form of plural at the pattern identification stage. Table 1 shows the results that have been obtained in terms of noun enrichment in AWN.

---

[7] YAGO is an ontology with a great amount of individuals (around 3 millions NEs) written in English. It is built from WordNet and Wikipedia and is connected with the SUMO ontology.

Table 1. Results of the noun content enrichment in AWN

Measures	Using BP form	Using regular plural form	Overall/Total
#AWN hypernym synsets	700	700	1,400
#Succeeded Patterns	17 (73.91%)	9 (39.13%)	17 (73.91%)
#Candidate hyponyms	1,426	828	2,254
Avg. candidate hyponyms per AWN synset	2.04	1.22	1.61
#Correct hyponyms	458 (32.12%)	415 (50.12%)	832 (36.91%)
#AWN synset with correct hyponsyms	94 (13.43%)	191 (27.29%)	284 (40.57%)
# New AWN hyponyms	265 (57.86%)	205 (49.40%)	459 (55.17%)
#New AWN hyponymy associations	193	196	359

The obtained results listed in the table above show also that 832 correct hyponyms have been identified (roughly 37% of the candidate hyponyms). About 55% of these correct hyponyms can be added in AWN as new synsets. Even though the remaining hyponyms exist already in the AWN ontology, they can be added in new hypernym/hyponym associations.

4.3 Current Step and Future Works: Complete Experiments of the Arabic Q/A Approach

The current step of the present research keeps on evaluating the impact of the whole AWN enrichment process on the Arabic Q/A task using the two first levels.

Future works aim at developing the semantic reasoning module in order to evaluate the performances reached by applying the three levels of our approach. The semantic reasoning process is divided into the following sub processes:

- Converting the enriched release of the AWN ontology into an Artificial Intelligence Platform (Amine Platform[8]);
- Enriching the verbs in the Amine AWN ontology with their semantic definition. These definitions have to be extracted from the Arabic Verbnet resource;
- Developing a module which converts a text into a CG. This module uses the Stanford Arabic Syntactical Parser[9] and a set of CG patterns;
- Developing a module which compares the CGs of the question and candidate passages in order to perform a new ranking on top of the one made by the two first levels of our approach.
- Conducting experiments for the whole three-levels Arabic Q/A approach.

Acknowledgements. I would like to thank my advisors Dr. Karim Bouzoubaa (Ecole Mohammadia d'Ingénieurs, Med Vth University-Agdal, Rabat, Morocco) and Dr. Paolo Rosso (Universidad Politécnica de Valencia, Spain) for their assistance and guidance.

---

[8] http://sourceforge.net/projects/amine-platform
[9] http://nlp.stanford.edu/software/lex-parser.shtml

References

1. Denicia-Carral, C., Montes-y-Gómez, M., Villaseñor-Pineda, L., Hernández, R.G.: A Text Mining Approach for Definition Question Answering. In: Salakoski, T., Ginter, F., Pyysalo, S., Pahikkala, T. (eds.) FinTAL 2006. LNCS (LNAI), vol. 4139, pp. 76–86. Springer, Heidelberg (2006)
2. Elkateb, Sabry, Farwell, D., Vossen, P., Pease, A., Fellbaum, C.: Arabic Wordnet and the Challenges of Arabic. In: The Challenge of Arabic for NLP/MT. International Conference at the British Computer Society, London, pp. 15–24 (October 23, 2006)
3. Abouenour, L., Bouzoubaa, K., Rosso, P.: On the extension of Arabic Wordnet Named Entities and its Impact on Question/Answering. In: Proceeding of the 2nd International Conference on Knowledge Engineering and Ontology Development, Valencia, Spain (October 2010)
4. Abouenour, L., Bouzoubaa, K., Rosso, P.: An evaluated semantic QE and structure-based approach for enhancing Arabic Q/A. The Special Issue on Advances in Arabic Language Processing for the International Journal on Information and Communication Technologies, IJICT (June 2010)
5. Abouenour, L., Bouzoubaa, K., Rosso, P.: Using the Yago ontology as a resource for the enrichment of Named Entities in Arabic WordNet. In: Workshop on Language Resources (LRs) and Human Language Technologies (HLT) for Semitic Languages Status, Updates, and Prospects, LREC 2010 Conference, Malta (May 2010)
6. Abouenour, L., Bouzoubaa, K., Rosso, P.: Three-level approach for Passage Retrieval in Arabic Question Answering Systems. In: Proc. Of the 3rd International Conference on Arabic Language Processing CITALA 2009, Rabat, Morocco (May 2009)
7. Brini, W., Ellouze, M., Hadrich Belguith, L.: QASAL : Un système de question-réponse dédié pour les questions factuelles en langue Arabe. In: 9ème Journées Scientifiques des Jeunes Chercheurs en Génie Electrique et Informatique, Tunisia (2009)
8. Suchanek, F.M., Kasneci, G., Weikum, G.: YAGO: a core of semantic knowledge unifying WordNet and Wikipedia. In: Proc. of the 16th WWW, pp. 697–706 (2007)
9. Benajiba, Y., Rosso, P., Lyhyaoui, A.: Implementation of the ArabiQA Question Answering System's components. In: Proc. Workshop on Arabic Natural Language Processing, 2nd Information Communication Technologies Int. Symposium, ICTIS-2007, Fez, Morroco, April 3-5 (2007)
10. Hammou, B., Abu-salem, H., Lytinen, S., Evens, M.: QARAB: A Question answering system to support the ARABic language. In: Proc. of the Workshop on Computational Approaches to Semitic languages, ACL, Philadelphia, pp. 55–65 (2002)
11. Mohammed, F.A., Nasser, K., Harb, H.M.: A knowledge-based Arabic Question Answering System (AQAS). ACM SIGART Bulletin, 21–33 (1993)
12. Kanaan, G., Hammouri, A., Al-Shalabi, R., Swalha, M.: A New Question Answering System for the Arabic Language. American Journal of Applied Sciences 6(4), 797–805 (2009) ISSN 1546-9239

Ontology Extension and Population: An Approach for the Pharmacotherapeutic Domain

Jorge Cruanes*

Department of Software and Computer Systems
University of Alicante, Mail Box 99, Alicante, Spain
jcruanes@dlsi.ua.es

Abstract. For several years ontologies have been seen as a solution to share and reuse knowledge between humans and machines. An ontology is a knowledge photography at the moment of its creation. Nevertheless, in order to keep an ontology useful throughout time, it must be expanded and maintained regularly, mainly in the pharmacotherapeutic domain. Drug-therapy needs up-to-date and reliable information. Unfortunately, achieving a systematic ontology updating has is an arduous and a tedious task that becomes a bottleneck. To limit this obstacle we need methods that expedite the process of extension and population.This proposal aims the designing and validating method able to extract, from a corpus of summary of product characteristics and a pharmacotherapeutic ontology, the relevant knowledge to be added to the ontology.

Keywords: Ontology Expansion, Ontology Extension, Ontology Population, Ontology Learning, Pharmacotherapeutic.

1 Introduction

Medication information is one of the most important types of clinical data. It is critical for healthcare safety and quality and, therefore, drug-therapy needs up-to-date and reliable information during the care time and in the shortest time. But it is still an obstacle for daily practice [1, 2]. This obstacle is tried to be overcome with the use of updated ontologies. In order to have an updated ontology, it must be expanded and maintained, but these tasks are arduous and tedious to be performed by humans. Therefore, this proposal focuses on an automatic method to expand ontologies (both extension and population) in the Spanish pharmacotherapeutic domain. Focus on Spanish is due to the limited terminology resources int the pharmacotherapeutic domain that cover only specific aspects (e.g., active ingredients). Although the English terminology is varied and broad, for clinical care resources are required in the native language.

---

* This paper has been partially supported by Ministerio de Ciencia e Innovación - Spanish Government (grant no. TIN2009-13391-C04-01), and Conselleria d'Educació - Generalitat Valenciana (grant no. PROMETEO/2009/119, ACOMP/2010/286 and ACOMP/2011/001).

R. Muñoz et al. (Eds.): NLDB 2011, LNCS 6716, pp. 342–347, 2011.
© Springer-Verlag Berlin Heidelberg 2011

One of the most referenced definitions of ontology in the literature is the one given by [3]: "An ontology is a specification of a conceptualization". Several authors clarified this definition, including [4], who stated that "[...] conceptualization refers to an abstract model of some phenomenon in the world to be identified by relevant concepts of that phenomenon". As a matter of fact, an ontology is an explicit representation of the ideas of the real world, where those are represented formally by their characteristics and relationships.

This proposal aims to help in the maintenance issues designing and validating a method able to expand a pharmacotherapeutic ontology from a corpus. Ontology expansion involves both extension (adding concepts and relationships) and population (adding instances) tasks. As Spanish pharmacotherapeutic ontology it is used a specific ontology called OntoFIS ([5]) and as corpus it is used a summary of product characteristics (written in natural language) from the AEMPS (Spanish Agency of Medicines and Health Products) of the Ministry of Health, Social Policy and Equality. OntoFIS has an intermediate granularity, with 23 concepts, 647 relationships and around 55000 instances.

This paper is divided in five sections, beginning with this introduction. Then, we attempt the state of the art, followed by the main objectives and the method used. Finally, there are showed the extracted conclusions.

2   State of the Art

In order to cover the ontology expansion task in the pharmacotherapeutic domain, we aim to use general and domain specific approaches as baseline for our research. According to the classification and definitions given in [6], there are two tasks to expand an ontology: (i) Ontology Learning (OL) and (ii) Ontology Population (OP). OL is to add new concepts and relationships to an existing ontology. However, OL is also used to create new ontologies, but this task is out of the scope of this project, so we are going to focus just on the Ontology Extension (OE). Otherwise, the task of OP is focused in adding new instances, i.e., different denotations of a concept. Although OE and OP have been traditionally addressed separately, this work is going to deal both. In order to deal with OE and OP in the pharmacotherapeutic domain, it is important to study existing general and specific domain techniques, establishing a base line for this research.

On the one hand, among the OE methods, the studies of [7–16] can be highlighted. These authors use documents in natural language which are structured ([7–10, 14–16]) or semi-structured ([11–13]) to extract the knowledge needed to help to extend the target ontology. The main approaches to determine the most relevant words in a text in these studies are the word frequency (e.g. in [8] or [11]) and patterns and rules (as in [10] or [12]). In [11] they do not just use word frequency, but they also use entropy to know how much information is related with a word, helping to determine the word relevance.

One of the problems to be tackled in OE is to determine the granularity of the target ontology. In this sense, [7] considers that the instances can only appear as the ontology leaves. According to the authors, this problem forces them to identify the most general concepts before the most specific ones (which will contain the instances later), so they do not lose knowledge in their learning process. Another issue found in those approaches based on pattern matching and rules is that they are static and they do not provide an automatic learning process, even when there are rule-based methods to solve this problem in the Machine Learning field. This learning process allows to learn new patterns and rules using big corpus, where the manual checking is not possible.

The best results achieved in this task have a recall value between 67.22% and 77.99% and a precision value between 24.66% and 33.61% ([11]). However, the study of [7] reports precision levels from 43% to 86%, but they do not give results of their recall which makes an objective comparison between both approaches difficult.

On the other hand, some of the most important studied methods of Ontology Population (OP) are [6, 17–24]. These studies basically use unstructured texts ([6, 20–24]), but there are also some of them that use queries and knowledge bases ([17–19]). As in OE, in OP most common methodologies involve patterns and rules (e.g., [22] or [6]) and word frequency (as in [20] or [21]). Other approaches can be found, as clustering techniques and Hidden Markov Models ([23]), the approach developed by [6] or the one developed by [18]. The approach of [6], called 'Class-Example', is based on the context of a word to determine its relevance in a domain. This last work uses a training set to increase their results determining the context of the word. The method showed in [17, 18] proposes a variation of the nearest neighbor, using an optimized weight committee for each domain. This weight committee allows to adapt the algorithm to each domain.

The problem of domain adaptation problem can be observed in the results of [17, 19, 18] where, despite their optimized weight committee, they have a high variability on their results. Their recall levels ranges in more than 55% and their precision levels in almost 60%. How to select the content in big ontologies it is also an important issue when human intervention is needed. One solution is given in [21], pruning the base ontology using a set of experts defined keywords. This approach has a limitation, because the proposed filter is very sensitive to mistake, oversight or neglect of the expert, causing noise or silent in the base ontology pruning. Another troubled point found is the need of too much previous information to identify new knowledge. For example, in the method of [23] requires to know the 75% of the whole instances in the corpus to reach their better results. In this same method, another problem is that the Hidden Markov Models used to discover new knowledge have to be manually set and they are domain dependent. Finally, as it was previously seen in OE task, there is also a lack of patterns and rules learning methods.

3   Objective and Method

The main objective of this proposal is to design and validate a method able to expand OntoFIS from a corpus of summary of product characteristics in Spanish. To achieve it, there is proposed a four phases method:

1. Phase of semantic tagging: the input corpus is processed using a NLP tool, with several knowledge sources, in two steps: (i) using EuroWordNet (in Spanish) to perform a general knowledge tagging and (ii) using OntoFIS for a pharmacotherapeutic domain tagging. The use of EuroWordNet will help to identify concepts and synonims. This phase ends with the text corpora tagged twice.
2. Phase of merging: it unifies the results of the previous phase, giving priority to the OntoFIS tags. If a token is only tagged with EuroWordNet, then it is needed to know its mapping to OntoFIS. To achieve that we will use semantic distance metrics ([18]). In order to consider a tag as equivalent, candidate or not related to a certain OntoFIS concept, relationship or instance, we will established thresholds experimentally. After this phase all the known knowledge is semantically tagged.
3. Phase of knowledge discovery: this is the main phase in our method and the main contribution of this proposal: discovering new knowledge from unknown tokens. This phase uses triplets as identification pattern (T1,R,T2), where T1 (subject) and T2 (predicate) are concepts or instances, and R is a relationship. The nouns in the corpus can be classified as concepts or instances candidates, while the named entities will be considered just as instances candidates. Verbs will be considered as relationships candidates. If there are verbs with prepositions those prepositions will be used to determine the relationship direction. To discover new knowledge there are considered four basic rules described below (the given examples use the figure 1 ontology as known ontology).

Fig. 1. OntoFIS extract to be used as known ontology example

- If T1 or T2 are known, and R is a known taxonomic relationship ([9]), the unknown token will be classified, according to its triplet position, as a hyponym, a hypernym or as an instance of the known class. E.g., considering the triplet "(Ibuprofeno,is,drug)", the token 'Ibuprofeno' will be considered as 'drug' instance, because it is a named entity and, as it was explained previously, it can only be an instance.
- If T1 or T2 are known, and R is a non taxonomic known relationship, the unknown token will be tagged using the domain or range of R. E.g., considering the triplet "(Ibuprofeno,ingest,oral route)" and we have learned that 'Ibuprofeno' is a 'drug' instance, then 'oral route' will be tagged as route instance, because it is related to an instance.
- If only R is unknown, its domain will be the semantic class of T1 and its range the semantic class of T2. E.g., considering the triplet "(drug,inject,route)", the relationship 'inject' will have 'drug' as domain and 'route' as range.

- If just R is known but it has only one domain and one range, T1 and T2 will be classified according to their triplet positions. E.g., considering the triplet "(Ibuprofeno,ingest,oral route)" and supposing that neither 'Ibuprofeno' nor 'oral route' are known, both will be considered as instances of 'drug' and 'route', respectively, due to the domain and range of 'ingest' and because 'Ibuprofeno' is a named entity.

This phase is completed using statistical values and a minimum appearance threshold to be processed. Moreover, Machine Learning techniques will be applied to automatically learn new patterns and rules.

4. Phase of quality control: this phase firstly checks the validity and consistency of the ontology, i.e., it checks if there are inheritance loops or instances which instantiate disjoint classes, using a reasoner. Secondly, if there are conflicts, they will be solved using a decision making algorithm, and performing the quality control again until no conflicts are found.

4   Conclusions

This paper makes a proposal of an automatic method for extending and populating a pharmacotherapeutic ontology, helping to perform the OntoFIS maintenance. This method uses NLP general and domain specific techniques (such as name entity recognition) and a combination of general (EuroWordNet) and specific (OntoFIS) knowledge sources. The existing approaches for general and specific domain ontology expansion will bring a baseline in recall and precision levels. Lately, new approaches will be adopted, such as using other specific knowledge sources (like UMLS) or applying Machine Learning methods for discovering new patterns and rules.

References

1. Slaughter, L., Soergel, D., Rindflesch, T.C.: Semantic representation of consumer questions and physician answers. International Journal of Medical Informatics 75, 513–529 (2006)
2. Gonzalez-Gonzalez, A., Dawes, M., Sanchez-Mateos, J., Riesgo-Fuertes, R., Escortell-Mayor, E., Sanz-Cuesta, T., Hernandez-Fernandez, T.: Information needs and information-seeking behavior of primary care physicians. Annals of Family Medicine 5, 345 (2007)
3. Gruber, T.: Toward principles for the design of ontologies used for knowledge sharing. International Journal of Human Computer Studies 43, 907–928 (1995)
4. Studer, R., Benjamins, V., Fensel, D.: Knowledge engineering: principles and methods. Data & Knowledge Engineering 25, 161–197 (1998)
5. Romá-Ferri, M., Cruanes, J., Palomar, M.: Quality Indicators of the OntoFIS Pharmacotherapeutics Ontology for Semantic Interoperability. In: Proceedings IADIS International Conference e-Health, pp. 107–114 (2009)
6. Tanev, H., Magnini, B.: Weakly supervised approaches for ontology population. In: Proceeding of the 2008 Conference on Ontology Learning and Population: Bridging the Gap between Text and Knowledge, pp. 129–143. IOS Press, Amsterdam (2008)

7. Alfonseca, E., Manandhar, S.: An unsupervised method for general named entity recognition and automated concept discovery. In: Proceedings of the First International Conference on General WordNet, pp. 1–9 (2002)
8. Avancini, H., Lavelli, A., Magnini, B., Sebastiani, F., Zanoli, R.: Expanding domain-specific lexicons by term categorization. In: Proceedings of the 2003 ACM symposium on Applied computing - SAC 2003, pp. 793–797. ACM Press, New York (2003)
9. Cimiano, P., Pivk, A., Schmidt-Thieme, L., Staab, S.: Learning taxonomic relations from heterogeneous sources of evidence, pp. 59–73. IOS Press, Amsterdam (2005)
10. Dean, M.: Semantic Web rules: Covering the use cases. In: Rules and Rule Markup Languages for the Semantic Web, pp. 1–5 (2004)
11. Jung, J., Oh, K., Jo, G.: Extracting Relations towards Ontology Extension. In: Agent and Multi-Agent Systems: Technologies and Applications, pp. 242–251 (2009)
12. Kietz, J., Maedche, A., Volz, R.: A method for semi-automatic ontology acquisition from a corporate intranet. In: Workshop Ontologies and text, Citeseer, pp. 2–6 (2000)
13. Maedche, A., Staab, S.: Ontology learning for the semantic web. IEEE Intelligent Systems 16, 72–79 (2001)
14. Valencia-Garcia, R.: Un entorno para la extracción incremental de conocimiento desde texto en lenguaje natural. PhD thesis, University of Murcia (2005)
15. Valencia-García, R., Castellanos-Nieves, D., Fernández-Breis, J.T., Vivancos-Vicente, P.J.: A Methodology for Extracting Ontological Knowledge from Spanish Documents. In: Gelbukh, A. (ed.) CICLing 2006. LNCS, vol. 3878, pp. 71–80. Springer, Heidelberg (2006)
16. Valencia-Garcia, R., Ruiz-Sánchez, J.M., Vivancos-Vicente, P.J., Fernández-Breis, J.T., Martínez-Béjar, R.: An incremental approach for discovering medical knowledge from texts. Expert Systems with Applications 26, 291–299 (2004)
17. D'Amato, C., Fanizzi, N.: Query answering and ontology population: An inductive approach. In: Proceedings of the 5th European, pp. 288–302 (2008)
18. d'Amato, C., Fanizzi, N., Esposito, F.: Distance-based classification in OWL ontologies. In: Lovrek, I., Howlett, R.J., Jain, L.C. (eds.) KES 2008, Part II. LNCS (LNAI), vol. 5178, pp. 656–661. Springer, Heidelberg (2008)
19. D'Amato, C., Fanizzi, N., Esposito, F., Lukasiewicz, T.: Inductive Query Answering and Concept Retrieval Exploiting Local Models. In: 2009 Ninth International Conference on Intelligent Systems Design and Applications, pp. 1209–1214 (2009)
20. Makki, J., Alquier, A., Prince, V.: An NLP-based ontology population for a risk management generic structure. In: Proceedings of the 5th International Conference on Soft Computing as Transdisciplinary Science and Technology, pp. 350–355. ACM, New York (2008)
21. Novalija, I., Mladenic, D.: Content and structure in the aspect of semi-automatic ontology extension. In: 2010 32nd International Conference on Information Technology Interfaces (ITI), pp. 115–120. IEEE, Los Alamitos (2010)
22. Shi, L., Sun, J., Che, H.: Populating CRAB Ontology Using Context-Profile Based Approaches. Context, 210–220 (2007)
23. Valarakos, A., Paliouras, G., Karkaletsis, V., Vouros, G.: Enhancing ontological knowledge through ontology population and enrichment. In: Engineering Knowledge in the Age of the Semantic Web, pp. 144–156 (2004)
24. Volkova, S., Caragea, D., Hsu, W.H., Drouhard, J., Fowles, L.: Boosting Biomedical Entity Extraction by Using Syntactic Patterns for Semantic Relation Discovery. In: 2010 IEEE/WIC/ACM International Conference on Web Intelligence and Intelligent Agent Technology, pp. 272–278 (2010)

Author Index

GPSR Compliance

*The European Union's (EU) General Product Safety Regulation (GPSR)
is a set of rules that requires consumer products to be safe and our
obligations to ensure this.*

*If you have any concerns about our products, you can contact us on
ProductSafety@springernature.com*

In case Publisher is established outside the EU, the EU authorized
representative is:

Springer Nature Customer Service Center GmbH
Europaplatz 3
69115 Heidelberg, Germany

Batch number: 09478804

Printed by Printforce, the Netherlands